The Hanle Effect and
Level-Crossing Spectroscopy

PHYSICS OF ATOMS AND MOLECULES

Series Editors

P. G. Burke, *The Queen's University of Belfast, Northern Ireland*
H. Kleinpoppen, *Atomic Physics Laboratory, University of Stirling, Scotland*

Editorial Advisory Board

R. B. Bernstein *(New York, U.S.A.)*
J. C. Cohen-Tannoudji *(Paris, France)*
R. W. Crompton *(Canberra, Australia)*
Y. N. Demkov *(Leningrad, U.S.S.R.)*
J. N. Dodd *(Dunedin, New Zealand)*
W. Hanle *(Giessen, Germany)*

C. J. Joachain *(Brussels, Belgium)*
W. E. Lamb, Jr. *(Tucson, U.S.A.)*
P.-O. Löwdin *(Gainesville, U.S.A.)*
H. O. Lutz *(Bielefeld, Germany)*
K. Takayanagi *(Tokyo, Japan)*

Recent volumes in the series:

ATOMIC INNER-SHELL PHYSICS
Edited by Bernd Crasemann

ATOMIC PHOTOEFFECT
M. Ya. Amusia

ATOMIC SPECTRA AND COLLISIONS IN EXTERNAL FIELDS
Edited by K. T. Taylor, M. H. Nayfeh, and C. W. Clark

ATOMS AND LIGHT: INTERACTIONS
John N. Dodd

COHERENCE IN ATOMIC COLLISION PHYSICS
Edited by H. J. Beyer, K. Blum, and R. Hippler

COLLISIONS OF ELECTRONS WITH ATOMS AND MOLECULES
G. F. Drukarev

ELECTRON-MOLECULE SCATTERING AND PHOTOIONIZATION
Edited by P. G. Burke and J. B. West

THE HANLE EFFECT AND LEVEL-CROSSING SPECTROSCOPY
Edited by Giovanni Moruzzi and Franco Strumia

ISOTOPE SHIFTS IN ATOMIC SPECTRA
W. H. King

MOLECULAR PROCESSES IN SPACE
Edited by Tsutomu Watanabe, Isao Shimamura, Mikio Shimizu, and Yukikazu Itikawa

PROGRESS IN ATOMIC SPECTROSCOPY, Parts A, B, C, and D
Edited by W. Hanle, H. Kleinpoppen, and H. J. Beyer

QUANTUM MECHANICS VERSUS LOCAL REALISM: The Einstein-Podolsky-Rosen Paradox
Edited by Franco Selleri

RECENT STUDIES IN ATOMIC AND MOLECULAR PROCESSES
Edited by Arthur E. Kingston

THEORY OF MULTIPHOTON PROCESSES
Farhad H. M. Faisal

ZERO-RANGE POTENTIALS AND THEIR APPLICATIONS IN ATOMIC PHYSICS
Yu. N. Demkov and V. N. Ostrovskii

A Continuation Order Plan is available for this series. A continuation order will bring delivery of each new volume immediately upon publication. Volumes are billed only upon actual shipment. For further information please contact the publisher.

The Hanle Effect and Level-Crossing Spectroscopy

Edited by
Giovanni Moruzzi and Franco Strumia
University of Pisa
Pisa, Italy

PLENUM PRESS • **NEW YORK AND LONDON**

Library of Congress Cataloging-in-Publication Data

The Hanle effect and level-crossing spectroscopy / edited by Giovanni
  Moruzzi and Franco Strumia.
       p.   cm. -- (Physics of atoms and molecules)
    Includes bibliographical references and index.
    ISBN 0-306-43630-2
    1. Hanle effect. 2. Level-crossing spectroscopy. I. Moruzzi,
  Giovanni. II. Strumia, F. III. Series.
  QC467.H36 1990
  543'.0858--dc20   90-47813
                                                               CIP

ISBN 0-306-43630-2

© 1991 Plenum Press, New York
A Division of Plenum Publishing Corporation
233 Spring Street, New York, N.Y. 10013

All rights reserved

No part of this book may be reproduced, stored in a retrieval system, or transmitted
in any form or by any means, electronic, mechanical, photocopying, microfilming,
recording, or otherwise, without written permission from the Publisher

Printed in the United States of America

CONTRIBUTORS

JÁNOS A. BERGOU • *Center for Advanced Studies, The University of New Mexico, Albuquerque, New Mexico 87131, USA*

NICOLÒ BEVERINI • *Diparimento di Fisica dell'Università di Pisa, 56126 Pisa, Italy*

DANUTA GAWLIK • *Sektion Physik, Universität München, and Max-Planck-Institut für Quantenoptik, D-8046 Garching, Federal Republic of Germany. Permanent address: Institute of Physics, Jagellonian University, Reymonta 4, PL-30-059 Kraków, Poland*

WOJCIECH GAWLIK • *Sektion Physik, Universität München, and Max-Planck-Institut für Quantenoptik, D-8046 Garching, Federal Republic of Germany. Permanent address: Institute of Physics, Jagellonian University, Reymonta 4, PL-30-059 Kraków, Poland*

GIOVANNI MORUZZI • *Dipartimento di Fisica dell'Università di Pisa, 56126 Pisa, Italy*

G. E. PIKUS • *A.F. Ioffe Physico-Technical Institute, USSR Academy of Sciences, 194021 Leningrad, USSR*

MARLAN O. SCULLY • *Center for Advanced Studies, The University of New Mexico, Albuquerque, New Mexico 87131, USA*

JAN OLOF STENFLO • *Institute of Astronomy, ETH-Zentrum, CH-8092 Zürich, Switzerland*

FRANCO STRUMIA • *Dipartimento di Fisica dell'Università di Pisa, 56126 Pisa, Italy*

A. N. TITKOV • *A.F. Ioffe Physico-Technical Institute, USSR Academy of Sciences, 194021 Leningrad, USSR*

HERBERT WALTHER • *Sektion Physik, Universität München, and Max-Planck-Institut für Quantenoptik, D-8046 Garching, Federal Republic of Germany*

H. G. WEBER • *Heinrich-Hertz-Institut and Optisches Institut der Technischen Universität Berlin, D-1000 Berlin 10, Federal Republic of Germany*

Foreword

I am most pleased and, in a way, I feel honored to write the Foreword for the book *The Hanle Effect and Level-Crossing Spectroscopy*, which covers such a very wide range of applications not only in the initial areas of atomic and molecular physics, but also in solid state physics, solar physics, laser physics, and gravitational metrology. To link these fields together in a *coherent* way has been the merit of the editors of the book, who attracted most distinguished authors for writing the chapters.

In retrospect to Hanle's discovery of *quantum mechanical coherence* between two quantum states about 65 years ago, this book demonstrates the enormous impact and central importance the effect has had, and most vividly still has, on modern physics. On the other hand, the concept of quantum mechanical coherence, which is an outgrowth of the linear superposition principle of quantum states, has been evident through a considerable number of experimental methods beyond the original Hanle effect; some of these methods were only recently discovered or applied and they have indeed revolutionized research fields such as atomic collision physics. For example, measurements of interfering magnetic sublevel scattering amplitudes in impact excitation of atoms opened up a new area of research in which so-called *complete* or *perfect atomic collision experiments* are applied. The famous Franck–Hertz experiment of electron impact excitation of atoms which proved energy conservation between colliding atoms and electrons and the emitted photons has been extended by the electron–photon coincidence technique during the last two decades. The analysis of such modern Franck–Hertz experiments in which the electron, having excited the atom, and the photon are detected in coincidence is based upon coherence effects for the excitation amplitudes of magnetic sublevels. Allow me just to mention that these "dynamical coherence effects" of impact excitation of atoms link a considerable part of my own research to that initiated through Professor Hanle in the Institute of James Franck at Göttingen. I am using the word *dynamics* because the coherence effects in impact excitation of atoms depend both on the energy of the incoming electrons and on the angle of the inelastically scattered electrons. Only very recently have quantum chromodynamical coherence effects been discovered in jets of high-energy particle reactions.

Hanle's life was not free from dramatic experiences which were, particularly in connection with the discovery of the Hanle effect itself, highly exciting. When he was a student in his fifth semester at Heidelberg and 21 years old, he presented a seminar on the energy–mass equivalence and the deflection of light from the sun that had just been detected by a British research group. This seminar led to a massive conflict with Noble prize winner Professor Lenard, who did not accept Einstein's theory of relativity. As a consequence of this severe clash with Lenard, Hanle was advised to leave Heidelberg so he went to James Franck and Max Born at Göttingen, where he encountered a completely different, humane attitude and a highly stimulating scientific environment. Göttingen was developing into a world center of research in physics and Hanle's investigation on fluorescence polarization was very highly regarded. However, distinguished theoretical physicists did not initially agree with Hanle on the interpretation of his effect. Max Planck personally said to Hanle, "your interpretation cannot be correct, it contradicts quantum theory." Of course, modern quantum mechanics was not yet born and most theoreticians considered his effect as a kind of Faraday effect. However, the decisive breakthrough came after Arnold Sommerfeld's visit to Göttingen when the great master of atomic theory asked him for an explanation of the magnetic effects on the polarization of resonance fluorescence detected by Wood and Ellet and Hanle himself. Next day, after having worked the whole night in the laboratory and checking experimental data, he presented his classical explanation and the suggestion of coherence of the two σ-components of the resonance light.

After gaining his PhD in Göttingen in 1924 Hanle stayed on as a research assistant for one year, and went then to Gerlach in Tübingen for half a year. J. Franck subsequently recommended Hanle to Gustav Hertz in Halle; there Hanle pioneered another piece of work, namely, the measurement of optical excitation functions of atoms by electron impact. Again, this work opened up a whole new field of research leading to cross-section measurements and, in the 1970s, to angular correlation experiments of coincident electrons and photons in impact excitations of atoms. Two and a half years after his PhD, Hanle received his Habilitation and was promoted to "Privatdozent," the first step toward a successful University career in Germany. In 1929 Hanle became "ausserplanmässiger Professor" in the Institute of Geheimrat Max Wien in Jena/Thuringia; at that time he was the youngest professor of experimental physics in Germany. During his stay at Jena University, Hanle was also asked to administer the chair of the distinguished Professor P. Debye and also another extraordinary chair of radiation physics. However, owing to clashes with the Nazis, Hanle lost his appointment in Jena in 1937. His friend, Professor Joos, saved Hanle's

University career by appointing him as an assistant back at Göttingen University subject to the special condition that he should refrain from any political statement. In 1939 Hanle was drafted for military service with responsibility as head of the weather station for a military airport. Afterward he was delegated to research on important war activities, such as radar and infrared technologies. In 1941 Hanle succeeded Professor Gerthsen of Giessen University. Of course research and scientific work was almost impossible during the war and in 1944 the building of his Physics Institute was bombed and destroyed to a large degree.

Immediately after the war, Hanle and his colleagues were responsible for traffic and the allocation of the needs of the population in the region. Convinced of the importance of science and physics, Hanle put an enormous effort into rebuilding Giessen University, the science faculty, and the Physics Institute. In 1953 Hanle was offered the distinguished chair of Ramsauer at the Technical University in Berlin. He was tempted to accept the offer and to help the City of West Berlin in its fight for survival and to reestablish Berlin once again as a world center of scientific research. As a diploma student of Professor Hanle in Giessen, I remember his many journeys between Berlin and Giessen and his concern for promoting what he had built up at the Justus-von-Liebig-University in Giessen. He finally decided to remain in Giessen and also turned down a chair at Saarbrücken University in 1957.

A large number of publications have resulted from Hanle's work with many research associates and research students at Giessen. Polarization of fluorescence and phosphorescence radiation, in a way related to his famous PhD thesis at Göttingen, continued to be one of his favorite and beloved fields of research. Continuous studies of his well-known excitation and ionization functions of atomic, molecular, and polymer systems and the development and extension of the ultrasonic fluorescence meter by Hanle and Maercks (1938) showed his preference for research in light emission. The development of scintillation counters linked his studies on the production of light by energetic particles to nuclear physics.

In 1954 Hanle became chairman of the committee of radiation measurements of the atomic ministry with the Government at Bonn for a full decade. As a result of his duties at the ministry and his two offers of full professor at Berlin and Saarbrücken, Hanle's Institute at Giessen was provided with a generous new building and also a new Institute on Radiation Physics ("Strahlenzentrum") was provided as one of the first interfaculty institutions in the Federal Republic of Germany (FRG). Hanle was a member or chairman of many national and international committees, notably of the German-French-Rector Conference, chairman of the Röntgen Museum, chairman of the education committee on the diploma in physics, and editorial adviser of the *Physikalische Blätter*, which is the official journal of the German Physical Society. A series of honors were bestowed on Hanle: Dr. Ing. E.h (honorary doctor of the faculty of engineering) by the University of Stuttgart in 1970, the Grosses Verdienstkreuz (an important medal) of the FRG in 1973, the Röntgen prize of the Röntgen Museum in 1975, and, last but not least, in the same year, the University of Göttingen renewed his doctorate to commemorate the 50th anniversary of his famous PhD award in 1925.

When he became Professor Emeritus, Hanle continued to be active in research on level-crossing spectroscopy. He still takes a lively interest in this field of research and attempts to record every known and new activity of his effect. He also devoted much work to editing the first two volumes of *Progress in Atomic Spectroscopy*, which, together with the subsequent two volumes, have developed into a most comprehensive survey of the achievements in atomic physics research.

I was always highly impressed by Hanle's clear and direct approach in understanding and analyzing the fundamental physics of a given research problem. I highly treasure the many and fruitful discussions I had with Professor Hanle as my teacher and highly regarded mentor.

Not only the scientific community of physicists but also laymen and people of our generations will be in debt and grateful to Professor Hanle for his achievements.

H. Kleinpoppen
Stirling

PREFACE

The Hanle effect was presented by Wilhelm Hanle in *Zeitschrift für Physik* in 1924. During the following years, the efforts to provide a consistent interpretation of this effect played a dominant role in the development of quantum theory. Several works on this subject by N. Bohr, G. Breit, R. Oppenheimer, V. Weisskopf, and W. Hanle himself contributed to the definite supersession of the old theory, according to which quantum systems could exist only in what are now called stationary states, in favor of the present quantum theory, which regards the possible states of a system as vectors of a Hilbert space. Experimentally, the Hanle effect provided a method for measuring the natural linewidth, and thus the lifetime, of an excited level independently of the Doppler broadening.

In more recent years, since the revival of atomic and molecular spectroscopy which followed the introduction of double resonance spectroscopy by Bitter, Brossel, and Kastler in 1949, of optical pumping by A. Kastler in 1950, and of (nonzero field) level crossing by Colegrove, Franken, Lewis, and Sands in 1959, the Hanle effect has constantly found novel applications in pure and applied physics.

The aim of the present book is to provide a complete physical insight into the Hanle effect and level-crossing spectroscopy, and to present an overview, as complete as possible, of their applications in many different fields of physics. The introduction provides a brief historical survey, an English translation of Prof. Hanle's original paper, and a general discussion of the effect. In the subsequent chapters, appropriate authors present the applications of the Hanle effect and level crossing in several fields. These cover atomic, molecular, and solid state physics, where physical properties like, for instance, the quantum state lifetimes can be determined by measuring the Hanle effect at known external fields; solar physics where, conversely, the observation of the effect on states of known lifetime can lead to the determination of the field intensities; gravitation, where the Hanle laser is proposed as a possible gravitational probe; and laser physics and technology, where the nonlinear Hanle effect, apart from its theoretical interest, can lead to large enhancement in laser output power. The chapter dedicated to laser physics provides the first available extensive survey on the effects of an external field on the output power of gas lasers. Special emphasis is dedicated to far-infrared molecular lasers and to noble-gas ion lasers; the combined action of magnetic plasma confinement and the nonlinear Hanle effect in a noble-gas ion laser are also thoroughly discussed, in view of their important commercial applications. Several new results are presented for the first time in this book.

We hope that the scientific community will appreciate the efforts made by the authors, who have undertaken the difficult task of a clear and, as far as possible within the limits of the available space, complete description of the aspects of an exciting subject, which has played extremely important roles in several branches of physics for more than 60 years since its discovery.

Giovanni Moruzzi
Franco Strumia
Pisa

CONTENTS

CHAPTER 1
THE HANLE EFFECT AND LEVEL-CROSSING SPECTROSCOPY—AN INTRODUCTION
Giovanni Moruzzi

1. Historical Survey	1
2. Classical Interpretation of the Hanle Effect	9
3. Quantum Mechanical Interpretation of the Hanle Effect	13
4. The Density Matrix Formalism for the Hanle Effect (Broad-Band Excitation)	17
5. Laser Excitation and Pressure-Induced Coherences	23
6. Nonzero-Field Level Crossing	28
7. Conclusions	30
References	32
Appendix. Magnetic Effects on the Polarization of Resonance Fluorescence (*original work by Wilhelm Hanle, translated by G. Moruzzi*)	34

CHAPTER 2
THE HANLE EFFECT AND ATOMIC PHYSICS
Wojciech Gawlik, Danuta Gawlik and Herbert Walther

1. Introduction	47
1.1. General Expression for the Hanle Signal in Terms of the Density Matrix	47
2. Spectroscopic Applications	49
2.1. Determination of Atomic Constants	49
2.2. Measurements of Laser-Level Populations	51
2.3. Increasing Resolution, Subnatural Linewidth Effects	51
2.4. Forward Scattering, Line Crossing	54
2.5. Technical Applications	55

3. Collisions 56
   3.1. Hanle Effect with Collisional Excitation 56
   3.2. Hanle Effect and Optogalvanic Detection 57
   3.3. Collision-Induced Hanle Resonances 59
   3.4. Fluctuation-Induced Hanle Resonances 64
4. Hanle Effect in Strong Laser Fields 65
   4.1. General Characteristic 65
   4.2. Specific Situations 69
   4.3. Hanle Effect and Nonlinear Optics 76
5. Hanle Effect in Quantum Optics 77
   5.1. Dressed-Atom Model 77
   5.2. Hanle Effect with Fluctuating Fields 78
   5.3. Squeezing in the Hanle Effect 79
   References 79

CHAPTER 3

THE HANLE EFFECT AND LEVEL-CROSSING SPECTROSCOPY ON MOLECULES

H. G. Weber

1. Introduction 87
2. Molecular Level-Crossing Signal 88
3. Comparison with Quantum Beat Experiments 91
4. Excitation of Molecules 92
5. Lifetime Investigations 95
6. Landé g-Factors 98
7. Electric-Field Level Crossing 100
8. Stark–Zeeman Recrossing and High-Field Level Crossing . . 102
9. Hanle Effect on NO_2 104
   9.1. The Influence of Detection Geometry 104
   9.2. Details of the Hanle-Effect Signal 109
   9.3. Collisions 113
   9.4. Discussion of Hanle-Effect Experiments on NO_2 115
10. Conclusion 118
    References 118

CHAPTER 4

THE NONLINEAR HANLE EFFECT AND ITS APPLICATIONS TO LASER PHYSICS

Giovanni Moruzzi, Franco Strumia, and Nicolò Beverini

1. The Nonlinear Hanle Effect and Its Experimental Observation   123
2. Saturation Intensity and Saturated Linewidth 132
3. The Three-Level Case: Homogeneously Broadened Lines . . . 139

4. The Three-Level Case: Doppler-Broadened Lines and the Rate Equations 147
5. The General Case 151
6. The Rate-Equation Approach to the Nonlinear Hanle Effect in Inhomogeneously Broadened Transitions 155
7. The Nonlinear Hanle Effect with a Gaussian Laser Beam . . 163
8. The Nonlinear Hanle Effect in Absorption 166
9. The Nonlinear Hanle Effect in Laser-Active Media 183
   9.1. The He–Ne Laser 192
   9.2. The Xe Laser 200
   9.3. The He–CdII and He–ZnII Lasers 201
   9.4. The Noble-Gas Ion Lasers 206
   9.5. Optically Pumped Far-Infrared Lasers 218
   9.6. Other Lasers 228
   9.7. Conclusions 228
   References 229

CHAPTER 5

APPLICATIONS OF THE HANLE EFFECT IN SOLAR PHYSICS

Jan Olof Stenflo

1. Introduction 237
2. Brief Review of the Properties of Solar Magnetic Fields . . . 238
3. Overview of the Diagnostic Possibilities and Limitations of the Hanle Effect 240
4. Basic Theoretical Concepts for Applications in Astrophysics . 243
5. Diagnostics of Magnetic Fields in Solar Prominences 254
6. Survey of Scattering Polarization on the Solar Disk 262
7. Diagnostics of Turbulent Magnetic Fields 269
8. Diagnostics of Magnetic Fields in the Chromosphere-Corona Transition Region and Above 275
9. Concluding Remarks 277
   References 279

CHAPTER 6

APPLICATIONS OF THE HANLE EFFECT IN SOLID STATE PHYSICS

G. E. Pikus and A. N. Titkov

1. Introduction 283
2. The Hanle Effect on Free Electrons 284
   2.1. Optical Orientation of Electron Spins 285

2.2. Occurrence of Electron–Nucleus Interaction in Polarized Luminescence 299
2.3. Optical Alignment of Electron Momenta in a Magnetic Field 304
3. The Hanle Effect on Excitons 309
3.1. The $\Gamma_8 \times \Gamma_6$ and $\Gamma_7 \times \Gamma_6$ Excitons in Cubic Crystals . . . 310
3.2. The $\Gamma_9 \times \Gamma_7$ and $\Gamma_7 \times \Gamma_7$ Excitons in Hexagonal II–VI Crystals with Wurtzite Structure 316
3.3. The $\Gamma_7 \times \Gamma_8$ Excitons in III–VI Crystals with Symmetry Class D_{3h} 317
3.4. The Influence of Reemission on the Hanle Effect 324
3.5. Hot Excitons and Polaritons 327
4. The Hanle Effect on Impurity Centers 329
4.1. The Hanle Effect on Sm^{2+} in CaF_2 330
4.2. The Hanle Effect on Eu^{2+} in CaF_2 and SrF_2 331
5. Summary 332
References 334

CHAPTER 7

QUANTUM THEORY OF THE HANLE LASER AND ITS USE AS A METRIC GRAVITY PROBE

János A. Bergou and Marlan O. Scully

1. Introduction 341
2. The Model 343
3. Linear Theory 344
4. Nonlinear Theory 351
5. Vanishing of Diffusion Constant for the Relative Phase . . . 357
6. Applications as a Metric Gravity Probe 359
7. Discussion and Summary 365
References 367

INDEX 369

CHAPTER 1

THE HANLE EFFECT AND LEVEL-CROSSING SPECTROSCOPY— AN INTRODUCTION

GIOVANNI MORUZZI

1. HISTORICAL SURVEY

The Hanle effect has played a major role in the development of atomic physics and, more generally, of quantum mechanics. As we shall see later in this section, it contributed much to the replacement of the old quantum theory, which was unable to provide a satisfactory explanation of the effect. Moreover, for decades the Hanle effect together with level crossing, which is its extension to nonzero fields, has been the only available Doppler-free spectroscopic technique. Besides its historical relevance, the Hanle effect is still very often the most accurate, if not the only, experimental technique for determining lifetimes of excited states. In fact, owing to their inherent simplicity, these experiments are affected by extremely few sources of error.

The milestone of this research field is the "Doktorarbeit" (Ph.D. thesis) of Wilhelm Hanle, elaborated in James Franck's laboratory in Göttingen and discussed in 1922. The subject was "Magnetic field influence on the polarization of the resonance fluorescence of mercury." In 1924 Hanle's thesis was published in *Zeitschrift für Physik*.[1] The English translation of this paper comprises the Appendix at the end of this chapter. One year later Hanle published his ideas in a more precise and detailed form in *Ergebnisse der exakten Naturwissenschaften*.[2]

The experiment, like so many other experiments of relevant importance for the development of atomic physics, was performed on the 253.7 nm resonance line of mercury. The problem had arisen in 1912, when Wood, investigating the polarization of scattered light,[3] observed that light scattered by gases and vapors in nonresonant conditions was almost completely polarized (as, for example, in the blue sky), while the 253.7 nm resonance radiation of mercury, "scattered" by mercury vapor, was unpolarized. In 1919, Lord Rayleigh observed that the 253.7 nm fluorescent radiation reemitted by a mercury vapor could be slightly linearly polarized, the polarization degree changing significantly from one experiment to the other without any apparent cause. He also observed that white light, scattered at right angles by a dense mercury vapor, was practically completely polarized. Lord Rayleigh's results were published three years later.[4] The matter was reconsidered in 1923 by Wood and Ellett,[5] who noted that the different polarization degrees observed by Lord Rayleigh depended on the orientation of the experimental setup with respect to the earth's magnetic field. It turned out that the polarization degree could be strongly increased by neutralizing the earth's magnetic field with a large solenoid. These authors performed resonance fluorescence experiments on mercury with the experimental geometry sketched in Figure 1, which corresponds to the simplest experimental arrangement for the detection of the Hanle effect. The vapor contained in the cell is excited by 253.7 nm resonance light propagating along the y axis, and the resonance fluorescence emitted along the z axis is monitored. Wood and Ellett observed that, when the excitation light was linearly polarized parallel to the x axis, the resonance fluorescence light

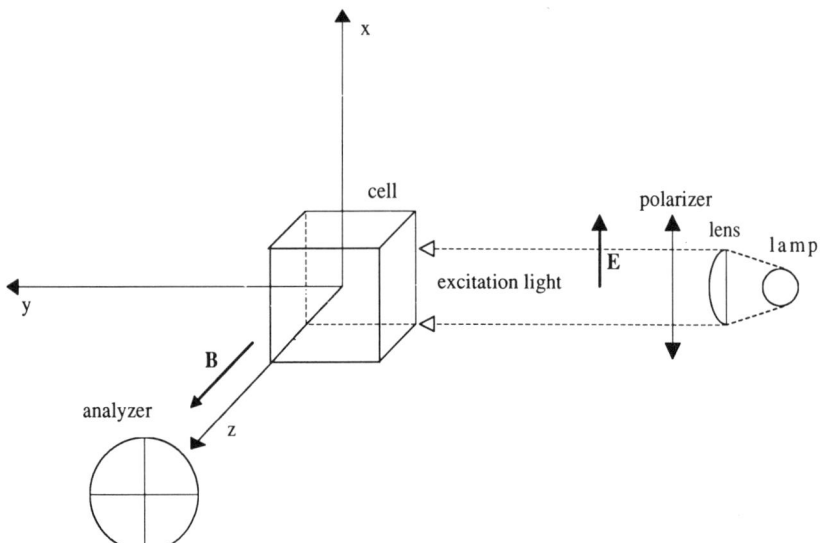

FIGURE 1. Experimental arrangement for the detection of the Hanle effect. The figure shows the simplest geometry.

was itself almost completely linearly polarized along the same direction in the absence of external magnetic fields. They also noted that the presence of a weak magnetic field parallel to z was sufficient to strongly affect the polarization degree of the fluorescence light. The polarization degree is defined as $P = (I_x - I_y)/(I_x + I_y)$, where I_x and I_y are the intensities of the x and y linearly polarized components. The effect was already clearly observable for field intensities smaller than 10^{-4} T (1 G). Wood and Ellett realized that this behavior could not be interpreted as a Zeeman effect, because the Zeeman separation in a field of less than one gauss is a small fraction only of the Doppler linewidth.[5] Moreover, the Zeeman effect could not account for the high polarization degree observed for mercury at zero field. The authors concluded that they were probably dealing with a new magneto-optic effect. The first correct interpretation of the phenomenon was published by Hanle in 1923[6] but, for a couple of years, Hanle's work was either ignored or not believed, and different interpretations were presented by several authors. Eldridge interpreted Wood and Ellett's results as a Zeeman effect, with the further assumptions that only linearly polarized radiation could be absorbed or reemitted by the atom, and that both absorption and reemission were sudden phenomena separated by a certain time interval τ during which the atom executed a Larmor precession.[7]

In Eldridge's work we find the first quantitative description of the dependence of the polarization degree on the magnetic field:

$$P = \frac{1}{1 + (eB\tau/m_0)^2}$$

where e is the electron charge, m_0 the electron mass, and B the magnetic field intensity. This Lorentzian shape fits quite well the experimental data by Wood and Ellett[5,8] (apart from the value of the maximum polarization degree observed at zero field). Eldridge's model was the first to realize that the signal shape depends on the *relaxation time* (lifetime of the excited level). It also predicted a rotation of the polarization plane of the fluorescence radiation induced by the magnetic field; this rotation was later observed by Hanle.[1] However, Eldridge's model presented two discrepancies with the experimental results. A 100% polarization degree was predicted at zero field, while Wood and Ellett had not been able to observe a polarization degree higher than 90% even at a temperature of −50 °C.[8] This first discrepancy actually arose from the use of natural mercury in the resonance cell. Now we know that a 100% polarization degree would have been observed in an experiment on a pure even Hg isotope, but separated Hg isotopes were unavailable in the 1920s. Thus the natural abundances of

16.84% for ^{199}Hg and 13.22% for ^{201}Hg were present in the mercury vapor used for the experiments. For odd isotopes, the presence of nuclear spin must be taken into account in a correct treatment. The second discrepancy between Eldridge's model and the experimental results was greater. Wood and Ellett[8] had observed a significant difference in the polarization degrees at zero field for mercury (90%) and sodium (6%), and Eldridge's model could not account for this difference. Attempts to interpret the phenomenon as a Zeeman effect, supplemented by *ad hoc* assumptions in order to account for the different behaviors of sodium and mercury, were presented by Pringsheim,[9] Breit,[10] and Joos.[11] Of these, Joos was able to obtain the correct value for the maximum polarization degree observable for the D_2 line of sodium by assuming that the atoms were forced to orient themselves relative to the excitation light beam. Hanle's interpretation of the phenomenon, which had already been published,[6] was not quoted in any of the above works. We are going to discuss it in the following.

The simplest experimental and theoretical conditions are sketched in Figure 1. The cell, containing Hg vapor, is located in a uniform magnetic field **B** directed along the z axis. For the sake of clarity, we shall assume in all our theoretical approaches that the cell contains an even Hg isotope, so that the nuclear spin I is zero (as reminded above, pure isotopes were not used in Hanle's original experiments; this accounts for the 90% polarization degree observed at zero field, while 100% polarization is predicted not only by Eldridge's, but also by Hanle's treatment). The excitation light propagates along the y axis and is linearly polarized along the x axis.

As a classical model for the Hg atom Hanle used a quasi-elastically bound electron. The incoming resonance light excites oscillations of the electron in the x direction. We detect the fluorescence resonance light emitted along z, whose polarization is determined by the electron motion in the xy plane. In the absence of external fields, the electron performs damped oscillations along x, so that linearly polarized fluorescence light is emitted, in agreement with Wood and Ellett's experimental result. The experiment in the presence of a nonzero magnetic field is easily interpreted if we decompose the electron oscillation into two counterrotating circular motions, each providing the origin (if observed along **z**) to a circularly polarized wave. At zero magnetic field, the superposition of the two counterrotating circular waves results in the observed linearly polarized fluorescence wave. If a uniform static magnetic field is now introduced parallel to z, the two circular components of the motion will occur at slightly different frequencies because of the normal Zeeman effect. If the electron oscillations were not damped, the electron would thus describe a rosette motion, as in Figure 2, and we would observe completely unpolarized fluorescence light.

Actually, and this was Hanle's point, the oscillations are damped because mechanical energy is continuously transformed into radiated energy

FIGURE 2. The rosette motion of an undamped charged oscillator in a magnetic field.

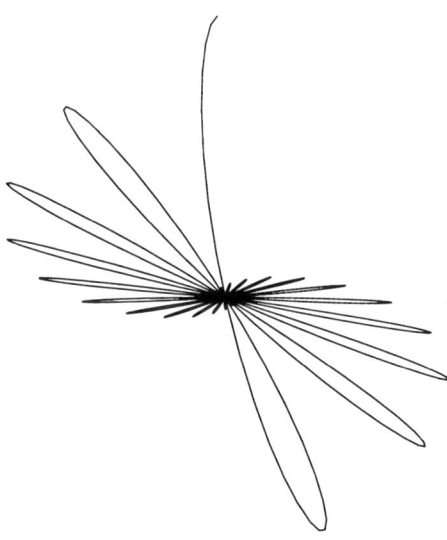

FIGURE 3. The damped rosette motion obtained by taking radiative decay into account.

(for the sake of simplicity, we are neglecting collisional relaxation at this stage). If the damping is fast enough, the resulting motion is represented in Figure 3, which describes a weak depolarization of the fluorescence light and a small rotation of its polarization plane. This rotation, predicted also by Eldridge's model, was observed experimentally by Hanle. All intermediate situations are possible, ranging from linear polarization, corresponding to an oscillator damping much faster than the precession motion about the z axis, up to the complete depolarization of Figure 2, corresponding to a precession motion much faster than damping.

The classical interpretation of the effect proposed by Hanle was thus straightforward. We shall consider its mathematical formulation with some more details in the next section. It is interesting to note that the classical equations correctly describe the experimental signals as long as no hyperfine structure, or nearly degenerate fine structure multiplets, are present.

On the contrary, the quantum mechanical interpretation presented serious difficulties at the time of Hanle's thesis. The necessity to override these difficulties played an important role in the subsequent development of the quantum theory. The first formulations of the quantum theory assumed that a quantum system could exist only in pure states (i.e., in eigenstates of its Hamiltonian, in current terminology). According to this assumption, the only available quantum analog for the superposition of the two counter-rotating circular waves, which we have just met in the classical interpretation, was the interference between a σ^+ and a σ^- photon emitted by two different atoms. The first atom should decay from the $|m = +1\rangle$ Zeeman sublevel of the excited state to the $|\mu = 0\rangle$ sublevel of the ground state, while the second should decay from $|m = -1\rangle$ to $|\mu = 0\rangle$ (throughout this chapter we shall label with m the Zeeman sublevels of the excited state, and with μ the sublevels of the ground state). Thus the two interfering photons would be emitted by atoms located at an average distance much larger than the wavelength. An interference in these conditions is, of course, not conceivable. Hanle argued that the only possible way out was the assumption that the directional quantization is not always rigorous. When the separation between a group of energy levels of a quantum system is of the order of, or smaller than, h/τ, where h is Planck's constant and τ the lifetime of the levels, the system can exist in a superposition of these levels.

This was a first step toward the interpretation of the states of a quantum system as vectors of a Hilbert space, and of the eigenstates of the Hamiltonian as one of the possible bases for this Hilbert space. This viewpoint was not easily accepted, and Franck himself felt compelled to withdraw Hanle's contribution to the 1924 Joint Conference of the German and Austrian Physical Societies in Innsbruck. All the theoreticians he had heard were convinced that Hanle's interpretation of the phenomenon could not be true, and that a Faraday rotation had actually been observed.

Back in Göttingen, Hanle was soon able to prove that the observed rotation of the polarization plane could not possibly be due to a Faraday effect. Almost simultaneously, Bohr[12] developed an interpretation of the Hanle effect which was a further important step for the development of the quantum theory. Bohr noted that the directional quantization in a magnetic field is related to a precessional motion, and suggested that an atom can be in a pure directionally quantized state only if the corresponding precession period is shorter than the lifetime of the state. For a precession period of the order of, or longer than, the lifetime, degeneration must occur. For Zeeman sublevels at zero field, which corresponds to an infinitely long precission period, the degeneration must be complete. Bohr's argument is extremely well visualized by the vector model for the angular momentum, which subsequently played a primary role in the development of atomic quantum mechanics.

Bohr's argument was combined (by Hanle himself[2]) with Hanle's assumption that the quantization of the energy levels is not rigorous when the spacing is of the order of, or smaller than, h/τ. Hanle noted that, because of their finite lifetime, the energy levels of a quantum system are not rigorously sharp, but display a certain width h/τ due to Heisenberg's uncertainty principle (natural width). The distance between the $|m = +1\rangle$ and $|m = -1\rangle$ levels can be varied arbitrarily by adjusting the external magnetic field. The quantization of the levels, and the corresponding directional quantization, are affected by degeneration when the levels overlap within their natural widths. The case when the $|m = +1\rangle$ and $|m = -1\rangle$ Zeeman sublevels of the excited state are shifted with respect to each other by an external magnetic field, but are still degenerate within their natural linewidths, is illustrated in Figure 4, which is taken from Ref. 13. The two crossing stripes represent the energies of the $|m = +1\rangle$ and $|m = -1\rangle$ levels; the stripe width is the natural linewidth (about h/τ) of the level. As long as the stripes overlap, coherence is possible. Thus, the interference of the circularly polarized components σ^+ and σ^- can be observed as long as the spacing between the levels is smaller than h/τ.[2]

Considerations on the experiments of Wood and Ellett,[5] Gaviola and Pringsheim,[14] and of Hanle himself[1] led Heisenberg[15] to formulate an extension of the correspondence principle, which allows one to predict the intensity and polarization of the radiation emitted in the case of a transition from a degenerate to a nondegenerate quantum system.

Soon after Bohr and Heisenberg, also Oppenheimer,[16] Weisskopf,[17] and Breit[18] considered the theoretical aspects of the problem, and the state coherence proposed by Hanle became a well-established quantum mechanical effect. The work by Breit[18] implicitly contained also the theoretical treatment of level crossing. The designation *Hanle effect* was used for the first time in 1925 by Heisenberg in two letters to Pauli[19]; this expression, however, entered current use only later, with the revival of atomic spectroscopy which followed, in the early fifties, the introduction of optical pumping by Kastler.[20] In 1974, Kastler himself wrote a commemorative paper on the 50th anniversary of the Hanle effect.[21]

With our experimental geometry, mercury atoms are excited into a coherent superposition of the two eigenstates of the atomic Hamiltonian with $m = +1$ and $m = -1$. If the separation between these levels is small compared to h/τ, i.e., if their energies overlap within their natural widths, coherent effects will be observable in the emitted fluorescence radiation. If we sweep the magnetic field intensity from negative through zero to positive values, we observe the bell-shaped Lorentz curves of Figure 5 for I_x and I_y. It will be seen in the next sections that one can evaluate the lifetime of the excited level from the widths of these curves, when the Landé factor of the level is known.

This coherence effect can occur whenever two levels overlap, within their natural widths, for a particular value of an external applied field. Typically, this overlap can occur when Zeeman sublevels, originating from different fine or hyperfine structure levels, separated at zero field, cross at a certain field intensity. This phenomenon is at the origin of the name *level crossing*, which is now well established. However, Hanle[13] has pointed out that this name is not rigorous, because the important phenomenon is not the *crossing*, but the *overlapping* of the levels. The expression *level overlapping* is found already in Ref. 2. The Hanle effect is also called *zero field level*

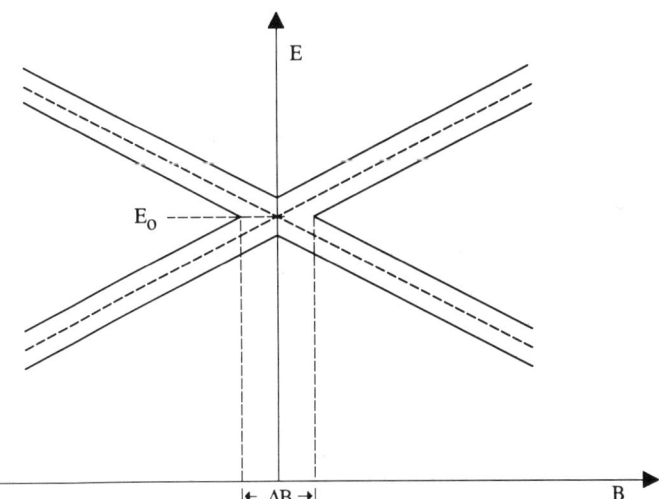

FIGURE 4. The case of two energy levels shifted with respect to each other by an external magnetic field. Coherence effects are possible as long as the spacing between the levels is smaller than, or of the order of, the natural width.[2,13]

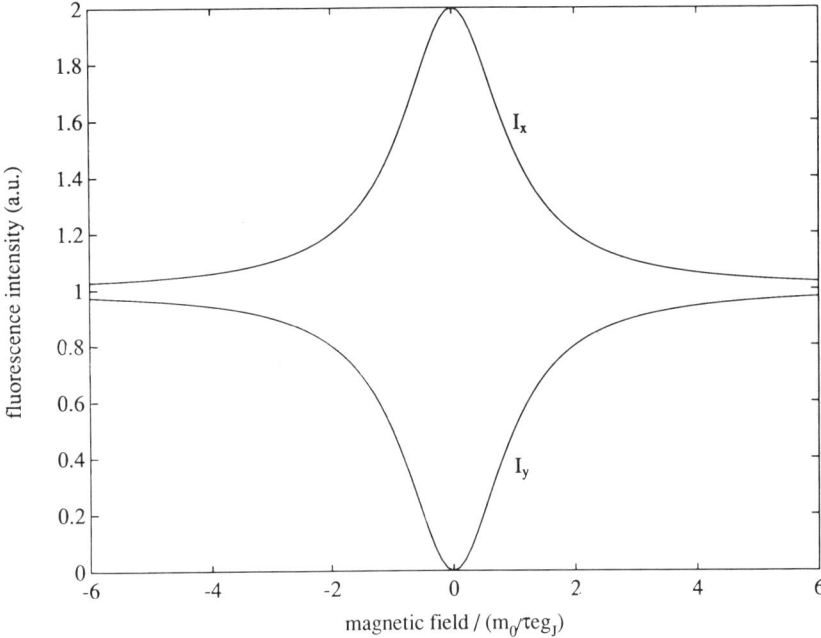

FIGURE 5. The Hanle signal. Intensities of the x and y linearly polarized components of the fluorescence light observed along the z axis. See Figure 1 for the experimental arrangement.

crossing because it is due to the crossing (and overlapping) at zero field of Zeeman sublevels, which are separated at positive and negative values of the magnetic field.

(Nonzero field) Level crossing was first used by Colegrove, Franken, Lewis, and Sands in 1959[22] to determine the 2^3P_1-2^3P_2 fine structure separation in helium. The physics underlying level crossing is the same as for the Hanle effect, but while the signal width is the only measured parameter for the latter, in a level-crossing experiment further information is gained from the field intensity at which the signal occurs. The level-crossing technique has allowed measurement of several atomic fine and hyperfine structures with great accuracy.

2. CLASSICAL INTERPRETATION OF THE HANLE EFFECT

In this section we shall develop the classical interpretation of the Hanle effect, carrying out the calculations explicitly only for the simple experimental geometry of Figure 1. The excitation light propagates along the y

axis and is linearly polarized with its electric field parallel to x. The x and y polarized components of the resonance fluorescence light emitted along z are monitored. We shall denote the unit vectors along the coordinate axes by $\hat{\mathbf{x}}, \hat{\mathbf{y}}$, and $\hat{\mathbf{z}}$. Throughout this chapter we shall use circumflexes to denote unit vectors.

The oscillating electric field of the incoming light excites oscillations of one of the optical electrons of an Hg atom along the x axis. We start by decomposing these oscillations into two counterrotating circular motions in the xy plane. Owing to the presence of the magnetic field $\mathbf{B}$ along z, the frequencies of the two circular motions are different (the classical Zeeman effect):

$$\omega_1 = \omega_0 + \omega_L \quad \text{and} \quad \omega_2 = \omega_0 - \omega_L \qquad (1)$$

where $\omega_L = g_J eB/(2m_0)$, e being the electron charge, m_0 the electron rest mass, g_J the Landé factor for the excited 6^3P_1 state of mercury, and B the intensity of the external magnetic field. Actually, in a strictly classical treatment we should have $\omega_L = eB/(2m_0)$, since the Landé factor g_J for a classical orbit is 1. However, we shall keep the factor g_J in our classical formulas, since it gives quantum validity to the formulas as long as hyperfine structures and nearly degenerate fine structures are absent.

If we denote by $1/\tau$ the damping rate of the fluorescence radiation, the damping rate of the electron oscillations is $\gamma = 1/2\tau$ because of the quadratic relation between electric field and radiation intensity, and the electron motion is described by the equations

$$x = R_0 \, e^{-2\gamma(t-t_0)} \cos[\omega_L(t-t_0)] \cos[\omega_0(t-t_0)]$$

$$y = R_0 \, e^{-2\gamma(t-t_0)} \sin[\omega_L(t-t_0)] \cos[\omega_0(t-t_0)] \qquad (2)$$

where t_0 is the time at which the electron was excited. These equations describe the *damped rosette* motion of an electron oscillating at the optical (actually UV for Hg) frequency ω_0, decaying with a time constant τ, and whose oscillation axis rotates at frequency ω_L in the xy plane. Quantity R_0 is the initial oscillation amplitude. The electric dipole moment $\mathbf{p}$ is obtained by multiplying the electronic charge e by the x and y components of the motion. The electric field generated by a dipole $\mathbf{p}$ in the wave zone $(r \gg \lambda)$ is

$$\mathbf{E} = \frac{1}{4\pi\varepsilon_0 c^2 r} \left[\frac{d^2 \mathbf{p}(t - r/c)}{dt^2} \times \hat{\mathbf{r}} \right] \times \hat{\mathbf{r}}$$

with the usual assumptions that the electron velocity is much smaller than c and that the linear dimensions of the electron trajectory are much smaller than the radiated wavelength. When we develop the second derivative of **p** with respect to time, we may neglect the terms in γ^2, ω_L^2, $\gamma\omega_L$, $\gamma\omega_0$, and $\omega_0\omega_L$ with respect to the much higher term in ω_0^2. This is equivalent to writing $\mathbf{p}(t)$ in the form $\mathbf{p}(t) = \mathbf{p}'(t)\cos[\omega_0(t-t_0)]$, with

$$\mathbf{p}'(t) = p_0 e^{-\gamma(t-t_0)}\{\hat{\mathbf{x}}\cos[\omega_L(t-t_0)] + \hat{\mathbf{y}}\sin[\omega_L(t-t_0)]\}$$

where $p_0 = eR_0$, and then considering $\mathbf{p}'(t)$ as a constant when performing the derivatives. Thus the electric field observed along z in the wave zone is obtained in the form

$$\mathbf{E} = \frac{\omega_0^2}{4\pi\varepsilon_0 c^2}\mathbf{p}'(t-z/c)\frac{\cos[kz-\omega_0(t-t_0)]}{z}$$

where $k = 2\pi/\lambda = \omega_0/c$ is the magnitude of the wave vector.

The intensity of the fluorescence light emitted along z, and observed through a polarizer which selects the linearly polarized component whose electric field forms an angle α with the x axis, is proportional to the square of the corresponding component of the electric field. If we consider the contribution of a single oscillator we get

$$I(\alpha, t) = T\frac{\omega_0^4 p_0^2}{32\pi^2\varepsilon_0 c^3} e^{-2\gamma(t-t_0)}\cos^2[\omega_L(t-t_0) - \alpha]$$

$$= T\frac{\omega_0^4 p_0^2}{64\pi^2\varepsilon_0 c^3} e^{-2\gamma(t-t_0)}\{1 + \cos[2(\omega_L(t-t_0) - 2\alpha)]\} \quad (3)$$

where T is the transmittance of the polarizer. This damped modulation of the fluorescence intensity at twice the Larmor frequency was later observed in time-resolved experiments.[23-25] In a steady-state experiment, the atoms can be considered as having been excited at a constant rate R since $t_0 = -\infty$ up to the time of observation t. Thus, in order to obtain the steady-state signal, we must multiply equation (3) by R and integrate it over the excitation time t_0 from $-\infty$ to t. The result is

$$I(\alpha) = NRT\frac{p_0^2\omega_0^4}{128\pi^2\varepsilon_0 c^3}\left[\frac{1}{\gamma} + \frac{\gamma\cos(2\alpha)}{\gamma^2 + \omega_L^2} + \frac{\sin(2\alpha)}{\gamma^2 + \omega_L^2}\right] \quad (4)$$

where N is the number of atoms whose radiation reaches the detector. Equation (4) gives the shape of the Hanle signal for the experimental geometry of Figure 1 as a function of the angle α between the axis of the polarizer and the x axis, and of the applied static field, since $\omega_L = g_J eB/(2m_0)$. According to equation (4), if we sweep the static field B from negative to positive values we observe a constant bias plus (minus) a bell-shaped Lorentzian curve with its maximum (minimum) at $B = 0$ (Figure 4) if $\alpha = \pm\pi/4$ or $\alpha = \pm 3\pi/4 [\cos(2\alpha) = 0]$. For $\alpha = 0$ or $\alpha = \pm\pi/2$ $[\sin(2\alpha) = 0]$ we observe the same bias plus (minus) a dispersion-shaped Lorentzian curve centered at $B = 0$. For angles which are not integer multiples of $\pi/4$ superpositions of bell-shaped and dispersion-shaped Lorentzians are observed. The polarization degree is

$$P = \frac{I_x - I_y}{I_x + I_y} = \frac{1}{1 + (g_J e\tau/m_0 c)^2 B^2} \qquad (5)$$

i.e., an absorption-shaped Lorentz curve as a function of B. If we use an arbitrary experimental geometry we shall observe a dispersion-shaped or an absorption-shaped Lorentz curve, a superposition of the two, or no field dependence at all, according to the particular geometry. Usually the geometry of the experiment is arranged in order to observe absorption-shaped signals. The full width at half the maximum intensity is $\Delta B = 2m_0/(eg_J\tau)$. From its measurement it is thus possible to determine the time damping constant of the oscillators which, in the quantum theory, corresponds to the lifetime of the excited state

$$\tau = 2m_0/eg_J\Delta B$$

provided that the Landé factor g_J is known. An important feature of the Hanle effect is its independence of the Doppler width of the optical line. In fact, the detection system is equally sensitive to all frequencies within the line profile and, since the global fluorescence intensity is detected as a function of the static field intensity, no attempt is made to resolve the shape of the optical line.

Thus, classically, the Hanle effect observed with the experimental geometry of Figure 1 is due to the precessional motions of the damped oscillating dipoles in a static magnetic field B. The fluorescence light observed along the z axis is linearly polarized parallel to the excitation light polarization when $B = 0$ (linear oscillations of the dipoles), the polarization degree decreases monotonically for increasing absolute values of B, and becomes completely unpolarized for $B \gg m_0/(g_J e\tau)$. If one of the linearly polarized fluorescence light components, or the polarization degree, is measured as a function of the applied field intensity, the damping constant of the oscillators is easily obtained.

3. QUANTUM MECHANICAL INTERPRETATION OF THE HANLE EFFECT

In this section, and in the next two, we shall sketch a semiclassical, rather than a full quantum mechanical, treatment of the Hanle effect. It is known that the semiclassical treatment of radiation-matter interactions practically always leads to qualitatively and quantitatively correct results. The quantization of the electromagnetic field is indispensable only for the treatment of a few special cases like, for instance, spontaneous emission.[26] Of course, spontaneous emission plays a determinant role in the Hanle effect but, within the frame of our semiclassical treatment, it is possible to deal with it by means of phenomenological rate equations. For the sake of clarity, in this section we shall again restrict ourselves to the case of the 253.7 nm line of an even isotope of mercury ($I = 0$), and to the experimental geometry of Figure 1. The more general case of an arbitrary transition and an arbitrary experimental geometry is left to Section 4.

The Heisenberg diagram of the $6\,^1S_0$-$6\,^3P_1$ transition, corresponding to the 253.7 nm resonance line, is shown in Figure 6. Let us choose the field direction z as quantization axis. Atoms excited into the $|m = +1\rangle$ ($|m = -1\rangle$) level decay into the $|\mu = 0\rangle$ ground state by emitting, along z, circularly polarized σ^+ (σ^-) radiation. The probability per unit time for an atom in an excited state $|m\rangle$ to decay into a sublevel $|\mu\rangle$ of the ground state by emitting a photon with polarization vector $\hat{\mathbf{e}}$ in the direction z within the solid angle $d\Omega$ is

$$W_{m\mu}(\hat{\mathbf{e}})\, d\Omega = \frac{\omega_0^3}{8\pi^2 \varepsilon_0 \hbar c^3} |\langle m|\hat{\mathbf{e}} \cdot \mathbf{P}|\mu\rangle|^2 \, d\Omega \tag{6}$$

where $\mathbf{P}$ is the atomic dipole moment operator.

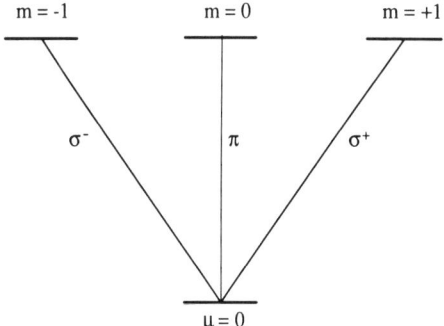

FIGURE 6. The Heisenberg diagram of the $6\,^1S_0$-$6\,^3P_1$ transition of mercury.

In order to understand the Hanle effect, we shall consider two different cases of excitation of the $|m = +1\rangle$ and $|m = -1\rangle$ levels. In the first case, which we shall label as *incoherent excitation*, the two levels are excited independently of each other, and each excited atom is either in $|m = +1\rangle$ or in $|m = -1\rangle$. In the second case, which corresponds to the experimental arrangement of Figure 1 and which we shall label as *coherent excitation*, the atoms are excited into a coherent superposition of the two levels.

For the incoherent case, we assume the presence of an excitation mechanism which maintains constantly $N/2$ atoms in the pure $|m = +1\rangle$ and $N/2$ atoms in the pure $|m = -1\rangle$ excited states (we are speaking of dynamical equilibrium, of course). The fluorescence intensity emitted along z and detected by a photomultiplier located after a polarizer which selects the polarization vector $\hat{\mathbf{e}}$ is

$$I(\hat{\mathbf{e}}) = \frac{A\omega_0^4}{2\pi c^3}[|\langle m = +1|\hat{\mathbf{e}} \cdot \mathbf{P}|\mu = 0\rangle|^2 + |\langle m = -1|\hat{\mathbf{e}} \cdot \mathbf{P}|\mu = 0\rangle|^2] \quad (7)$$

where A is a constant depending on the density of the excited atoms in the cell, on the transmittance of the polarizer, and on the solid angle covered by the photomultiplier. The intensity has been obtained by multiplying the rate of incoming photons by the energy $\hbar\omega_0$ of the single photon. The fluorescence light observed in the z direction is completely unpolarized. If we apply an external magnetic field in this case, no field dependence of the polarization of the fluorescence light (no Hanle effect) is observed.

Let us now consider the coherent excitation case. If we choose the direction of $\mathbf{B}$ as the quantization axis, the linearly polarized excitation light excites the Hg atoms, initially in the $|\mu = 0\rangle$ ground state, into a coherent superposition $a|m = -1\rangle + b|m = +1\rangle$. For our experimental geometry, we have $a = b = 1/\sqrt{2}$. The probability per unit time for an excited atom to decay into the ground state $|\mu = 0\rangle$ by emitting a photon of polarization $\hat{\mathbf{e}}$ along the z axis within the solid angle $d\Omega$ is

$$W(\hat{\mathbf{e}})\, d\Omega = \frac{\omega_0^3}{8\pi^2\varepsilon_0 \hbar c^3}|[a\langle m = -1| + b\langle m = +1|]\hat{\mathbf{e}} \cdot \mathbf{P}|\mu = 0\rangle|^2\, d\Omega$$

or, if we develop the square modulus,

$$W(\hat{\mathbf{e}})\, d\Omega = \frac{\omega_0^3}{8\pi^2\varepsilon_o \hbar c^3}[|a\langle m = -1|\hat{\mathbf{e}} \cdot \mathbf{P}|\mu = 0\rangle|^2 + |b\langle m = +1|\hat{\mathbf{e}} \cdot \mathbf{P}|\mu = 0\rangle|^2$$

$$+ 2\text{Re}\{ab^*\langle m = -1|\hat{\mathbf{e}} \cdot \mathbf{P}|\mu = 0\rangle$$

$$\times \langle\mu = 0|\mathbf{P} \cdot \hat{\mathbf{e}}^*|m = +1\rangle\}]\, d\Omega \quad (8)$$

The last term within the square brackets of equation (8) has no equivalent in equation (7), and is responsible for the Hanle effect. Let us now denote by γ the common decay rate of the $|m = +1\rangle$ and $|m = -1\rangle$ excited states, by E_{+1} and E_{-1} their respective energies, and by E_g the energy of the $|\mu = 0\rangle$ ground state. We have

$$E_{\pm 1} = E \pm \mu_0 g_J B = E \pm \hbar \omega_L$$

where E and g_J are the zero field energy and the Landé factor of the $6\,^3P_1$ state, respectively, and B is the external static field. The average contribution of a single atom to the signal becomes

$$W(\hat{\mathbf{e}})\,d\Omega = \frac{e^{-2\gamma(t-t_0)}\omega_0^3}{8\pi^2\varepsilon_0 \hbar c^3}\left[\tfrac{1}{2}|\langle m = -1|\hat{\mathbf{e}}\cdot\mathbf{P}|\mu = 0\rangle|^2 + \tfrac{1}{2}|\langle m = +1|\hat{\mathbf{e}}\cdot\mathbf{P}|\mu = 0\rangle|^2\right.$$

$$+ 2\,\mathrm{Re}\{\tfrac{1}{2}e^{i2\omega_L(t-t_0)}\langle m = -1|\hat{\mathbf{e}}\cdot\mathbf{P}|\mu = 0\rangle$$

$$\left.\times \langle \mu = 0|\mathbf{P}\cdot\hat{\mathbf{e}}^*|m = +1\rangle\}\right]\,d\Omega \qquad (9)$$

The quantum analog of equation (3) is obtained by remembering the relations between the unit vectors $\hat{\mathbf{x}}, \hat{\mathbf{y}}$, and $\hat{\mathbf{z}}$ parallel to the Cartesian axes, and the spherical unit vectors $\hat{\varepsilon}_{-1}, \hat{\varepsilon}_0$, and $\hat{\varepsilon}_{+1}$:

$$\hat{\varepsilon}_{-1} = (\hat{\mathbf{x}} - i\hat{\mathbf{y}})/\sqrt{2}, \qquad \hat{\varepsilon}_{+1} = -(\hat{\mathbf{x}} + i\hat{\mathbf{y}})/\sqrt{2}, \qquad \hat{\varepsilon}_0 = \hat{\mathbf{z}} \qquad (10)$$

In spherical coordinates, the scalar product between two vectors $\mathbf{A}$ and $\mathbf{B}$ is written

$$\mathbf{A}\cdot\mathbf{B} = \sum_{m=-1}^{+1}(-1)^m A_m B_{-m} \qquad (11)$$

With our choice of the quantization axis, waves with $\hat{\varepsilon}_{-1}, \hat{\varepsilon}_0$, and $\hat{\varepsilon}_{+1}$ polarization are related to $\Delta m = -1\ (\sigma^-)$, $\Delta m = 0\ (\pi)$, and $\Delta m = +1\ (\sigma^+)$ transitions, respectively.

If our polarizer is arranged so that the photomultiplier detects the linear polarized component of the fluorescence light whose electric vector forms an angle α with the x axis, the polarization unit vector $\hat{\mathbf{e}}$ which must be introduced into equation (9) is

$$\hat{\mathbf{e}} = -\hat{\varepsilon}_{+1}(\cos\alpha - i\sin\alpha)/\sqrt{2} + \hat{\varepsilon}_{-1}(\cos\alpha + i\sin\alpha)/\sqrt{2} \qquad (12)$$

The nonzero matrix elements of the spherical components of the vector operator **P** are particularly simple for the case of an even isotope of mercury ($J = 1$ for the excited state, $J = 0$ for the ground state), i.e.,

$$\langle \mu = 0|P_0|m = 0\rangle = -\langle \mu = 0|P_{-1}|m = +1\rangle = -\langle \mu = 0|P_{+1}|m = -1\rangle = \langle \mathbf{P}\rangle \quad (13)$$

where $\langle \mathbf{P}\rangle$ is the reduced matrix element $\langle 6\,^1S_0||\mathbf{P}||6\,^3P_1\rangle$. We can now evaluate the matrix elements appearing in equation (9):

$$|\langle m = -1|\hat{\mathbf{e}} \cdot \mathbf{P}|\mu = 0\rangle|^2 = |\langle m = +1|\hat{\mathbf{e}} \cdot \mathbf{P}|\mu = 0\rangle|^2 = |\langle \mathbf{P}\rangle|^2/2$$

$$\langle m = -1|\hat{\mathbf{e}} \cdot \mathbf{P}|\mu = 0\rangle\langle \mu = 0|\mathbf{P} \cdot \hat{\mathbf{e}}^*|m = +1\rangle = |\langle \mathbf{P}\rangle|^2\, e^{-2i\alpha}/2$$

Thus the contribution of a single atom to the signal is

$$W(\alpha)\,d\Omega = \frac{e^{-2\gamma(t-t_0)}\omega_0^4}{16\pi^2\varepsilon_0 c^3}|\langle \mathbf{P}\rangle|^2\{1 + \cos[2\omega_L(t - t_0) - 2\alpha]\}\,d\Omega \quad (14)$$

which is the quantum analog of equation (3). In a time-resolved experiment (23-25) a short light pulse (short compared to $1/\omega_L$) simultaneously excites a certain number N of Hg atoms into the $(|m = -1\rangle + |m = +1\rangle)/\sqrt{2}$ state at time t_0, so that all the excited atoms subsequently evolve in phase. In these conditions the observed collective signal is a damped light pulse modulated at twice the Larmor frequency ω_L. This is a particular case of a quantum beat signal. It will be shown in Chapter 7 that the *quantum beat laser* and the *Hanle laser* are actually two sides of the same coin.

For a steady-state experiment, the quantum analog of equation (4) is again obtained, as in the classical treatment, by remembering that now the atoms are continuously excited into the $(|m = -1\rangle + |m = +1\rangle)/\sqrt{2}$ state at a certain rate R, and integrating over t_0 from $-\infty$ to the time of observation t. In these conditions no phase relations exist between the time evolutions of the excited atoms, and no modulation of the signal is observed.

Thus, quantum mechanically, the Hanle effect is due to the excitation of the atom (or molecule) into a coherent superposition of two or more sublevels of the excited state. The observed signal is due to the presence of the coherent term in the development of the square modulus of the matrix element for the fluorescence emission. The coherent superposition can be considered an approximate eigenstate of the atomic Hamiltonian as long as the superposed sublevels overlap within their natural linewidths. Since the separation between the Zeeman sublevels depends on the intensity of the applied magnetic field **B**, and the natural linewidth is the reciprocal of the lifetime of the level, the determination of the dependence of the Hanle signal on B leads to the determination of the lifetime of the excited state.

4. THE DENSITY MATRIX FORMALISM FOR THE HANLE EFFECT (BROAD-BAND EXCITATION)

The use of the density matrix formalism is the most convenient approach to the treatment of the Hanle effect in arbitrary experimental conditions.[27] Actually, the density matrix formalism can be regarded as a general approach to the prediction of the result of any experiment performed on a quantum mechanical system. Any observable physical quantity is represented, in quantum mechanics, by an operator $\mathscr{G}$. If the density matrix ρ describing the system is known, the expectation value for an experimental measurement of the physical observable is $\text{Tr}(\rho\mathscr{G})$. Thus, in principle, the result of any measurement can be predicted, provided that one is able to evaluate ρ. Very often this is, however, absolutely not a simple task.

In arbitrary experimental conditions, the cell containing the atomic vapor will be located in a uniform static field $\mathbf{B}$ along which we choose the quantization axis z (Figure 7). The vapor will be excited by light of arbitrary polarization $\hat{\mathbf{e}}_e$, whose propagation direction $\hat{\mathbf{r}}_e$ is defined by the Euler angles θ_e and φ_e. A photomultiplier detects the fluorescence light emitted along a direction $\hat{\mathbf{r}}_d$ defined by the Euler angles θ_d and φ_d. In front of the photomultiplier a polarizer selects the polarization vector $\hat{\mathbf{e}}_d$. The excited

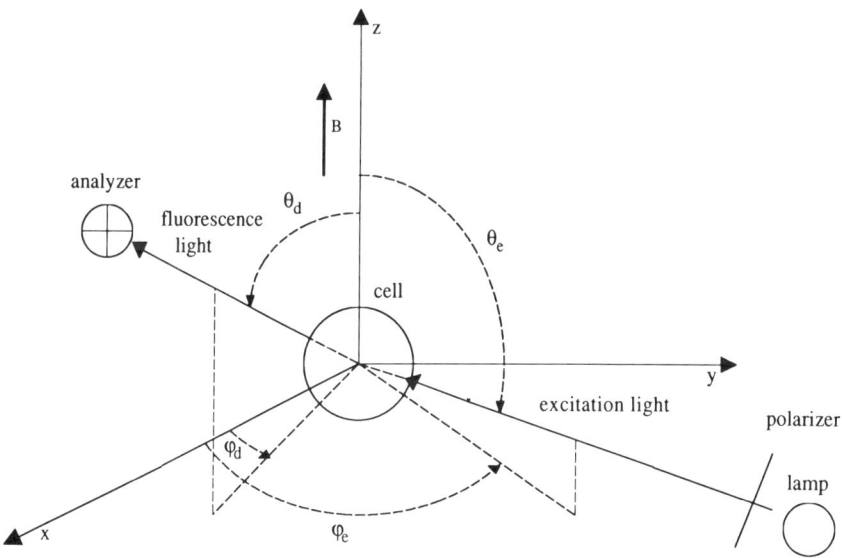

FIGURE 7. Hanle experiment with arbitrary geometry of the experimental arrangement. The excitation lines propagates toward the cell along a direction defined by the Euler angles θ_e and φ_e, while the fluorescence light emitted along (θ_d, φ_d) is detected. The analyzer consists of a photomultiplier preceded by a polarizer which selects the polarization vector $\hat{\mathbf{e}}$.

state will have an arbitrary electronic angular momentum J_e with Zeeman sublevels $|m\rangle$, the ground state an arbitrary electronic angular momentum J_g with Zeeman sublevels $|\mu\rangle$; of course $|J_e - J_g| \leq 1$. Our treatment will be valid also in the presence of a nonzero nuclear spin I; in this case the quantum number J should be replaced by the quantum number F, and the magnetic quantum numbers m, m' by m_F, m'_F throughout all the formulas.

Much of the following will be valid for experiments on both atomic and molecular vapors, so that the words *molecule* and *molecular* can be substituted for the words *atom* and *atomic* practically everywhere throughout the text. Of course, some care must be taken. For instance, while usually only the ground state and the state connected to it by the excitation radiation are populated appreciably in an atomic system, this is usually not the case for a molecular vapor. In fact, the average spacing between neighboring molecular levels is usually much smaller than kT. Another difference is that molecules can have permanent electric dipole moments, so that the level splitting in a polar molecule can be originated by an electric field, while the Stark effect is very small for all atoms but hydrogen. However, the effect of the Stark splitting on the polarization of the resonance fluorescence light of mercury was already observed by Hanle himself in 1925.[28] More recently, the observation of the *electric Hanle effect* on the Lyman-α line has been regarded by Favati, Landi degl'Innocenti, and Landolfi[29] as a possible method for measuring electric fields in astrophysical plasmas. These authors propose the name *second Hanle effect* for the Hanle effect in an electrostatic field. This subject will be treated more extensively in Chapter 5.

The Hamiltonian of our system can be decomposed into the sum of a static Hamiltonian $\mathcal{H}_0$, and two operators $\mathcal{H}_1$ and $\mathcal{H}_2$ describing the couplings with the electromagnetic field and with the nonradiative relaxation mechanisms, respectively. The Liouville equation for the density matrix ρ describing the system is

$$\frac{d\rho}{dt} = \frac{1}{i\hbar}[\mathcal{H}_0, \rho] + \frac{1}{i\hbar}[\mathcal{H}_1, \rho] + \frac{1}{i\hbar}[\mathcal{H}_2, \rho] \qquad (15)$$

Quantity $\mathcal{H}_0$ is itself the sum of the unperturbed atomic Hamiltonian $\mathcal{H}_{at}$, which describes the unperturbed atom, and of the Zeeman Hamiltonian $\mathcal{H}_Z = -\boldsymbol{\mu} \cdot \mathbf{B}$, which describes the interaction with the external static magnetic field. The difficulties for an exact solution of equation (15) arise from the fact that the atomic system is not closed with respect to the Hamiltonians $\mathcal{H}_1$ and $\mathcal{H}_2$. The quantity $\mathcal{H}_1$ is responsible for *excitation* and *induced emission*, which can both be described accurately by the semiclassical theory, and for *spontaneous emission* which, being due to coupling of the atomic system with the zero-point fluctuations of the vacuum radiation fields, requires second quantization for a rigorous treatment; $\mathcal{H}_2$ represents the various collisional processes. Usually, it is not possible to express $\mathcal{H}_2$ in a rigorous way, since the closed system should include, for instance, the cell walls. However, it is normally possible to describe accurately the effects of the nonradiative relaxation mechanisms by means of rate equations.

A rigorous treatment of the effects due to the presence of $\mathcal{H}_1$ is possible by quantizing the field and then writing a global Hamiltonian for the system *atoms + photons*, as first proposed by Dirac,[30] and later developed by Haroche in the dressed-atom formalism for the treatment of the interaction between atoms and radio-frequency photons.[31] This method has been used for treating the Hanle effect excited by laser radiation. A dressed-atom approach to the atom in the laser radiation field can be found in a paper[32] which, however, is essentially devoted to the problem of the atomic motion. A discussion of the Hanle effect excited by laser light is found in Ref. 33. In the next section we shall briefly discuss the laser excitation of the Hanle effect for the simple three-level case at low laser intensities, while nonlinear phenomena are left to Chapter 4.

In the present section, we shall limit ourselves to the case of excitation by means of a resonance lamp (broad-band excitation). A resonance lamp has very small temporal and spatial coherences, so that it cannot give origin to an ensemble electric dipole moment of the sample oscillating at the optical frequency of the incident radiation. Thus all off-diagonal elements of the density matrix $\rho_{\mu m}$ connecting ground-state Zeeman sublevels $|\mu\rangle$ with excited-state Zeeman sublevels $|m\rangle$ are identically zero, and we say that the density matrix possesses no optical coherence. In these conditions, both excitation and spontaneous emission are accurately described by the rate equations which we shall introduce below, and induced emission can be neglected with respect to spontaneous emission.

Since we have no optical coherence ($\rho_{\mu m} \equiv 0$), and since, for an atomic vapor, all states other than the ground state and the excited state connected to it by the excitation radiation are practically unpopulated, the density matrix is factorized into two principal minors $\rho_{\mu \mu'}$ and $\rho_{mm'}$, describing the ground and excited states, respectively.

In our experiment we detect the fluorescence light, which is due to spontaneous emission. We shall thus consider only the evolution of the principal minor $\rho_{mm'}$, describing the excited state. In the most common experimental conditions for a Hanle experiment, the nonradiative relaxation mechanisms which affect $\rho_{mm'}$ are negligible with respect to spontaneous emission. The time evolution of $\rho_{mm'}$ can thus be described by a Liouville equation analogous to equation (15) where, however, the commutator with $\mathcal{H}_2$ is absent and the commutator with $\mathcal{H}_1$ is replaced by two phenomenological terms,

$$\frac{d\rho}{dt} = \frac{1}{i\hbar}[\mathcal{H}_0, \rho] + \frac{d^{(1)}}{dt}\rho + \frac{d^{(2)}}{dt}\rho \tag{16}$$

The first phenomenological derivative on the right-hand side of equation (16) describes the rates at which the elements of the excited-state density matrix are created. Neglecting coherence transfer from the ground to the excited state, these rates are

$$\frac{d^{(1)}}{dt}\rho_{mm'} = \frac{\pi u(\omega_0)}{\varepsilon_0 \hbar^2} \sum_\mu \langle m|\hat{\mathbf{e}}_e \cdot \mathbf{P}|\mu\rangle\langle\mu|\mathbf{P}\cdot\hat{\mathbf{e}}_e^*|m'\rangle \rho_{\mu\mu} \tag{17}$$

where the sum is extended to all the ground-state sublevels μ; $u(\omega_0)$ is the spectral density of the incident radiation, $\mathbf{P}$ is the atomic electric dipole operator, and $\hat{\mathbf{e}}_e$ is the polarization unit vector of the excitation light. The matrix elements appearing in equation (17) can be easily evaluated by expanding $\hat{\mathbf{e}}_e$ in terms of the spherical unit vectors $\hat{\varepsilon}_{-1}$, $\hat{\varepsilon}_0$, and $\hat{\varepsilon}_{+1}$, which were encountered in the preceding section. Transitions with $m - \mu = -1$, 0, and +1 are respectively excited by the $\hat{\varepsilon}_{-1}$, $\hat{\varepsilon}_0$, and $\hat{\varepsilon}_{+1}$ components.

We shall soon see that the Hanle effect is observed only in the presence of nonzero off-diagonal elements $\rho_{mm'}$, the diagonal elements ρ_{mm} giving origin only to a field-independent background of resonance fluorescence. Thus the excitation light must excite the atoms into a superposition of different Zeeman sublevels. This is possible only if the polarization vector $\hat{\mathbf{e}}_e$ of the excitation light is a superposition of at least two of the spherical unit vectors $\hat{\varepsilon}_{-1}$, $\hat{\varepsilon}_0$, nad $\hat{\varepsilon}_{+1}$, i.e., the polarization of the excitation light must be coherent, and a coherence is transferred from radiation to the atoms. If we label E_m and $E_{m'}$ the energies of the $|m\rangle$ and $|m'\rangle$ Zeeman sublevels of the excited state, $\rho_{mm'}$ evolves at the angular frequency $(E_m - E_{m'})/\hbar$. Since $(E_m - E_{m'})/\hbar$ lies in the radio-frequency spectral range, we speak of *radio-frequency* or *Hertzian* coherence. A shorthand form for equation (17) is

$$\frac{d^{(1)}}{dt}\rho_{mm'} = \frac{\pi u(\omega_0)}{\varepsilon_0 \hbar^2} \sum_\mu F_{\mu\mu mm'} \rho_{\mu\mu} \tag{18}$$

where, according to Corney,[27] we have defined the excitation matrix F by

$$F_{\mu\mu mm'} = \langle m|\hat{\mathbf{e}}_e \cdot \mathbf{P}|\mu\rangle\langle\mu|\mathbf{P}\cdot\hat{\mathbf{e}}_e^*|m'\rangle \tag{19}$$

The second phenomenological derivative on the right-hand side of equation (16) describes the effect of spontaneous emission. Since spontaneous emission is a random process, it can be represented as a rate process,

$$\frac{d^{(2)}}{dt}\rho_{mm'} = -2\gamma \rho_{mm'} \tag{20}$$

where $2\gamma = 1/\tau$ is the common spontaneous decay rate for all the $|m\rangle$ sublevels due to all possible decay channels.

Equation (16) now becomes

$$\frac{d\rho}{dt}\rho_{mm'} = \frac{1}{i\hbar}[\mathcal{H}_0, \rho]_{mm'} + \frac{\pi u(\omega_0)}{\varepsilon_0 \hbar^2} \sum_\mu F_{\mu\mu mm'}\rho_{\mu\mu} - 2\gamma\rho_{mm'} \qquad (21)$$

Time-resolved experiments have been considered in the preceding section. Here we are interested in steady-state experiments, and must therefore look for a steady-state solution of equation (21), under the assumption that the excitation light is time-independent. We remember that $\mathcal{H}_0$ is diagonal on the $|m\rangle$ basis and, in the absence of hyperfine structure, its eigenvalues for the Zeeman sublevels of interest are

$$\langle m|\mathcal{H}_0|m\rangle = E_m = E + g_J\mu_0 B J_z = E + \hbar\omega_L m \qquad (22)$$

where E is the unperturbed energy of the excited state. If we set $d\rho/dt = 0$ in equation (21), the solution is

$$\rho_{mm'} = \frac{\pi u(\omega_0)}{\varepsilon_0 \hbar^2} \sum_\mu \frac{F_{\mu\mu mm'}\rho_{\mu\mu}}{2\gamma + i\omega_L(m - m')} \qquad (23)$$

As already stated in this section, equation (23) shows that the diagonal elements ρ_{mm} do not depend on the magnetic field, and thus do not contribute to the Hanle signal. The off-diagonal elements $\rho_{mm'}$, which display a magnetic-field dependence and thus contribute to the Hanle signal, can be significantly different from zero only when their angular frequency $\omega_L(m - m')$ is comparable to or smaller than the relaxation rate 2γ. According to the uncertainty principle, $\gamma = 1/\tau$ determines the width of the excited-state sublevels, and equation (23) states that there can be appreciable Hertzian coherence only between overlapping levels, i.e., between levels whose separation is of the order of, or smaller than, their homogeneous linewidth.

In order to evaluate the observed signal we must build the quantum mechanical operator $\mathcal{L}$ corresponding to the observed physical quantity, multiply it by the density matrix ρ, and evaluate the trace of their product. We are monitoring the fluorescence light emitted in the (θ_d, φ_d) direction within the solid angle $d\Omega$. A polarizer in front of the detector selects the fluorescence light whose polarization vector is $\hat{\mathbf{e}}_d$. The operator $\mathcal{L}$ is thus given by

$$\mathcal{L} = \frac{\omega_0^4 \, d\Omega}{8\pi^2 \varepsilon_0 c^3} \sum_{\mu'} \hat{\mathbf{e}}_d \cdot \mathbf{P}|\mu'\rangle\langle\mu'|\mathbf{P} \cdot \hat{\mathbf{e}}_d^* \qquad (24)$$

where, as usual, the photon rate has been multiplied by $\hbar\omega_0$ in order to obtain the energy radiated per unit time. The matrix elements of $\mathscr{L}$ are

$$\mathscr{L}_{mm'} = \frac{\omega_0^4 \, d\Omega}{8\pi^2 \varepsilon_0 c^3} \sum_{\mu'} G_{mm'\mu'\mu'} \tag{25}$$

where $G_{mm'\mu'\mu'} = \langle m|\hat{\mathbf{e}}_d \cdot \mathbf{P}|\mu'\rangle\langle\mu'|\mathbf{P} \cdot \hat{\mathbf{e}}_d^*|m'\rangle$ is the emission matrix. The observed signal $\mathrm{Tr}(\rho\mathscr{L})$ is

$$dI = \frac{u(\omega_0)\omega_0^4}{8\pi\varepsilon_0^2\hbar^2 c^3} \sum_{\substack{mm'\\ \mu\mu'}} \frac{F_{\mu\mu mm'} G_{m'm\mu'\mu'}}{2\gamma + i(m-m')\omega_\mathrm{L}} \rho_{\mu\mu} \, d\Omega \tag{26}$$

which is known as the Breit–Franken formula, from the works by Breit[18] and Franken.[34] If the appropriate values are inserted for the various matrix elements, equation (26) describes the Hanle signal, i.e., the intensity of a certain polarized component of the fluorescence light, as a function of the applied static magnetic field, for any experimental geometry. We should remember, however, that this equation has been derived under the assumptions of no optical coherences, of no coherence transfer between the lower and upper states, and of a common decay rate for all the sublevels of the upper state. From the study of an experimental Hanle signal it is thus possible to determine γ, i.e., the inverse of the lifetime of the excited state. As already discussed in the classical treatment, this measurement is not affected by the Doppler broadening.

Equation (26) remains valid in the presence of hyperfine structure. In this case the quantum numbers m and μ must be understood as the projections of the total angular momentum $\mathbf{F}$ along the quantization axis, and the Larmor frequency ω_L is

$$\omega_\mathrm{L} = g_F \mu_B B / \hbar \tag{27}$$

where, to a good approximation,

$$g_F = g_J \frac{F(F+1) + J(J+1) - I(I+1)}{2F(F+1)} \tag{28}$$

Thus, except for the case $J = I$, the different hyperfine levels have different g_F values, and the widths ΔB of the corresponding signals are different. Usually it is not possible to excite the atoms to a single hyperfine level, and the observed signal is a superposition of the signals corresponding to the single hyperfine levels. Since it is difficult to derive an accurate lifetime value in these conditions, experiments on isotopes with $I = 0$ are to be preferred whenever possible.

5. LASER EXCITATION AND PRESSURE-INDUCED COHERENCES

The Hanle effect in the presence of laser excitation requires a more rigorous density matrix treatment than the one developed in the preceding section for broad-band excitation. Laser excitation, in fact, does induce optical coherences, and the corresponding matrix elements $\rho_{\mu m}$ cannot be neglected. As a simple example, let us consider the case of the V-shaped three-level system of Figure 8. The two excited sublevels are labeled $|+\rangle$ and $|-\rangle$, and their energies above the ground state are denoted by $\hbar\omega_+$ and $\hbar\omega_-$, respectively. The ground state is labeled $|0\rangle$. The laser frequency is ω, and $\Delta\omega = \omega - \omega_0$ is its detuning from the center of the doublet $\omega_0 = (\omega_+ + \omega_-)/2$. The spacing between $|+\rangle$ and $|-\rangle$ is denoted by $2\delta\omega$. Assuming a linear Zeeman effect, $\delta\omega$ is proportional to the strength of the applied magnetic field, and ω_0 equals the absorption frequency at zero field.

For simplicity, we shall assume again the experimental geometry of Figure 1, and we shall first consider homogeneously broadened lines. The

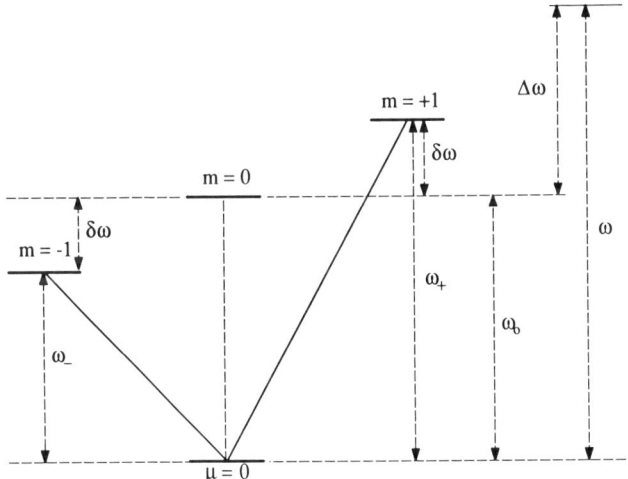

FIGURE 8. Laser excitation of the Hanle effect. For the sake of simplicity, a transition with a Heisenberg diagram analogous to Figure 6 is considered. The exciting laser beam is assumed linearly σ polarized, so that the system behaves as a three-level system. $\omega_+ = \omega_0 + \delta\omega$ and $\omega_- = \omega_0 - \delta\omega$ are the energies (in units of $\hbar$) of the two excited levels. In this case $\omega_0 = (\omega_+ + \omega_-)/2$ coincides with the energy of the unperturbed, and unaffected, level $|m = 0\rangle$. The laser frequency is ω and $\Delta\omega = \omega - \omega_0$.

time evolution of the density matrix is described by the Liouville equation (15), where the density matrix ρ and $\mathcal{H}_0$ now have the forms

$$\rho = \begin{bmatrix} \rho_{++} & \rho_{+0} & \rho_{+-} \\ \rho_{0+} & \rho_{00} & \rho_{0-} \\ \rho_{-+} & \rho_{-0} & \rho_{--} \end{bmatrix} \quad \text{and} \quad \mathcal{H}_0 = \begin{bmatrix} \hbar\omega_+ & 0 & 0 \\ 0 & 0 & 0 \\ 0 & 0 & \hbar\omega_- \end{bmatrix} \quad (29)$$

In a semiclassical treatment we can split $\mathcal{H}_1$, which in equation (15) is responsible for excitation, and both induced and spontaneous emission, into the sum of two operators. The first one is the interaction Hamiltonian $\mathcal{V}$, which describes the interaction with the electromagnetic field of the incoming wave. The second operator is responsible for the spontaneous emission and will be incorporated into a phenomenological term $\Gamma\rho$, where Γ is a superoperator (an operator operating on operators), describing both radiational and collisional decays. Thus, equation (15) becomes

$$\frac{d\rho}{dt} = -\frac{i}{\hbar}[\mathcal{H}_0 + \mathcal{V}, \rho] + \Gamma\rho \quad (30)$$

For the experimental geometry of Figure 1, $\mathcal{V}$ is written as (see Section 2 in Chapter 4)

$$\mathcal{V} = \begin{bmatrix} 0 & \hbar W e^{-i\omega t} & 0 \\ \hbar W e^{i\omega t} & 0 & \hbar W e^{i\omega t} \\ 0 & \hbar W e^{-i\omega t} & 0 \end{bmatrix} \quad (31)$$

where W is a real quantity defined by

$$W = -\frac{\sqrt{I}}{\hbar\sqrt{2c\varepsilon_0}}\langle +|\hat{\mathbf{e}} \cdot \mathbf{P}|0\rangle = -\frac{\sqrt{I}}{\hbar\sqrt{2c\varepsilon_0}}\langle -|\hat{\mathbf{e}} \cdot \mathbf{P}|0\rangle = -\frac{\sqrt{I}}{2\hbar\sqrt{c\varepsilon_0}}\langle\mathbf{P}\rangle \quad (32)$$

In equation (32) I and $\hat{\mathbf{e}}$ are the intensity and the unit polarization vector of the excitation light beam, respectively, and $\langle\mathbf{P}\rangle$ is the reduced matrix element of the transition. The term $\Gamma\rho$ can be written in the general form

$$\Gamma\rho = \begin{bmatrix} -2\gamma_p\rho_{++} & -\gamma_o\rho_{+0} & -2\gamma_c\rho_{+-} \\ -\gamma_o\rho_{0+} & 2\gamma_p(\rho_{++} + \rho_{--}) & -\gamma_o\rho_{0-} \\ -2\gamma_c\rho_{-+} & -\gamma_o\rho_{-0} & -2\gamma_p\rho_{--} \end{bmatrix} \quad (33)$$

where subscripts p, o, and c stand for *population, optical coherence,* and (Hertzian) *coherence.* The factors 2 have been inserted into equation (33) so that, when the decays from $|+\rangle$ and $|-\rangle$ to $|0\rangle$ are due to spontaneous emission only, the equation remains valid with $\gamma_p = \gamma_o = \gamma_c = \gamma$. The factor 2 does not appear in the coherences between the upper states and the ground state because the decay rate of the ground state has been assumed to be zero. The general solution of equation (30), which includes the treatment of saturation effects, will be discussed in Chapter 4.

In the present section we shall assume that the intensity of the excitation light is low enough not to alter in a sensible way the population distribution corresponding to thermal equilibrium. In these conditions it is possible to obtain a steady-state solution $\rho = \rho^{(0)} + \rho^{(1)} + \rho^{(2)} + \cdots$ for the density matrix in ascending powers of W, by solving the following sequence of equations[35]:

$$\frac{d\rho^{(0)}}{dt} = -\frac{i}{\hbar}[\mathcal{H}_0, \rho^{(0)}] + \Gamma\rho^{(0)}$$

$$\frac{d\rho^{(1)}}{dt} = -\frac{i}{\hbar}[\mathcal{H}_0, \rho^{(1)}] + \Gamma\rho^{(1)} - \frac{i}{\hbar}[\mathcal{V}, \rho^{(0)}]$$

$$\cdots$$

$$\frac{d\rho^{(n)}}{dt} = -\frac{i}{\hbar}[\mathcal{H}_0, \rho^{(n)}] + \Gamma\rho^{(n)} - \frac{i}{\hbar}[\mathcal{V}, \rho^{(n-1)}]$$

$$\cdots \qquad (34)$$

One obtains the density matrix at thermal equilibrium from the first equation, and $\rho_{00}^{(0)} = 1$ is its only nonzero element. The linear response of the system is described by $\rho^{(1)}$, whose nonzero elements are proportional to W and oscillate at the frequency of the excitation radiation:

$$\rho_{+0}^{(1)} = -\frac{iW e^{-i\omega t}}{\gamma_o - i(\Delta\omega - \delta\omega)} = \rho_{0+}^{(1)*}$$

$$\rho_{0-}^{(1)} = +\frac{iW e^{i\omega t}}{\gamma_o + i(\Delta\omega + \delta\omega)} = \rho_{-0}^{(1)*} \qquad (35)$$

The elements of $\rho^{(2)}$, which are stationary and proportional to W^2, provide the lowest nonvanishing perturbation order for the populations of the two upper states $|+\rangle$ and $|-\rangle$, and for the Hertzian coherence between them. These are the lowest-order terms which contribute to the fluorescence signal:

$$\rho_{++}^{(2)} = W^2 \frac{\gamma_o}{[\gamma_o^2 + (\Delta\omega - \delta\omega)^2]\gamma_p}$$

$$\rho_{--}^{(2)} = W^2 \frac{\gamma_o}{[\gamma_o^2 + (\Delta\omega + \delta\omega)^2]\gamma_p}$$

$$\rho_{+-}^{(2)} = W^2 \frac{(\gamma_o + i\delta\omega)}{(\gamma_c + i\delta\omega)[\gamma_o - i(\Delta\omega - \delta\omega)][\gamma_o + i(\Delta\omega + \delta\omega)]} = \rho_{-+}^{(2)}{}^* \quad (36)$$

We are assuming that the intensity of the excitation laser beam is low enough to allow one to disregard the terms in W^3 and higher. For much higher intensities it should be noted that the sequence of equations (34) actually leads to a power development in W/γ, where γ is of the order of γ_p, γ_o, and γ_c. This development is not applicable when W/γ is of the order of, or larger than, 1, in which case the exact solution of equation (3) is required. We shall consider this exact solution in Chapter 4.

Returning to the case when the powers of W higher than the second can be disregarded, the fluorescence signal observed along the z axis within the solid angle $d\Omega$ is obtained by combining equations (13) and (24):

$$\text{Tr}(\mathscr{L}\rho) = \frac{\omega^4 \langle \mathbf{P} \rangle^2 \, d\Omega}{8\pi^2 \varepsilon_0 c^3} [\rho_{++} + \rho_{--} + \rho_{+-} e^{i2\alpha} + \rho_{-+} e^{-i2\alpha}] \quad (37)$$

where $\langle \mathbf{P} \rangle$ is the reduced matrix element of the transition, and α is the angle between the x axis and the polarization vector of the detected fluorescence radiation.

It is apparent from equations (36) and (37) that, for a homogeneously broadened line, the fluorescence signal can depend on the strength of the external field even in absence of coherence. Let us assume that $\Delta\omega = 0$ and assume that coherence between $|+\rangle$ and $|-\rangle$ is immediately destroyed by some quenching mechanism, so that $\rho_{-+}^{(2)} = 0$. The laser is thus in resonance with both transitions at zero field, so that the $|+\rangle$ and $|-\rangle$ levels are equally populated and a maximum of fluorescence radiation is observed. As the field intensity is increased, the two absorption frequencies are shifted out of resonance, and the fluorescence radiation is correspondingly decreased. If fluorescence is monitored, the observed signal is a Lorentzian curve of width γ_o.

This effect is washed out if a large Doppler broadening is present. If the absorption lines are Doppler broadened and we sweep the magnetic field around zero, different velocity groups are brought into resonance with the laser radiation according to the field intensity. If the Zeeman shift is kept well below the Doppler width, the velocity distribution is practically flat in the region of interest. Thus, as long as $\delta\omega$ is much smaller than the Doppler width $\Delta\omega_D$, both ρ_{++} and ρ_{--} are practically constant. With a calculation analogous to those which we shall discuss in Chapter 4, we can easily obtain for large Doppler widths

$$\rho_{++} = W^2 \frac{\sqrt{\pi \log 2}}{\gamma_p \Delta\omega_D} = \rho_{--}$$

$$\rho_{+-} = W^2 \frac{\sqrt{\pi \log 2}}{\Delta\omega_D} \frac{\gamma_o + i\delta\omega}{\gamma_c + i\delta\omega} \frac{1}{\sqrt{\gamma_o^2 + 2i\gamma_o\delta\omega + \delta\omega^2}} \qquad (38)$$

Thus the resonant behavior of the matrix elements ρ_{+-} and ρ_{-+}, which are responsible for the actual Hanle effect, is not destroyed by the integration over the velocity distribution, and a Hanle signal of width γ_c is visible also in the case of Doppler broadening.

An interesting phenomenon occurs when $\Delta\omega \gg \delta\omega$ and $\Delta\omega \gg \gamma_o$, i.e., when the laser frequency is strongly detuned from both absorption frequencies.[36] Under these conditions equations (36) become

$$\rho_{++}^{(2)} = \frac{W^2 \gamma_o}{\Delta\omega^2 \gamma_p} = \rho_{--}^{(2)}$$

$$\rho_{+-}^{(2)} = \frac{W^2}{\Delta\omega^2} \frac{(\gamma_o + i\delta\omega)}{(\gamma_c + i\delta\omega)} = \rho_{-+}^{(2)*} \qquad (39)$$

Thus, even for homogeneously broadened lines, no magnetic-field dependence is displayed by the populations, while the Hanle effect, due to Hertzian coherence, is observable only if $\gamma_o \neq \gamma_c$. We have already stated that $\gamma_p = \gamma_o = \gamma_c$ when spontaneous emission is the only cause of decay. Thus a collisional relaxation mechanism, such as to introduce different decay rates for the optical and for the Zeeman coherences, is needed in order to observe the Hanle signal in the presence of a large detuning between excitation light and absorption. This phenomenon is known as the *pressure-induced Hanle effect* (*PIHE*) and has been observed, for instance, by Lange for the $^1S_0$-$^1P_1$ transition of barium using argon as buffer gas.[36]

Another kind of PIHE was recently found in absorption by Gong and Zou.[37] It is based on collision-enhanced Zeeman coherence in the ground state in transverse optical pumping (for instance, on Na). The resonance width is very small (about 100 mG). The intensity of the signal is proportional to $\Delta\omega^{-4}$ and to p^2, where p is the buffer gas pressure.

A third kind of PIHE was observed also in Na in four-wave mixing.[38] It is a nonlinear effect with a very small resonance width (approximately 20 mG); the intensity is proportional to $\Delta\omega^{-6}$ and, within a certain approximation, to p^2. PIHE in four-wave mixing is discussed more extensively in Section 3.3 of Chapter 2.

6. NONZERO-FIELD LEVEL CROSSING

An atomic term may be split into different sublevels already in the absence of external fields. This splitting can be due to fine structure (sublevels with different J values resulting from coupling of the same L and same S) or hyperfine structure (sublevels with different F values resulting from coupling of the same J and same I) interactions. Each of the sublevels with nonzero total angular momentum may, in its turn, be split into Zeeman sublevels by an external magnetic field. A Zeeman sublevel, whose energy decreases with the magnetic field and which starts from a higher-lying (hyper)fine level, may cross, at a particular value B_{LC} of the field, another Zeeman sublevel whose energy increases with the field and which starts from a lower-lying (hyper)fine level (Figure 9). In this case we speak of (nonzero field) *level crossing*.

A first theory of the level crossing was presented by Breit in 1936,[18] but the first experimental measurement was performed by Colegrove et al. in 1959[22] and provided a high-precision determination of the helium $2\,^3P_1$-$2\,^3P_2$ fine structure separation. A theoretical treatment was presented by Franken in 1961.[34]

A level-crossing signal in the fluorescence intensity is observed by sweeping the external field around the crossing value B_{LC}, while the vapor is excited by a light beam with a polarization such as to excite the atoms into a coherent superposition of the crossing levels. The physics underlying the level crossing is the same underlying the Hanle effect, which actually is a level crossing occurring at zero field. The main difference between the Hanle effect and the nonzero-field level crossing is that in the latter case not only the width of the signal, but also the value of B_{LC}, is measured. From the value of B_{LC} it is then possible to determine, for instance, the HFS or the FS constant. An expression for the intensity of the fluorescence light of a given polarization emitted in a given direction is simply obtained by replacing the term $(m - m')\omega_L$ in the denominator of equation (26) by the more general expression $(\omega_m - \omega_{m'}) = (E_m - E_{m'})/\hbar$, where E_m and $E_{m'}$ are the energies of the Zeeman sublevels of the hyperfine levels under consideration. This substitution is necessary because, usually, crossings occur at intermediate field values, where the partial decoupling between L and S (fine structure, transition to the Paschen-Back effect), or between J and I (hyperfine structure, transition to the Back-Goudsmit effect), gives origin to a nonlinear dependence of the level energies on the external field. In the case of hyperfine structure m refers to the F projection, rather than to the J projection, over the quantization axis. The resulting expression, known as the *Franken equation*, is

$$dI = \frac{u(\omega_0)\omega_0^4}{8\pi\varepsilon_0\hbar^2 c^3} \sum_{mm'} \frac{F_{\mu\mu mm'} G_{m'm\mu'\mu'}}{2\gamma + i(E_m - E_{m'})/\hbar} \rho_{\mu\mu}\, d\Omega \qquad (40)$$

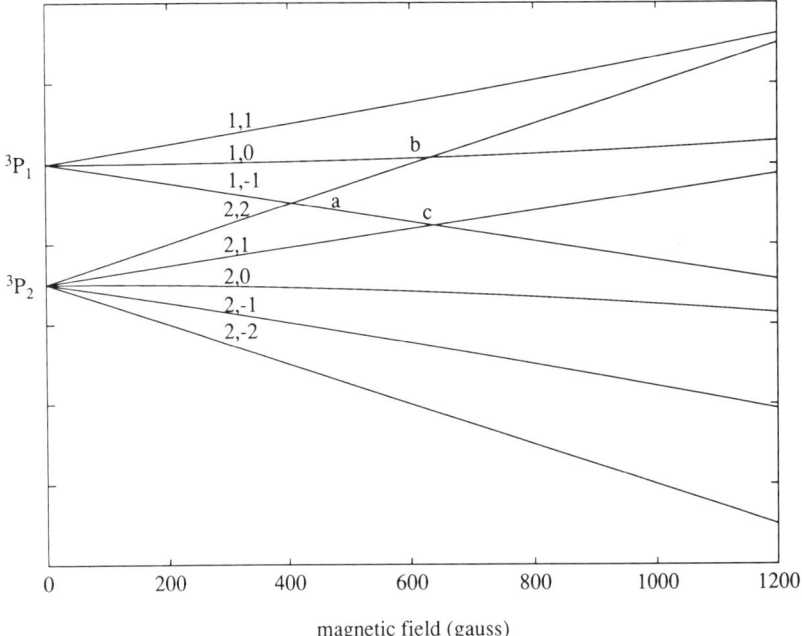

FIGURE 9. The nonzero-field level-crossing experiment of Colegrove et al.[22] on the $2\,^3P_1$ and $2\,^3P_2$ fine structure separation of helium. Level-crossing signals are observed, for instance, at locations b and c. No signal is observed at location a, since the selection rule $\Delta m = 0, \pm 1$ for an electromagnetic transition does not allow the creation of coherence between states with $\Delta m > 2$.

Of course, not only the correct level energies must be used in the denominator of equation (40), but also the correct intermediate field wave functions must be used for the evaluation of the matrix elements appearing in the expressions for $F_{\mu\mu mm'}$ and $G_{m'm\mu'\mu'}$.

A limit to the accuracy in the determination of the HFS constants from a level-crossing experiment is due, in some cases, like the case of the HFS of the $3\,^2P_{3/2}$ of sodium, to the incomplete resolution of the level-crossing signal. This problem has been solved, for instance, by Deech, Hannaford, and Series[39] by performing an experiment with resolution beyond the natural linewidth, and by Figger and Walther[40] by time-resolved level-crossing spectroscopy. A review on the experimental determination of the hyperfine structure in the alkali atoms, including the level-crossing technique, can be found elsewhere.[41]

The nonlinear dependence on the external field of the energies E_m and $E_{m'}$ appearing in the denominator of the last term in equation (40) is particularly relevant for the case of level crossings in electric fields for nonhydrogenic atoms. In this case the lowest nonvanishing order of the Stark shift is quadratic. The level crossing in an electric field was first applied by Khadjavi, Lurio, and Happer[42] to the measurement of the electric tensor polarizability of the $^2P_{3/2}$ states of rubidium and cesium. It was then applied to potassium by Schmieder, Lurio, and Happer.[43] Level-crossing experiments have also been performed in combined electric and magnetic fields. This technique was introduced by Volikova in order to investigate the electric polarizabilities.[44] Level-crossing spectroscopy in combined electric and magnetic fields has also been applied, for instance, by Kollath and Kleinpoppen to the measurement of the nonlinear Stark effect in the $2\,^2P_{1/2}$ state of hydrogen.[45] A mathematical formalism for evaluating the crossings of sublevels with the same Zeeman quantum number in superimposed electric and magnetic fields has been presented by von Oppen.[46] A review on the Stark effect in atomic spectroscopy, including level crossing in a pure electric field and in superimposed electric and magnetic fields, is found in Ref. 47.

7. CONCLUSIONS

We have seen that the Hanle effect has played an important role in the history of quantum mechanics. The correct interpretation of the experimental results required assumptions in contradiction to some of the postulates of the early quantum theory, and the necessity to override these difficulties leads, in particular, to the concepts of state superposition and coherence.

From an experimental point of view, the Hanle effect has allowed precise measurements of the lifetimes of excited atomic states for decades. Its extension to nonzero fields, or the level-crossing technique, had allowed precise sub-Doppler measurements of fine and hyperfine structure separations. Since the development of laser sub-Doppler methods, the Hanle effect and the level crossing are no longer the only available spectroscopic sub-Doppler techniques.

It is astonishing, however, to see how these methods, and particularly the Hanle effect, which was born in 1924, continue to play an irreplaceable role in modern spectroscopy. The problem of determining the excited-state lifetimes, and the atomic transition probabilities, maintains its importance in atomic spectroscopy. This problem has also gained astrophysical relevance, since in relatively cold stars, where ion lines are weak or absent, atomic resonance lines become very important for diagnosis. The determination of excited-state lifetimes is also important in plasma physics and space research. These measurements also provide a quite sensitive check of the reliability of approximate solutions to the many-electron problem of atomic structure.

Some alternative methods for measuring the excited-state lifetimes have been developed in more or less recent years. A few examples follow. In the *beam-foil technique* one monitors the spatial variation of the intensities of the spectral lines emitted by a fast atomic beam. This is the basis of the canal ray method developed by Wien in 1927.[48] In the modern version an accelerated ion beam is converted into an excited neutral atomic beam by passing through a thin carbon foil, as observed independently by Kay[49] and Bashkin.[50] A description of the method can be found elsewhere.[27,51] Lifetimes can also be measured by direct measurement of the exponential decay following a brief excitation pulse,[52] by measuring the phase shift between the modulated exciting radiation and the emitted fluorescence radiation,[53,54] or by natural linewidth determinations using optically or electronically excited double resonance.[55-57]

Because of its inherent simplicity, the Hanle effect is affected by very few sources of error and, whenever applicable, its accuracy compares favorably with more recent techniques. The limits to its accuracy come from radiation trapping and collisional mechanisms, which are negligible at low vapor densities and can be corrected by zero density extrapolations. A typical uncertainty is 3%. The main limit to applicability is the requirement of strong resonance lines, unaffected by self-reversal, which limits the method to the first few resonance lines of an element. This limit can be overcome by measuring the Hanle effect in the presence of electron excitation rather than optical excitation, as demonstrated by the group of Pebay-Peyroula[58,59] for atomic lifetimes and by Kahan, Lucatorto, and Novick for ion lifetimes.[60] These experiments must overcome the difficulties of a small zero-field polarization degree (the polarization is parallel to the direction of the exciting electron beam) and of the influence of the magnetic field on the electron beam. The Hanle effect with collisional excitation will be discussed more extensively in Section 3.1 of Chapter 2. The Hanle-effect technique with optical excitation has been extended to the measurement of the lifetimes of excited states of ions by Smith and Gallagher.[61]

Novel important applications, covering such a wide range of fields as atomic and molecular physics, solid state physics, laser physics, and gravitational metrology, are continuously found by the experimenters, as shown in the subsequent chapters of this book. Unfortunately it has not been possible to find in time an author to cover the applications in nuclear physics. This would have further widened the already wide coverage of physical fields of this book.

ACKNOWLEDGMENTS. I am indebted to N. Beverini, G. E. Pikus, M. O. Scully, J. O. Stenflo, F. Strumia, and A. N. Titkov for helpful comments and suggestions, and to Brenda P. Winnewisser for revising my translation of Hanle's original paper. Extremely pleasant conversations with Professor Hanle, scattered in time over the past thirteen years, have been a strong stimulus to participate in the editing and writing of this book, and have been of great help for writing the historical survey.

REFERENCES

1. W. Hanle, *Z. Phys.* **30**, 93 (1924).
2. W. Hanle, *Ergebnisse der exakten Naturwissenschaften* **4**, 214 (1925).
3. R. W. Wood, *Philos. Mag.* **23**, 689 (1912).
4. Lord Rayleigh, *Proc. R. Soc. London, Ser. A* **102**, 190 (1922).
5. R. W. Wood and A. Ellett, *Proc. R. Soc. London, Ser. A* **103**, 396 (1923).
6. W. Hanle, *Naturwissenschaften* **11**, 690 (1923).
7. J. A. Eldridge, *Phys. Rev.* **24**, 234 (1924).
8. R. W. Wood and A. Ellett, *Phys. Rev.* **24**, 243 (1924).
9. P. Pringsheim, *Naturwissenschaften* **12**, 247 (1924).
10. G. Breit, *Philos. Mag.* **47**, 832 (1924).
11. G. Joos, *Phys. Z.* **25**, 130 (1924).
12. N. Bohr, *Naturwissenschaften* **12**, 1115 (1924).
13. W. Hanle, *Oberhess. Naturwiss. Z.* **48**, 57 (1984).
14. E. Gaviola and P. Pringsheim, *Z. Phys.* **25**, 367 (1924).
15. W. Heisenberg, *Z. Phys.* **31**, 617 (1925).
16. R. Oppenheimer, *Z. Phys.* **43**, 27 (1927).
17. V. Weisskopf, *Ann. Phys. (Leipzig)* **9**, 23 (1931).
18. G. Breit, *Rev. Mod. Phys.* **5**, 91 (1936).
19. W. Pauli, *Wissenschaftlicher Briefwechsel mit Bohr, Einstein, Heisenberg u.A.*, Band I: 1919–1929, edited by A. Hermann, K. v. Meyenn, and V. F. Weisskopf (Springer-Verlag, Berlin, Heidelberg, New York, 1979), pp. 219–220.
20. A. Kastler, *J. Phys. Radium* **11**, 255 (1950).
21. A. Kastler, *Phys. B.* **30**, 394 (1974).
22. F. D. Colegrove, P. A. Franken, R. R. Lewis, and R. H. Sands, *Phys. Rev. Lett.* **3**, 420 (1959).
23. J. N. Dodd, R. D. Kaul, and D. M. Warrington, *Proc. Phys. Soc. London* **84**, 176 (1964).
24. E. B. Aleksandrov, *Opt. Spectrosc. (USSR)* **17**, 522 (1964).
25. J. N. Dodd, W. J. Sandle, and D. Zisserman, *Proc. Phys. Soc. London* **92**, 497 (1967).
26. M. O. Scully and M. Sargent III, *Phys. Today* **25**, No. 2, 38 (1972).
27. A. Corney, *Atomic and Laser Spectroscopy*, Chapter 15, *The Hanle Effect* (Clarendon Press, Oxford, 1977).
28. W. Hanle, *Z. Phys.* **35**, 346 (1925).
29. B. Favati, E. Landi degl'Innocenti, and M. Landolfi, *Astronaut. Astrophys.* **179**, 329 (1987).
30. P. A. M. Dirac, *The Principles of Quantum Mechanics*, 4th ed., Chapter X, *Theory of Radiation* (Oxford University Press, Oxford, 1958).
31. S. Haroche, *Ann. Phys. (Paris)* **6**, 189, 327 (1971).
32. J. Dalibard and C. Cohen-Tannoudji, *J. Opt. Soc. Am. B* **2**, 1707 (1985).

33. B. Decomps, M. Dumont, and M. Ducloy, *Linear and Nonlinear Phenomena in Laser Optical Pumping*, in *Topics in Applied Physics*, Vol. 2, *Laser Spectroscopy of Atoms and Molecules*, edited by H. Walther (Springer-Verlag, Berlin, Heidelberg, New York, 1976), p. 283.
34. P. A. Franken, *Phys. Rev.* **121**, 508 (1961).
35. N. Bloembergen and Y. R. Shen, *Phys. Rev.* **133**, A37 (1964).
36. W. Lange, *Optics Commun.* **59**, 243–248 (1986).
37. Q. H. Gong and Y. H. Zou, *Opt. Commun.* **65**, 1 (1988).
38. Y. H. Zou and W. Bloembergen, *Phys. Rev. A* **33**, 1730 (1986).
39. J. S. Deech, P. Hannaford, and G. W. Series, *J. Phys. B* **7**, 1131 (1974).
40. H. Figger and H. Walther, *Z. Phys.* **267**, 1 (1974).
41. E. Arimondo, M. Inguscio, and P. Violino, *Rev. Mod. Phys.* **49**, 31 (1977).
42. A. Khadjavi, A. Lurio, and W. Happer, *Phys. Rev.* **167**, 128 (1968).
43. R. W. Schmieder, A. Lurio, and W. Happer, *Phys. Rev.* **173**, 76 (1968).
44. L. A. Volikova, V. N. Grigorieva, G. I. Khvostenko, and M. P. Chaika, *Opt. Spektrosk.* **30**, 170 (1971) [*Opt. Spectrosc.* **30**, 88 (1971)].
45. K. J. Kollath and H. Kleinpoppen, *Phys. Rev. A* **10**, 1519 (1974).
46. G. von Oppen, *Physica* **80**, 553 (1975).
47. K. J. Kollath and M. C. Standage, *Stark Effect*, in *Progress in Atomic Spectroscopy*, Part B, edited by W. Hanle and H. Kleinpoppen (Plenum Press, New York and London, 1979).
48. W. Wien, *Ann. Phys. (Leipzig)* **83**, 1 (1927).
49. L. Kay, *Phys. Lett.* **5**, 36 (1963).
50. S. Bashkin, *Nucl. Instrum. Methods* **28**, 88 (1964).
51. S. Bashkin, *Appl. Opt.* **7**, 41 (1968).
52. S. Heron, R. W. P. McWhirter, and E. H. Rhoderick, *Proc. R. Soc. London, Ser. A* **234**, 565 (1956).
53. O. Osberghaus and K. Ziock, *Z. Naturforsch., A* **11**, 762 (1956).
54. W. Demtröder, *Z. Phys.* **166**, 42 (1962).
55. J. Brossel and F. Bitter, *Phys. Rev.* **86**, 308 (1952).
56. J. C. Pebay-Peyroula, *J. Phys. Radium (Paris)* **20**, 669 (1959).
57. J. C. Pebay-Peyroula, *J. Phys. Radium (Paris)* **20**, 721 (1959).
58. A. Faure, O. Nédelec, and J. C. Pebay-Peyroula, *C. R. Acad. Sci. Paris* **256**, 5088 (1963).
59. M. Chenevier, J. Dufayard, and J. C. Pebay-Peyroula, *Phys. Lett.* **25A**, 283 (1967).
60. W. Kahan, T. Lucatorto, and R. Novick, *Bull. Am. Phys. Soc.* **9**, 451 (1964).
61. W. W. Smith and A. Gallagher, *Phys. Rev.* **145**, 26 (1966).

APPENDIX*

MAGNETIC EFFECTS ON THE POLARIZATION OF RESONANCE FLUORESCENCE[1]

WILHELM HANLE, GÖTTINGEN
[*Translation by Giovanni Moruzzi (1988)*]

With six figures. (Received 1 November 1924)

1. Historical background of the magnetic effects on resonance fluorescence polarization and their interpretation as a Zeeman effect. 2. Determination of the lifetime of the excited state of mercury from the rotation of the resonance fluorescence polarization plane due to weak magnetic fields. 3. Experiments on the excitation of mercury resonance fluorescence by light of different polarizations and directions.

R. W. Wood and A. Ellett[2] have shown that, upon excitation of the resonance fluorescence of mercury by the linearly polarized 2536.7 Å line, the resonance fluorescence light is also almost completely linearly polarized. The degree of polarization of the fluorescence resonance is, however, significantly affected by very weak magnetic fields (of the order of 1 gauss). For instance, if the fluorescence perpendicular to the propagation direction of the exciting light is observed and if the magnetic field lines are parallel to the direction of observation, complete depolarization is observed already at a field strength as low as 2 gauss. The author interpreted these and similar results by Wood and Ellett at first as Zeeman effects,[3] in which the expected splitting is negligibly small compared to the linewidths. Only the polarizations typical for the Zeeman effect can be detected. This rather qualitative interpretation as a Zeeman effect has been developed further in several

---

* This appendix is a translation, by Giovanni Moruzzi (1988), of "Über magnetische Beeinflussung der Polarisation der Resonanzfluoreszenz," *Z. Phys.* **30**, 93–105 (1924). It is reprinted by permission of the author and the publisher, Springer-Verlag. The reference citations within the appendix are presented as numbered footnotes.

[1] Dissertation, University of Göttingen.
[2] R. W. Wood and A. Ellett, *Proc. R. Soc.* **103**, 396 (1923). *Translator's note.* Here and subsequently the name of Alexander Ellett is misspelled "Ellet" in the original work. The correct spelling is used in the translation.
[3] W. Hanle, *Naturwissenschaften* **11**, 690 (1923).

works by P. Pringsheim,[4] G. Joos,[5] G. Breit,[6] and E. Gaviola and P. Pringsheim,[7] leading to good agreement with the quantum interpretation of the Zeeman effect, at least within the present experimental accuracy.[8] Serious difficulties, however, persist in the interpretation of the high polarization level observed at zero field. An extended and more detailed treatment of the problem, carried out by Bohr,[9] and presumably by now in print in *Naturwissenschaften*, will be discussed below.

The present work presents some further experimental contributions to the problem of the polarization of mercury resonance fluorescence and its behavior under the influence of magnetic fields.[10] We shall start from a description of the rotation of the polarization plane of the resonance fluorescence, which is to be expected according to the classical theory of the Zeeman effect sketched below.

According to this theory, a quasi-elastically bound electron in a magnetic field performs a precession motion round about the field direction. In the special case of an electron oscillating perpendicularly to the magnetic field, it describes a rosette in a plane perpendicular to the field direction.

Because of the radiative damping of the electron, the rosette will decay with time. The resulting situation at different field intensities is shown in Figures 1 to 3. If the field is very weak the precession period is long compared to the decay time; the oscillation amplitude can decay practically completely after a rotation of 180° (Figure 1). The rosette motion becomes more and more complete if the field intensity is increased (Figure 2). When, eventually, the precision period becomes short compared to the decay time, a complete precession of the rosette is described, as in the case of an undamped oscillating electron (Figure 3). Therefore, if the fluorescence light is polarized in the vertical plane, we expect that it will be weakly depolarized by very weak fields, and that its polarized part will be rotated out of the vertical plane (Figure 1). We also expect that the depolarization effect and the rotation of the oscillation plane will increase with increasing field intensities (Figure 2), and that the fluorescence will be completely depolarized at high fields (Figure 3). The increase of depolarization with increasing magnetic field has been observed by Wood and Ellett, as we stated above. It seemed

---

[4] P. Pringsheim, *Naturwissenschaften* **12**, 247 (1924); *Z. Phys.* **23**, 324 (1924).
[5] G. Joos, *Phys. Z.* **25**, 130 (1924).
[6] G. Breit, *Philos. Mag.* **47**, 832 (1924).
[7] E. Gaviola and P. Pringsheim, *Z. Phys.* **25**, 367 (1924).
[8] See also A. E. Ruark, P. Foote, and F. L. Mohler, *J. Opt. Soc. Am.* **7**, 415 (1923) and F. Weigert, *Naturwissenschaften* **12**, 38 (1924).
[9] The contents of this work were made available to us in advance through the courtesy of N. Bohr.
[10] The major part of this work was presented at a meeting of the Lower Saxony (Niedersachsen) regional section of the German Physical Society in Hamburg, June 22, 1924.

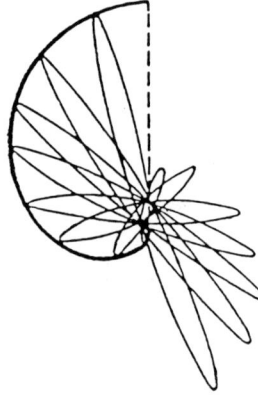

FIGURE A.1

important to investigate whether a rotation of the polarization plane also occurs at low fields. The experiment was performed with the following experimental setup (Figure 4).

A quartz mercury lamp Q is used as light source. The lamp is produced by Heraeus and operates at 220 V, 2.5 A. Self-reversal of the 2536.7 Å resonance line is prevented as far as possible by an electromagnet, which pushes the electric arc against the lamp walls, and by air cooling. The light emitted by the lamp is collimated by the quartz lens L_1, polarized by the ultraviolet transparent Nicol prism N_1, and focused on the resonance cell R by the lens L_2. L_2 is a feldspar lens, since quartz crystal lenses could not be used because of the birefringence and optical activity of quartz. The resonance cell simply consisted of a glass sphere with a thin window of

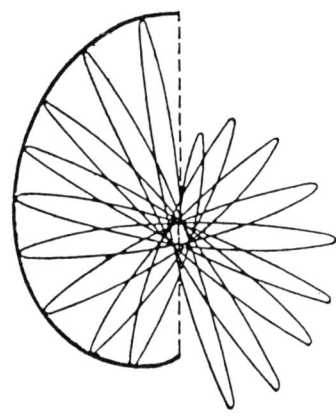

FIGURE A.2

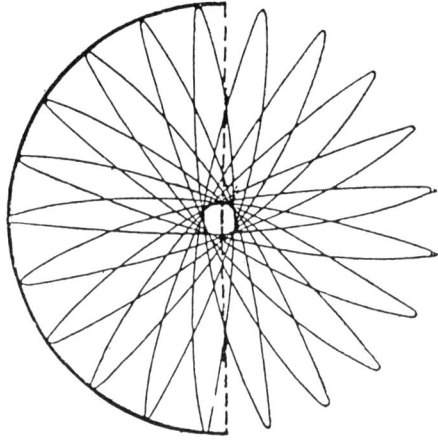

FIGURE A.3

fused quartz. A tube connects the cell to the vacuum pump through a Dewar flask. Experiments were usually performed at a mercury vapor pressure corresponding to room temperature, i.e., some 10^{-3} mm, and sometimes at a vapor pressure of 2×10^{-4} mm, corresponding to 0 °C. The resonance cell is located at the center of a large rotary coil in a Cardanic support which compensates the earth's field, in order to avoid its perturbative effect on the polarization of the resonance fluorescence.

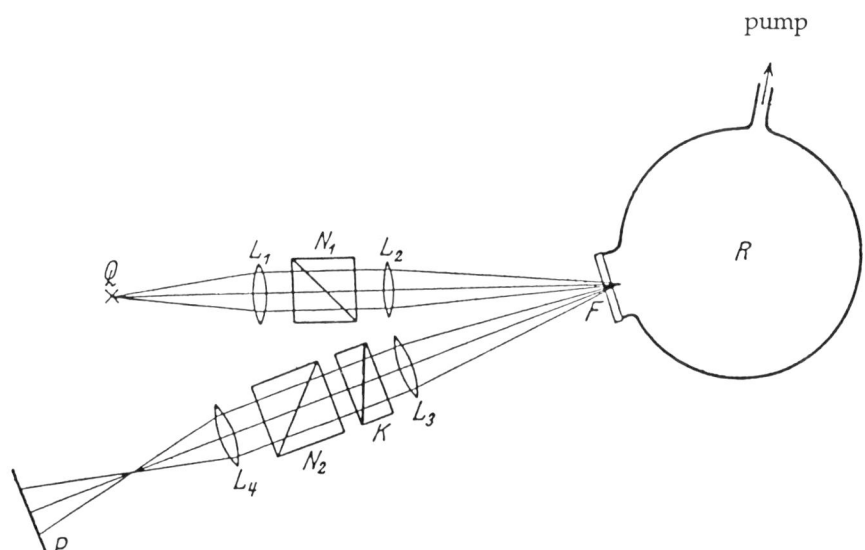

FIGURE A.4

The field compensation was checked by a ballistic measurement of the current induced in a coil which can be rotated about two mutually perpendicular directions. Compensation to about 1/100 gauss can be achieved. The fluorescence radiation is analyzed by means of a rotatory quartz prism. This consists of a combination of a right rotatory and a left rotatory quartz prism, both cut perpendicularly to their optical axis. The rotatory quartz prism, combined with a Nicol prism, produces fringes in polarized light, from whose intensities one can evaluate the polarization degree, and from whose shift one can evaluate the rotation of the polarization plane. Cross hairs on the crystal surface serve as reference. The observation direction forms an angle of about 20° with the direction of incidence.[11] Fluorescence light was collimated by a feldspar lens L_3, and traversed the rotatory quartz prism K and the Nicol prism N_2; the quartz lens L_4 formed the image of the cross hairs on the camera plate P.

The mercury vapor in the cell was first excited by linearly polarized light, with complete compensation of the earth's field. The electric vector of the excitation light was vertical. The fluorescent light was then also linearly polarized. We did not try to measure accurately the polarization degree, because we had been informed that analogous measurements on mercury are simultaneously being performed in another institute.[12] If a magnetic field of 0.5 gauss, parallel to the observation direction, was added, we could observe a fading and a shift of the fringes, corresponding to a certain depolarization and to a rotation of the polarization plane (of about 20°). If the field was increased to 1 gauss we could observe a greater depolarization and rotation. At 2 gauss the fringes were scarcely distinguishable, and therefore the fluorescence light was practically completely depolarized. If the magnetic field was reversed, a reversed rotation of the polarization plane was observed. The rotation of the polarization plane proved to be independent of the orientation of the polarization plane of the exciting light. The sense of rotation of the polarization plane corresponds to the sense of the precessional motion.

The measurement of both the polarization degree and the rotation of the polarization plane offer the possibility to evaluate the decay time (according to quantum theory, the lifetime of the upper level; see below). The first method, which can be very inaccurate because of the difficulties found in precise polarization measurements in the ultraviolet, has been used

---

[11] R. W. Wood and A. Ellett's observation angle was 90°. Since in this experiment the intensity and polarization of fluorescence light are distributed symmetrically around the vertical, the angle between excitation and observation direction is irrelevant, as long as the magnetic field, introduced in the second step of the experiment, is applied parallel to the observation direction.

[12] In a recent publication by Wood and Ellett [*Phys. Rev.* **24**, 243 (1924)] a polarization of 90% is again reported for the fluorescence of mercury in the absence of a magnetic field.

also by Breit[13] and Eldridge.[14] The second method seems to be suitable for more precise determinations of the decay time. The present measurements have been of a rather qualitative nature up to now. A preliminary evaluation, based on rotation measurements at known field strength, and taking into account the 3/2 factor by which the classical precession velocity must be multiplied due to the 3/2 normal splitting, leads to a value of 10^{-7} s (lifetime of the $2p_2$ state of mercury). This is in good agreement both with the value calculated by Breit and Eldridge from depolarization, and with the direct measurements by W. Wien on Hg anodic rays.

The rotation effect can prove a further consequence of the model which we developed at the beginning of this work. Namely, an increased quenching of the resonators induced by additional gases must lead to a smaller rotation and to a higher polarization degree. This effect is expected to be especially strong for gases which exert a particularly strong collisional quenching on the radiation process. On the contrary, one expects only a depolarization increase for gases which exert only a weak quenching on the oscillator, but which are able to rotate it from its direction. The latter property is typical of the noble gases (see Wood,[15] and the thesis of K. Donat, which is going to appear soon), while hydrogen strongly quenches fluorescence, as shown in the same references. The experimental results were in good agreement with these expectations. The fringes are strongly smeared out by the presence of argon at a pressure of 2 mm, which means depolarization. On the other hand, if hydrogen is added, the fringes are shifted in a direction indicating a partial cancellation of rotation, and become sharper, indicating an increase in the polarization. Since the presence of hydrogen decreases the fluorescence intensity, we had to increase the exposure time correspondingly in order to have equal blackening with and without gas addition, so that one would not be mistaken as to the sharpness of the fringes. With the addition of argon, for which both an extremely weak quenching and, on the other hand, a strong pressure broadening of the absorption line increase the fluorescence intensity, the exposure was correspondingly shorter.

Thus the experiment has confirmed the theoretical predictions. In our treatment we have considered the electron, which has undergone a single excitation, as performing damped oscillations. Actually, the electron performs forced oscillations in the radiation field. Since, however, the electron does not respond only to its own eigenfrequency, but also, correspondingly more weakly, to nearby frequencies because of damping, its behavior

---

[13] G. Breit, *Philos. Mag.* **47**, 832 (1924).
[14] J. A. Eldridge, *Phys. Rev.* **23**, 772 (1924). *Translator's note.* Here and subsequently the name of John A. Eldridge is misspelled "Eldrige" in the original work. The correct spelling is used in the translation.
[15] R. W. Wood, *Philos. Mag.* **44**, 1107 (1922).

remains practically the same as we have sketched above for the freely oscillating electron.

We have thus shown that the predictions of the classical theory are perfectly verified, and we now ask ourselves how these same phenomena can be interpreted by quantum theory. Quantum theory replaces an oscillating electron with transitions from excited states to ground states. Damping is replaced by the average lifetime of the excited states and collisional damping by the action of collisions of the second kind. Quantum theory provides an even clearer interpretation than the classical theory for the different behaviors of argon and hydrogen as foreign gases. While the measurement of the depolarization degree, or the measurement of the rotation of the polarization plane at known field strength, only allow conclusions concerning damping according to the classical theory, according to the quantum theory the same measurements allow a determination of the lifetime of the excited state. So far there is no difficulty in substituting the classical interpretation with a quantum theoretical one. On another point, however, we are forced by the experiment to add some assumptions to quantum theory in order to avoid difficulties.

Namely, there is a further essential difference between classical and quantum theory. The Zeeman splitting of a spectral line into several components is no longer interpreted as due to the complicated motion of a single oscillator, which can be decomposed into different frequencies by mathematical methods. Rather, the different frequencies are considered as emitted by separate atoms which differ from one another with respect to their energy content ($h\nu$ relation) and to their orientation relative to the field direction (directional quantization). These assumptions of quantum theory are both proved by the behavior at higher fields, where larger splittings are observed. The first assumption is proved, albeit only for natural splittings (doublets), by the fact that only one line is emitted when the energy needed for the emission of this line is provided to the atom (either by electron collisions or by light excitation). The second assumption was definitely proved by the deviation of atomic beams in Stern and Gerlach's experiment. In our case, the longitudinal Zeeman effect on the 2536.7 Å mercury line, two circularly polarized components (σ components) of slightly different frequency should be emitted by two different atoms. In order to justify the rotation of the polarization plane, one should assume an interaction between the polarized emissions of two atoms located at an average distance from each other which is large compared to the wavelength. This is absolutely inconceivable. The only possible assumption left is that neither the directional quantization nor the splitting of the excited atoms into two levels are completely maintained at low fields. On the contrary, we are forced to accept the existence of a special transition region between directionally quantized, and therefore energetically different, atoms at high fields and not directionally quantized atoms at zero field. This result is therefore in disagreement with the usual assumption that directional quantization is rigorous also at zero field, only the energy splitting being zero.

Recently, in *Naturwissenschaften*, Bohr has developed a special interpretation of the nature of the gradual disappearance of the directional quantization, and of the resulting almost 100% polarization at zero field which we have observed. The interpretation of the phenomenon is based on the consideration that directional quantization, i.e., the quantization of a precessional period introduced by the magnetic field, only has a meaning if this period is short compared to the lifetime of the excited state. In the field range where the period is equal to, or longer* than the lifetime, a degeneracy must appear. At zero field, where the precessional period is infinitely long, we have a completely degenerate system. In the transition region, directional quantization becomes less and less sharp, corresponding to a statistical distribution of the direction of the atomic axis around the direction which it would have for rigorous directional quantization. In our case, we are observing this transition from the nondegenerate to the degenerate system. On the occasion of the above-mentioned seminar of the author, Mr. H. A. Kramers from Copenhagen, whom we thank heartily also for some other advice, pointed out the necessity to check that the observed rotation effect was not a Faraday effect of the mercury vapor in the region of its resonance line. This point was particularly important, because Gerlach and Schütz[16] have clearly observed such a Faraday effect in sodium vapor absorption around the D line. Therefore we concentrated our efforts to verify if here we were also facing a Faraday effect. Anticipating the result, we have proved that this was not the case. Because of the importance of this conclusion, we describe here the experiments which led to this result.

It is known that the Faraday effect consists in the rotation of the polarization plane by a magnetic field when light travels through a dispersive body. Therefore the rotation has the same sense as the Larmor precession associated with the Zeeman effect in both the right and left wings of the absorption line, while the rotation sense is reversed for frequencies close to the center of the absorption line. This behavior alone is sufficient to make the hypothesis of a Faraday effect unlikely, since in the case of resonance fluorescence only the central part of the line is active, and the observed rotation has the same sense as the precessional motion. A further typical feature of the Faraday effect is that the rotation increases proportionally to the length of the path traveled by the light beam in the magnetic field. Here, on the contrary, the rotation effect reaches its maximum as soon as the resonance cell experiences the longitudinal magnetic field, and there is

---

* *Translator's note.* There is a misprint in the original text, which actually reads *shorter.*
[16] W. Gerlach and W. Schütz, *Naturwissenschaften* **11**, 637 (1923).

absolutely no dependence on the length of the path traveled by the exciting light in the magnetic field, nor on the path traveled by the fluorescence light in the magnetic field before detection. Both results have been obtained in the following experiments.

A quartz absorption cell was located between the light source and the resonance cell described above. The absorption cell was evacuated down to the vapor pressure of mercury; vapor pressures between 10^{-3} and 2×10^{-4} mm were used. The absorption cell was located in a longitudinal magnetic field with intensities between several tenths up to 100 gauss. The length of the absorption tube was 2 cm. A rotation effect for the fluorescence radiation excited in the resonance cell was observable only when a magnetic field was applied to the resonance cell itself. The magnitude of this rotation was completely independent of the presence of a magnetic field on the absorption cell. Even when the order of the resonance cell and absorption cell on the light beam path was reversed, the rotation effect displayed the same independence of the magnetic field on the absorption cell.

It was also not possible to observe a Faraday effect on the 2536.7 Å line analogous to the one observed by Gerlach and Schütz[17] for the D lines, although a replication of the Gerlach-Schütz experiment for the D lines reproduced their result perfectly. Clearly the Faraday effect is much smaller for the 2536.7 radiation than for the sodium D lines because the absorption coefficient of the former line is smaller by a factor of about 50.

The proof that the rotation effect was independent of the path traveled by the fluorescence light in the magnetic field was obtained as follows. As resonance lamp we used a 3-cm-long fused quartz tube terminated by plane windows. Field direction, irradiation direction, and observation direction as shown in Figure 5. An advantage of this geometry is the possibility of exciting the vapor with unpolarized light, thus gaining much in intensity.

[17] W. Gerlach and W. Schütz, *Naturwissenschaften* **11**, 637 (1923).

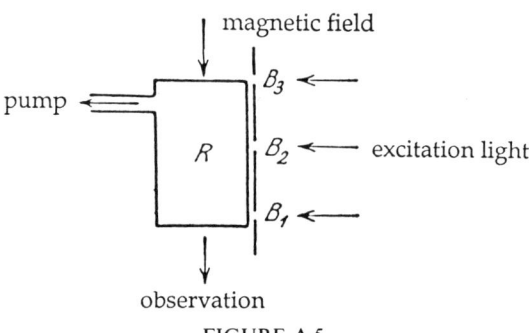

FIGURE A.5

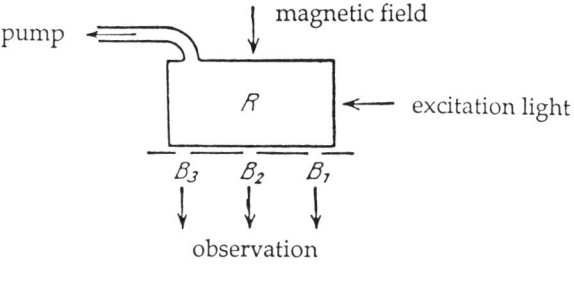

FIGURE A.6

(Exposition times are of some seconds rather than a quarter-hour.)[18] The observed rotation effect was independent of the diaphragm through which the excitation light had passed, which could be B_1, B_2, or B_3. Figure 6 shows the exchange of irradiation direction and observation direction. The experiments with this geometry led to the same result as above. Thus the observed rotation must be understood as described above.

The results of this experiment are so remarkable that it is naturally important to complete the work with further experiments on other monoatomic metal vapors. In particular we are planning to perform further experiments on sodium, since its anomalous Zeeman effect (splitting of the ground state in a magnetic field) promises particularly interesting results.[19]

A further point which we have investigated is related to an experiment by Wood and Ellett, which led to quantitative deviations from our theoretical model. Wood and Ellett excited resonance fluorescence by means of linearly polarized light and observed it perpendicularly to the oscillation plane. The fluorescence light was linearly polarized when the field was either horizontal or vertical. When the magnetic field formed an oblique angle with the vertical, polarization was not complete, and disappeared when the angle was 45°, while this angle should be 54° according to our theory, as can easily be verified. Even if this discrepancy could be limited by experimental inaccuracies, it seemed desirable to clear up the point by a more precise measurement. We chose therefore an experimental geometry which allowed a precise determination of the angle. The magnetic field was vertical in all experiments, while the electric vector E of the primary beam was rotated.

---

[18] The Nicol prisms, glued with Canada balsam, absorb the ultraviolet spectral region completely. Therefore one must use Nicol prisms separated by an air layer (Foucault or Glan prisms) which, however, are still strongly absorbing and require a good collimation of the light beam, otherwise the polarization is not complete and moreover these prisms, because of their dimensions, do not allow the full exploitation of the aperature of the lenses.

[19] Comment added in proof. In the meantime Heisenberg has extended Bohr's picture to the complicated Zeeman splitting of sodium, and has kindly provided us his results before publication; therefore we draw the reader's attention to this work here.

When E was vertical, the fluorescence light, observed at an angle of 20° with the incidence direction, was also linearly polarized in the vertical plane. When E was slowly rotated up to an angle of 50° with the vertical, the polarization plane of the fluorescence light was preserved, while the polarization degree decreased with increasing angles. On the other hand, if E was horizontal, the fluorescence light was polarized in the horizontal plane, and again the polarization plane was preserved, with changing polarization degree, when E was rotated from the horizontal up to an angle of 60° with the vertical. Since the minimum of the degree of polarization was always between 50° and 60°, and the jump from vertical to horizontal of the polarization plane of the polarized part of the light always occurred in the vicinity of 45°, the conclusions of our model can be regarded as proved, and the discrepancy with the experiment as removed.

Finally, we conducted experiments with excitation by elliptically polarized light. For this purpose a Soleil-Babinet compensator was inserted in the primary beam path between the Nicol prism and the fluorite lens, giving rise to elliptically polarized light. The Soleil wedges could be shifted relatively to each other, thus allowing variation of the ellipticity of the primary light beam.[20] The ellipticity of the fluorescence light was analyzed by substituting the rotary quartz wedge in the path of the secondary beam with a Babinet compensator. Again we observed the fluorescence emitted at an

---

[20] We mention here a difficulty encountered using the Soleil and Babinet compensators in the ultraviolet spectral region. Both consist of quartz wedges cut parallel to their optical axis. Thus a light beam propagating exactly perpendicularly to the crystal surface, and therefore perpendicularly also to the optical axis of the crystals, undergoes just one birefringence. On the contrary, a slightly obliquely incident beam will experience also a rotation of its polarization plane, due to the optical activity of quartz. This rotation is small for visible light, but very large for ultraviolet light (145°/mm for 2536.7 Å light propagating along the optical axis). For beams propagating at an angle α with the perpendicular to the crystal surface, the rotation will thus be 145° sin α per mm of layer depth. The beams are perfectly parallel only if (see the sketch of the experimental arrangement [Figure 4]) (i) the light source is located exactly at the focus of the first lens, and (ii) the surface area of the light source is small, or the focal length of the first lens is very long. The first condition can be obtained by careful adjustment, but limits are set to the second, since the total light intensity is proportional to the square of the source diameter, or, which amounts to the same thing, it is inversely proportional to the square of the focal length of the first lens. In our experiments the diaphragm in front of the light source had a 2 mm diameter, and the focal length of the first lens was 8 mm. The thickness of the Soleil available to us was 5 mm. The peripheral light rays thus experienced a rotation of their polarization plane of about 10°. The same difficulties had to be overcome in the analysis of the elliptical light by the use of a Babinet. Our Babinet analyzer consisted of a single very thin wedge (the thickness was some 0.5 mm in the center), therefore the rotation for the obliquely incident rays was not so large as for the Soleil in the path of the primary light beam. Less care is needed for the rotating quartz wedge, since although the oblique rays experience, in addition to the rotation of their polarization plane, also a double refraction, the latter is, however, much smaller than the rotation in the Soleil and Babinet compensators.

angle of 20° with the irradiation direction. It has not been possible to reduce this angle further because of technical difficulties. In a first experiment the vapor was excited by light of arbitrary elliptical polarization in the absence of magnetic fields (of course, the earth's field was compensated). In this case the fluorescence light was also elliptically polarized, and displayed the same ellipticity as the excitation light. If the ellipticity of the excitation beam is varied by shifting the Soleil wedges relatively to each other, the ellipticity of the fluorescence light is correspondingly varied. If the former was circularly polarized, so was the latter. The influence of a magnetic field of a few gauss along the observation direction was the following: When the vapor was excited by circularly polarized light the fluorescence light was itself circularly polarized. When the excitation light was linearly polarized the fluorescence light was completely depolarized. When the polarization of the excitation light was elliptical, with arbitrary phase difference, the fluorescence light was partly circularly polarized and partly unpolarized. In the first case only one of the circularly polarized components of the Zeeman splitting is excited, and therefore only circularly polarized light is emitted. In the second case, according to our model, two left- and right-polarized oscillating components are simultaneously excited, leading to unpolarized fluorescence light. In the third case we can think of the elliptically polarized light as a superposition of a circular component, which gives origin to circularly polarized fluorescence, and of a linear part, which is depolarized by the fluorescence emission. The experiment in the absence of a magnetic field is interpreted, classically, as the simultaneous oscillation of elliptical resonators. According to the new model developed by Bohr, we must look for a quantum theoretical interpretation in the degeneracy of a quantum system in the absence of, or at very low, magnetic fields. This degeneracy explains the phase relations between π and σ components and the fact that excitation by elliptical light always gives origin to fluorescence light of the same ellipticity.

The importance of the experiments on the excitation by circularly polarized light in a magnetic field is due to the fact that, while both σ components were used for excitation in the previous experiments (i.e., in our case linearly polarized light was used, which can be considered as the superposition of two counterrotating circular components), here only one component is used for excitation and, as is to be expected, only that component is observed in reemission. Of course, this is true only as long as no important collisional perturbations are present.

The experiments which we have described were essentially concluded last July; the formulation of their theoretical interpretation has delayed the publication till now. In the meantime further publications about the same theme, apart from the above-mentioned work by Bohr, have appeared. These include a more extensive experimental publication by Wood and

Ellett,[21] where their experiments published a year ago are completed, and a theoretical publication by Eldridge,[22] where the author, apparently without knowledge of most of the publications mentioned at the beginning of this work, presents a new discussion of the experiments by Wood and Ellett from the viewpoint of the classical theory. Eldridge, among other things, also predicts the rotation effect, and a footnote points out that Wood has observed this rotation effect last July. Eldridge does not present further conclusions; at the end of his work he just states briefly that difficulties are encountered in the quantum theoretical interpretation of the rotation effect.

Summary

It is shown that the presence of a magnetic field not only depolarizes the resonance fluorescence light of a mercury vapor excited by linearly polarized light, but also rotates its polarization plane. This effect is no Faraday effect. Its quantum theoretical interpretation is discussed.

The experiments by Wood and Ellett have been extended concerning one aspect of their work which seemed to indicate discrepancies with our theory.

Experiments on excitation of resonance fluorescence by means of elliptically polarized light are described.

This work has been performed with the encouragement and constant collaboration of Prof. Franck, to whom I wish to express my warmest gratitude here, in the Second Physical Institute of the Göttingen University. We are also particularly indebted to the Electrophysics Committee, which provided the means to buy some necessary instruments.

Göttingen, October 29, 1924

---

[21] R. W. Wood and A. Ellett, *Phys. Rev.* **24**, 243 (1924).
[22] J. A. Eldridge, *Phys. Rev.* **24**, 234 (1924).

CHAPTER 2

THE HANLE EFFECT AND ATOMIC PHYSICS

WOJCIECH GAWLIK, DANUTA GAWLIK, AND
HERBERT WALTHER

1. INTRODUCTION

This chapter presents a survey of applications of the Hanle effect to atomic physics. Emphasis is placed on the variety of phenomena studied, rather than on a complete survey of data obtained. If we take, for example, one of the classical applications of the Hanle effect, lifetime measurements, the discussion of all results obtained would fill a chapter of this book by itself. A concise survey of its abundant applications seems more useful at this point, all the more so since especially in connection with laser experiments new and exciting results have recently been obtained and new possibilities opened up. Examples of these are observation of squeezing, effects in connection with the correlated emission laser or applications in nonlinear optics. The more classical uses of the Hanle effect are described in previous reviews, e.g., Happer and Gupta,[1] Pokozan'ev and Skrotskii,[2] and Chaika.[3]

We start with some basic remarks on the Hanle effect and then turn to applications in spectroscopy, collision studies, experiments with strong laser fields, and finally discuss aspects of the Hanle effect in quantum optics.

1.1. General Expression for the Hanle Signal in Terms of the Density Matrix

To obtain the expression for the Hanle signal (see also Chapter 1), it is necessary to calculate the fluorescence intensity of an atom in an external

---

WOJCIECH GAWLIK, DANUTA GAWLIK, AND HERBERT WALTHER • Sektion Physik, Universität München and Max-Planck-Institut für Quantenoptik, D-8046 Garching, Federal Republic of Germany. *Permanent address* of Wojciech Gawlik and Danuta Gawlik: Institute of Physics, Jagellonian University, Reymonta 4, PL-30-059 Kraków, Poland.

magnetic field (see, for example, Cohen-Tannoudji[4]). The intensity of the fluorescence light polarized along $\hat{\varepsilon}_d$ is

$$L_F(t) \propto \langle \hat{\varepsilon}_d \cdot \mathbf{E}^{(-)}(\mathbf{r}, t) \hat{\varepsilon}_d \cdot \mathbf{E}^{(+)}(\mathbf{r}, t) \rangle \tag{1}$$

where angle brackets $\langle\ \rangle$ denote averaging over optical oscillations and $\mathbf{E}^{(\pm)}$ denote the positive- and negative-frequency parts of the electric field operator $\mathbf{E}(\mathbf{r}, t)$. By relating the electric field to the atomic dipole moment, one obtains

$$L_F(t) \propto \langle \hat{\varepsilon}_d \cdot \mathbf{D}^{(+)}(t) \hat{\varepsilon}_d \cdot \mathbf{D}^{(-)}(t) \rangle \tag{2}$$

where $D^{(+)} = D|e\rangle\langle g|$, $D^{(-)} = D|g\rangle\langle e|$, and $D = \langle e|\hat{D}|g\rangle$.

Equation (2) can be expressed in terms of the atomic density operator σ:

$$L_F(\hat{\varepsilon}_d, t) \propto \sum_{\mu m m'} \sigma_{mm'} D_{m'\mu} D_{\mu m} \tag{3}$$

where μ and m, m' denote the sublevels of the lower and upper atomic states, respectively.

The simplest atomic structure for discussion of the Hanle effect is $J_g = 0$ and $J_e = 1$. With the magnetic field directed along Oz, equation (3) yields the intensity of fluorescence light polarized along Ox and Oy:

$$L_F(\hat{\varepsilon}_x) \propto \sigma_{++} + \sigma_{--} - 2\,\text{Re}\,\sigma_{-+}$$

$$L_F(\hat{\varepsilon}_y) \propto \sigma_{++} + \sigma_{--} + 2\,\text{Re}\,\sigma_{-+} \tag{4}$$

The fluorescence intensity depends not only on the populations σ_{++}, σ_{--} of the excited state sublevels $|\pm\rangle$, with reference to $m = \pm 1$, but also on the Zeeman coherence σ_{-+} between these sublevels. This dependence on the coherent mixture is the essence of the Hanle effect. It reflects quantum interference between two light-scattering channels via the $|\pm\rangle$ Zeeman sublevels and represents a one-photon phenomenon.

For very weak and broad-band excitation of the $|\pm\rangle$ sublevels with appropriately polarized light beam, the populations do not depend on the magnetic field:

$$\sigma_{++} = \sigma_{--} = \gamma N_0/\Gamma$$

while the coherence $\sigma_{-+} = \gamma N_0/(\Gamma - 2i\omega_L)$ does so in a resonant way, where N_0 is the overall number of investigated atoms, γ and Γ are the excitation and decay rates, respectively, and ω_L is the Larmor frequency of the excited state (see also Chapter 1 of this book for density matrix formalism for the Hanle effect with weak broad-band light). In the case of unpolarized fluorescence detection the signal is given by

$$L_F = L_F(\hat{\varepsilon}_x) + L_F(\hat{\varepsilon}_y) = 2\gamma N_0/\Gamma = \text{const}(\omega_L)$$

and its intensity does not depend on the magnetic field. For observation of the Hanle effect appropriate polarization thus has to be chosen both for excitation and detection. In particular, with suitable differential detection the coherence effect can be isolated:

$$L_F(\hat{\varepsilon}_x) - L_F(\hat{\varepsilon}_y) \propto 2 \operatorname{Re} \sigma_{-+}$$

A convenient way of detecting only the coherent part of the signal is to use a rotating polarizer and lock-in detection.[2,5]

With increasing light intensities, atomic sublevels get more and more mixed and all density matrix elements, i.e., populations and coherences, are mutually coupled. This results in a resonance dependence of populations on the magnetic field, known as saturation resonances.

In more complex atomic structures, i.e., those with states of higher angular momenta, a general expression (3) describing fluorescence signals takes a more complicated form than equations (4), yet it still contains the dependence on the Zeeman coherences and their decoupling as a function of the magnetic field. For $J_e > 1$ there are more Zeeman coherences influencing the Hanle effect. Those which directly affect the fluorescence are between sublevels differing in their m values by $|\Delta m| \leq 2$, which results from $|\Delta m| \leq 1$ selection rules for the electric dipole matrix elements in equation (3). Nevertheless, with higher light intensities all density matrix elements are coupled and higher-order coherences with $|\Delta m| > 2$ might indirectly affect the fluorescence signal. These phenomena will be discussed in more detail later (see also Figure 9 below).

2. SPECTROSCOPIC APPLICATIONS

2.1. Determination of Atomic Constants

As is well known (see, for example, Chapter 1 of this book), the inverse width of the Hanle signal is a measure of the product of the Landé factor

g and the lifetime of the crossing levels τ. Thus, if one of these parameters is known, the other can be precisely determined. It should be pointed out that τ, which determines the width of the level-crossing resonance, is the "coherence lifetime," which is not always identical to the population lifetime, i.e., the decay constant of the excited state population. Besides spontaneous decay, other relaxation mechanisms (such as collisions or radiation trapping) also contribute to τ, making level-crossing spectroscopy an appropriate technique for studying these effects also. While collisions always destroy excited state coherence, making τ shorter, radiation trapping, i.e., reabsorption by one atom of a photon that has been spontaneously emitted by another one, can preserve the coherence and make τ longer. This is known as coherent diffusion of radiation or coherence narrowing.[6]

Extremely useful for hyperfine structure studies are level crossings occurring in higher magnetic fields because their position is uniquely determined by the strength of the magnetic-dipole and electric-quadrupole hyperfine interactions (see Chapter 1). Derivation of the hyperfine constants from level-crossing experiments in the case $J > \frac{1}{2}$ requires numerical diagonalization of a full (i.e., hyperfine and Zeeman) atomic Hamiltonian.[7]

When atoms with small hyperfine splittings and short-lived states are being studied, the resolution is often limited by the overlap of individual crossing resonances. Below (see Section 2.3 and also Section 5.1), several methods used to overcome this problem are described.

Level-crossing spectroscopy also allows precise studies of the Stark effect and determination of atomic scalar and tensor polarizabilities. First studies in this direction were performed by Hanle himself when, two years after the discovery of the Hanle effect (see Chapter 1), he studied the level-crossing effect in combined electric and magnetic fields.[8] Later, two different approaches were used: one of them employs an electric field to obtain the Hanle signal.[9,10] Though this is the most direct approach, it has the disadvantage that rather high electric fields are required. Much more popular therefore is the use of crossed electric and magnetic fields. The role of the electric field is to perturb the level structure which is studied by the standard level-crossing technique, i.e., by recording the fluorescence intensity as a function of the magnetic field. It is then possible to determine the Stark constants with good precision from the corresponding shifts and modifications of the level-crossing signals (see detailed reviews[11]).

It is worth mentioning that in a way similar to that in the Stark effect studies, level-crossing spectroscopy has been applied for studies of interaction of atoms with AC electromagnetic fields, in particular, when verifying predictions of the "dressed atom" model (see Section 5.1).

As already mentioned, it is not intended to give a complete survey of all atomic data obtained with the Hanle effect. They can be found in earlier reviews.[1-3,12,13]

Level-crossing spectroscopy is generally performed with allowed transitions. There are, however, very interesting applications of this technique to the forbidden $6S-7S$ transition in cesium. Bouchiat and co-workers performed a series of Hanle effect measurements for the purpose of testing and calibrating their atomic parity violation experiment.[14] These measurements determined the relative sign of the scalar and spin-dependent vector part of the polarizability,[15] depolarization cross-sections σ_{CsHe}, lifetime τ_{7S} and scalar polarizability of the $6S-7S$ transition[16] and, furthermore, tested the M1 nature of the transition.[17]

2.2. Measurements of Laser-Level Populations

The population of laser levels is the most important parameter for determining the efficiency of any laser. Its determination and its dependence on parameters such as gas pressure and discharge current in the case of gas lasers are of great importance when searching for new laser transitions or optimizing the efficiency.

There have been two methods allowing laser-level populations of gas lasers to be determined on the basis of the Hanle effect. The first one, devised by Winiarczyk,[18] is based on the "Doppler-broadened" or coherent Hanle effect, which is in fact a resonant Faraday effect observed in forward scattering in an external longitudinal magnetic field. The observed magneto-rotation effect is proportional to the population difference of the two levels of a transition and changes sign when population inversion is achieved. This allows a precise estimate of level population differences. The other method, according to Grigorieva et al.,[19] exploits the proportionality between the Hanle effect signal in a gas laser discharge and the population difference of the studied levels. Again, the change of the sign of the Hanle signal is used to determine the population difference between the levels involved.

2.3. Increasing Resolution, Subnatural Linewidth Effects

The width of level-crossing resonances is given by $(\mu_B \Delta m g \tau)^{-1}$, where Δm is the difference between the magnetic quantum numbers of the crossing levels, g and τ are the Landé factor and lifetime of the investigated level, respectively, and μ_B is the Bohr magneton. In the study of level-crossing signals of atoms with small hyperfine structures and short lifetimes it often happens that various high-field crossing resonances overlap and reduce resolution (Figure 1).

In such a case two different approaches can be applied to improve the resolution. The first is based on the fact that the resonance width decreases

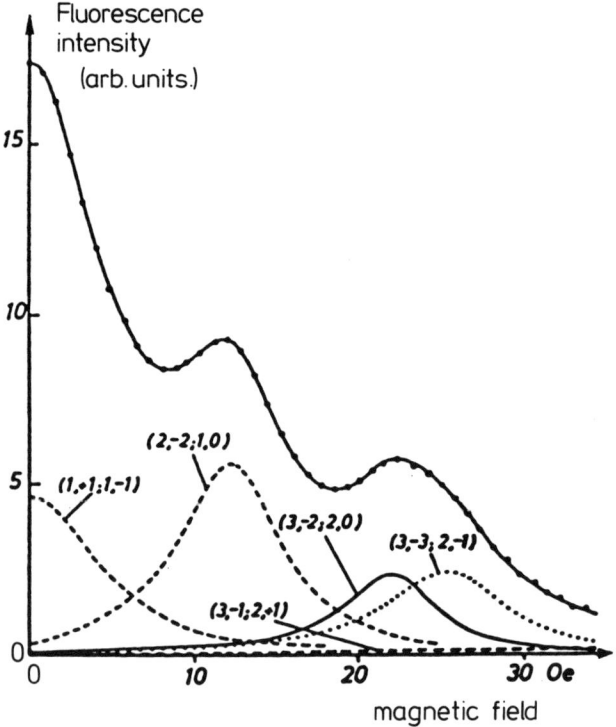

FIGURE 1. Level-crossing ($\Delta m = 2$) experimental signal from the $3\,^2P_{3/2}$ state of sodium (dots) fitted to the theoretical curve calculated with $A = 18.8$ MHz, $B = 2.9$ MHz, and $\tau = 1.59 \times 10^{-8}$ s. Contributions from individual crossings ($Fm; F'm'$) are also shown (from Ref. 20).

for longer lifetimes. Thus, with suitable time-differential and time-delayed fluorescence detection one can obtain narrower signals from those atoms which survive in the excited state longer than the average τ. In this way one can get "subnatural" signals, i.e., narrower than the natural linewidth. The other approach is the change of the relative amplitudes of selected crossing resonances by excitation from a lower state with nonisotropic population distribution, prepared by, for example, optical pumping.

The first experiment where delayed fluorescence detection was applied to level-crossing spectroscopy was performed by Copley et al.[21] for the $3\,^2P_{3/2}$ level in ^{23}Na. (The same idea of time-delayed detection was applied simultaneously in double-resonance experiments of Ma et al.[22,23]) In the experiments with time-delayed detection, abrupt excitation or its termination is necessary to set the time scale precisely. In the work by Ma et al.[22,23] it was easily accomplished by the use of an atomic beam with well-confined and appropriately separated excitation and detection regions. In the experiment of Copley et al. in a vapor cell[21] it was necessary to pulse the exciting light with a Kerr shutter. An effect of pulsed excitation and time-differential detection is not only to narrow the level-crossing and double resonance line shapes, but also to produce significant departures from a Lorentzian.[21,24,25] Such a line shape distortion can be minimized by an appropriate choice of a time-biasing function as was done in the experiment of Deech et al.[26] which resulted in an improvement of the hyperfine structure constants for the $3\,^2P_{3/2}$ state of Na.

The development of pulsed dye lasers allows time-differential experiments to be performed much more easily than with classical light sources. Laser excitation is by far more efficient than classical lamps, and so an exponential signal reduction after a long delay is a less severe problem than in earlier experiments. Figger and Walther[27] performed an experiment similar to those of Copley et al.[21] and Deech et al.,[26] but with a pulsed dye laser as light source. By means of an electronic apodization technique they resolved overlapping crossing signals and obtained accurate values for the hyperfine structure constants of the $3\,^2P_{3/2}$ state sodium.

Another approach to resolving overlapping level-crossing signals was developed by Kraińska-Miszczak[28] and by Baylis.[29] It relies on the use of optical pumping for redistribution of the population of the ground-state sublevels in such a way that some of the overlapping crossings are enhanced while the others are attenuated (Figure 2). The first such measurements were performed by Kraińska-Miszczak, who resolved overlapping crossings in fluorescence from the $4\,^2P_{3/2}$ state of potassium.[30] Later, similar experiments were performed with the $6\,P_1$ state of Yb by Bauer et al.[31] and Liening,[32] who also gave a general theoretical discussion of the level-crossing signals under the influence of optical pumping and buffer gas collisions.

Subnatural linewidth resolution of $2S$ and $2P$ level crossings in hydrogen was achieved by Hartmann and Oed[33] in an atomic beam experiment with pulsed electric fields. Their method is similar to the Ramsey method of spatially separated microwave fields[34] insofar as it also uses the interference-narrowing effect (see also Section 4.2.5).

As mentioned above, signals narrower than the natural (homogeneous) linewidth are sometimes termed "subnatural." Usually, they offer resolution within the natural linewidth and are very interesting for high-resolution spectroscopy. At this point it should be stressed, however, that not all "subnatural" structures allow increased resolution. An example is the ground-state Hanle signal, the width of which is determined by the ground-state relaxation (or transit) time, which is usually much longer than the excited state lifetime. Still, the ultranarrow ground-state Hanle signals are not useful for improving the resolution of the level-crossing spectroscopy,

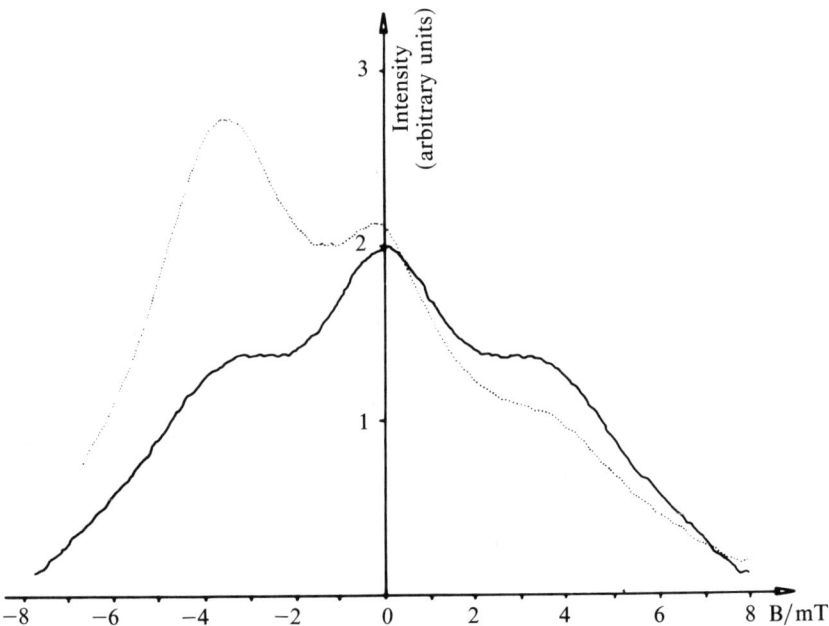

FIGURE 2. $\Delta m = 2$ level-crossing signal from the $4\,^2P_{3/2}$ state of potassium (solid line) and the same signal with additional optical pumping of the ground state (dotted line) (from Ref. 31).

since the fact that they only occur in zero magnetic field means that they have no influence on the excited-state crossings occurring in high magnetic fields.

2.4. Forward Scattering, Line Crossing

In the standard version of the Hanle effect with lateral scattering, the measured signal arises from interference between excitation and decay paths to different sublevels of a single atom. Averaging over the whole atomic ensemble cancels the phase relationship between different atoms and causes scattering from different atoms to be incoherent, resulting in an intensity proportional to the atomic density N. On the other hand, in forward scattering the phase relationship is preserved and all scattering atoms contribute coherently to the scattered intensity. Forward scattering signals thus arise from interference between transitions among sublevels of different atoms. As a result of this coherent character the intensity of forward scattering signals is proportional to N^2.

The theory of forward scattering was formulated in the context of level-crossing and double resonance by Corney et al.[35] According to this theory, the forward scattering signal as a function of the magnetic field can be interpreted as a Doppler-broadened Hanle effect. It has also been designated as the "coherent Hanle effect" by Flussberg et al. in their paper on magnetically induced three-wave mixing.[36]

First studies of the effect were performed with classical light sources and mainly contemplated the coherent character of forward scattering.[35,37-40] The mutual coherence between different scatterers was demonstrated explicitly in a beautiful experiment of Hackett and Series,[41] who observed a forward scattering analog of the high-field level-crossing effect which they termed "line crossing." Line crossings offer the possibility of determining the isotope shifts and were observed in both longitudinal[42,43] and transverse[44] fields. Church and Hadeishi[42] studied the forward scattering Hanle effect at foreign gas pressures so high that the lineshape of the zero-field crossing was a Lorentzian rather than a Doppler profile.

Though physically very interesting, line crossings with one-photon excitation do not seem to be very useful from a spectroscopic point of view, because the accuracy of isotope shift determination is obscured by the Doppler width. On the other hand, studies of forward scattering in the context of the Faraday rather than Hanle effect have been successfully applied for spectroscopic purposes, such as determination of laser level populations[18] or precise measurements of relative oscillator strengths.[45] Recently, studies of relaxation mechanisms with the help of forward scattering were also discussed theoretically by Giraud-Cotton et al.[46] Another interesting application of forward scattering is in trace element detection.[47]

It is worth mentioning here that backward scattering (selective or specular reflection) is of coherent character, too, and has also been used for Hanle effect and line-crossing studies.[48]

2.5. Technical Applications

2.5.1. Magnetometry. As pointed out above, the product $g\tau$ can be determined from the measurements of the width of the Hanle curve $\Delta B_{1/2}$. However, when g, τ, and the exact shape of the level-crossing curve are well known, the Hanle effect can also be used to determine an external magnetic field. The sensitivity of such a measurement is obviously limited by the width of the Hanle curve. It is therefore more appropriate to work with the ground rather than with the excited state. As in the former case, τ is now the relaxation time of the ground state, which can be orders of magnitude larger than the lifetime of an excited state. Indeed, with suitable paraffin coatings of the wall of alkali vapor cells, τ can be of the order of 0.1 s, which corresponds to $\Delta B_{1/2} \approx 10^{-6}$ G.

Possible magnetometric applications of the ground-state Hanle effect have been demonstrated by Dupont-Roc et al.[49] In an improved experiment with a parametric resonance technique,[50,51] magnetic fields of the order of 10^{-9} G were measured by the Paris group.[50,52] Application of the Hanle effect for measurements of weak magnetic fields has also been discussed[2,53–55] and the possibility of determining the different components of the field is considered elsewhere.[56]

2.5.2. Plasma Diagnostics. The Hanle effect has also found some applications in plasma diagnostics. A good example is the work of Kazantsev and Subbotenko,[57] who determined an anisotropic velocity distribution in a high-frequency capacitive discharge.

3. COLLISIONS

3.1. Hanle Effect with Collisional Excitation

There are many atomic systems for which application of the Hanle effect in its original version, i.e., excitation with polarized light, is impossible owing to the very small oscillator strength of the transition, too short a wavelength, or other specific physical or chemical properties. Fortunately, collisional excitation of atoms by an electronic beam can also result in alignment and coherent population in the excited states owing to the conservation of angular momentum.[58] If the particle beam is polarized, then orientation may also be obtained. Excitation at threshold causes, in most cases, a population of the magnetic substates equivalent to optical excitation with linear polarization parallel to the particle velocity.

The principle of Hanle experiments with electronic beam excitation is similar to that using optical excitation and only differs in the size of the atomic alignment. In order to obtain suitable alignment, it is necessary to orient the vector of the electron velocity perpendicular to the applied magnetic field B.[59,60] For that reason, the sample cell is equipped with appropriately situated electrodes. The electrons are produced by a heated cathode and are accelerated in the electric field between the grid and anode to the energies required for atomic excitation. In order to select a spectral line of interest, a monochromator or suitable interference filter must be used. This technique was used to measure the lifetimes of the levels in noble gases,[61–63] cadmium,[64] and potassium ions.[65] Experiments of this type were also performed on calcium atoms of an atomic beam[66] and on calcium, cadmium, and helium excited by electrons in a radio-frequency discharge.[67]

Although very efficient, electronic excitation in the magnetic field suffers from the fact that the electrons are subjected to the influence of the Lorentz force. It results in deflection of the electron trajectories and distortion of the Hanle curves, which sometimes limits the accuracy of the method by as much as 20%.[60] To eliminate the related systematic errors, very thorough analysis of the modification of electronic trajectories must be performed (see, e.g., work by Gorny et al.[63]).

Another problem connected with electron excitation is due to its poor energy selectivity, which is responsible for cascades of alignment from higher excited levels to the level of interest. They are particularly important when the electron energy is set well above threshold to obtain a sufficient signal/noise ratio. In extreme cases the cascades may lead to invertion of the Hanle signal.[68]

There are other collision experiments in which the Hanle signals were observed, such as measurement of the radiative decay rate of Sr ions coherently excited in Penning ionizing collisions[69] or of metal atoms generated in cathodic sputtering in hollow cathode lamps.[70,71] Another very important experimental method is the beam-foil spectroscopy, which in its special version with time-integrated detection allows the lifetimes of ionic states to be determined by means of the Hanle effect.[13] Atomic or ionic beams excited by collisions with a foil acquire a nonstatistical population of the m-sublevels; orientation, alignment, and coherent superposition may be produced. Application of a constant magnetic field destroys the coherence and allows Hanle signals to be observed. Many references to Hanle effect studies with this technique are listed in the review by Andrä.[13]

Another interesting phenomenon related to the Hanle effect is the so-called hidden (also termed latent, concealed, or implicit) alignment discovered by Chaika.[72] Owing to the Doppler effect and velocity anisotropy in beams or in cells with electric discharges, the atomic absorption and/or emission profiles undergo modification which depends in a characteristic way on the magnetic field and atomic lifetime. Hidden alignment occurs even in states which have no angular momentum ($J = 0$).

3.2. Hanle Effect and Optogalvanic Detection

The optogalvanic effect has been known since 1925,[73] but it became a widely used and sensitive spectroscopic method only after the development of tunable lasers. The first spectroscopic application of the optogalvanic effect was reported by Green et al.[74] Subsequently, numerous authors have combined optogalvanic detection with various spectroscopic methods, e.g., Doppler-free two-photon spectroscopy,[75] double resonance, level crossing, and the Hanle effect,[76,77] and also mode-crossing and saturation resonances.[78,79]

The optogalvanic effect is based on the change in the discharge parameters of a plasma exposed to radiation resonant with atomic transitions of the medium. Nonvolatile materials sputtered in a hollow cathode discharge or a heat-pipe oven can also be studied by this technique.[77] Theoretical description of the effect is not simple because many mechanisms contribute to it.[80] Still, it is clear that the probability of collisional ionization of a particle depends on the particular states which are populated. This population is changed by resonant excitation and must therefore be seen in the discharge current. The typical experimental arrangement for optogalvanic detection is shown in Figure 3.

First experimental evidence of the level-crossing effect monitored by the optogalvanic signal was reported by Hannaford and Series,[77] who sputtered zirconium atoms into a neon discharge. For observation of the level-crossing signal the experimental setup of Figure 3 required an additional static magnetic field ($\mathbf{B} \| Oz$) and linear polarization of the laser beam along the Oz direction. An increase in the optogalvanic current with increasing magnetic field B reflects the increase in population of the upper states, since the ionization probability is higher for the upper states than for the lower ones.

Hanle curves observed by optogalvanic detection show some important differences to those simultaneously observed by fluorescence[77]: (1) The signals always increase with the magnetic field, independently of the direction of the polarization vector with respect to the discharge current and also independently of the position of the light beam in the discharge. At the same time the Hanle resonances observed in the fluorescence depend sensitively on the relative orientations of the polarization vectors of the exciting and monitoring beams. (2) The resonance curves are more pointed than the Lorentzian ones. (3) The width of the optogalvanic Hanle resonances falls in the range 3-7 G, while the width of the Hanle resonance

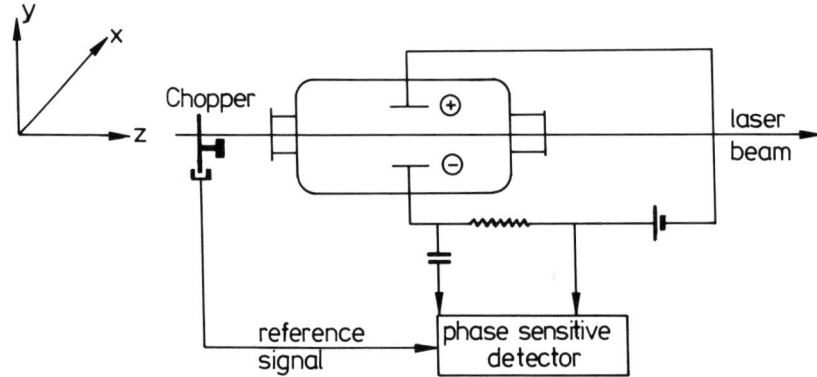

FIGURE 3. Experimental arrangement for optogalvanic detection. Observation of level-crossing effects requires an additional magnetic field along the Oz direction.

measured in the fluorescence under the same conditions is 0.7 G (the Doppler width in that case being 400 G).

3.3. Collision-Induced Hanle Resonances

A perturbation theory treatment of nonlinear effects in atoms, e.g., multiphoton transitions and the dynamic Stark effect, always leads to solutions depending resonantly on energy differences between the initially occupied state and various other states. When damping effects are taken into account in second- and higher-order perturbation theory, resonances may also occur in which the energy difference between two initially unoccupied states is involved.[81] At first, it appears strange that collisions, for example, can lead to new resonances in the coherence. There exists, however, a simple physical interpretation: the phase relationship between different pathways in the multiphoton process, usually interfering destructively, is changed by collisions so that a net collision-induced coherence contribution can show up.[82] The physical reality of extra resonances confirming the correct treatment of damping in nonlinear situations[83] was first demonstrated by Prior et al.[84] in a three-dimensional four-wave light-mixing experiment in Na vapor in helium buffer gas. Two light beams, with wave vectors $\mathbf{k}_1$ and $\mathbf{k}_2$, and frequencies ω_1 and ω_2, respectively, set up a coherence grating at $\omega_1 - \omega_2$ with a grating constant $|\mathbf{k}_1 - \mathbf{k}_2|^{-1}$. This coherence grating is probed by a third light beam at ω_1, with wave vector $\mathbf{k}'_1$, and leads to the generation of a new, fourth beam at $2\omega_1 - \omega_2$ in a new direction $\mathbf{k}_1 + \mathbf{k}'_1 - \mathbf{k}_2$. The frequencies were detuned by $\Delta \approx 1\,\text{cm}^{-1}$ away from the Na D doublet. The authors observed that the intensity of this new coherent beam showed a resonance when $\omega_1 - \omega_2 = 17\,\text{cm}^{-1}$ and the integrated intensity was proportional to the He buffer gas pressure.

Later Grynberg[85] noted that collision-induced coherence may also be described in terms of the dressed atom picture, which permits extension of the theory to very large detunings beyond the impact regime, in which case the line shape ceases to be Lorentzian. Grynberg also called attention to the fact that the two excited states may be Zeeman sublevels of the same electronic state.[86] In this case a resonance would occur when $\omega_1 = \omega_2$, i.e., in zero magnetic field. This level-crossing-type resonance can be probed by changing an external magnetic field. In accordance with these suggestions, experiments on this collision-induced Hanle effect were carried out by Scholz et al.[87,88] in Yb and Ba vapor with Ar as buffer gas to demonstrate these predictions.

While Zeeman resonances in excited manifolds are considerably broadened by collisions, collisions of Na atoms in the ground state with noble gas atoms do not affect the lifetime in the ground-state manifold, which requires a spin flip to be broadened. Collision-enhanced Hanle resonances can therefore be much better observed by using ground state atoms. Bloembergen and co-workers carried out extensive measurements using the four-wave mixing geometry.[89-91] Resonances can be obtained by varying the laser frequency ω_2 at fixed frequency ω_1 and magnetic field B, keeping $\omega_1 - \omega_2$ fixed and changing the magnetic field. For $B = 0$ the Zeeman resonances collapse to a single line occurring when $\omega_1 - \omega_2 = 0$. This resonance may also be observed by keeping $\omega_1 = \omega_2$ fixed and varying B through zero.

The latter case has the important experimental advantage that the laser beams in the four-wave mixing experiment can be derived from one single dye laser, and that the standard geometry for phase-conjugate reflection may be used.

The experimental configuration is sketched in Figure 4. A laser beam with wave vector $\mathbf{k}_1 = |k|\hat{\mathbf{x}}$ passes through a cell containing Na vapor at a temperature in the range of 250-300 °C, corresponding to a vapor pressure of about 1-10 mtorr. A buffer gas (He, Ne, Ar, Xe) is present at pressures of 10^3-10^4 torr (1.5-15 atm). The laser beam is linearly polarized in the $\hat{y}$ direction parallel to B_0 and is reflected back onto itself by a mirror. A second laser beam at the same frequency has a wave vector $\mathbf{k}_2$ which makes a small angle $\theta \approx 5 \times 10^{-3}$ to 5×10^{-2} radian with $\mathbf{k}_1$. This second beam is linearly polarized in the $\hat{z}$ direction perpendicular to B_0. The phase-conjugate beam created by the three incident light waves has a wave vector $-\mathbf{k}_2$.

Figure 5 shows the observed resonance when the magnetic field is varied through zero. The observed width (FWHM) is 35 mG or about 23 kHz

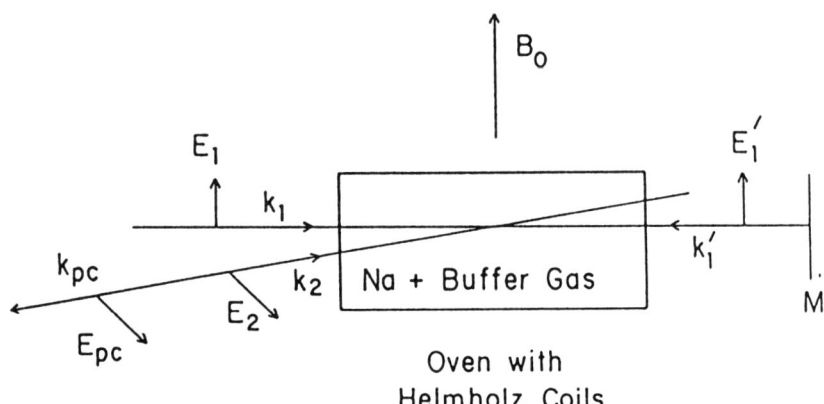

FIGURE 4. Geometry of a collision-induced Hanle experiment with degenerate four-wave mixing. All light beams have the same frequency. The intensity of the phase-conjugate backward wave $|E_{PC}|^2$ is observed as B_0 is varied through zero (from Ref. 91).

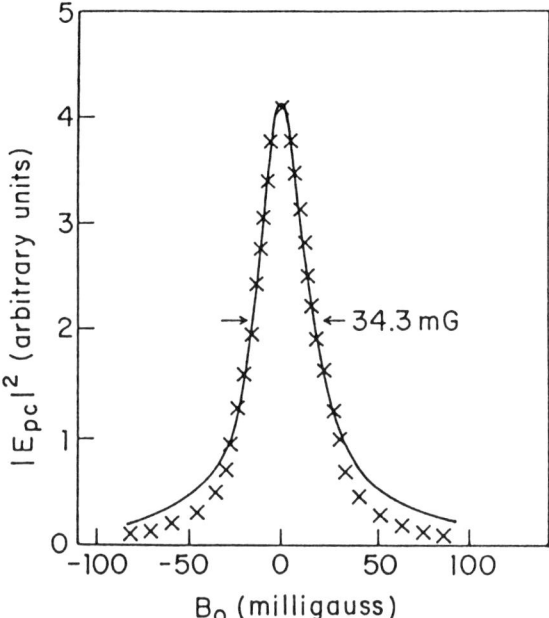

FIGURE 5. Hanle-type resonance in phase-conjugate four-wave light mixing. The solid curve is a Lorentzian fit to the experimental points (×). The sodium vapor was at 277 °C with 3050 torr (~4 atm) of argon buffer gas. Other experimental parameters are given in the text (from Ref. 91).

for the following experimental parameters: $p_{Na} \sim 1$ mtorr, $p_{Ar} \sim$ atm, detuning $\Delta \approx 50$ GHz below the D_1 resonance line, laser power in each incident beam about 0.1 W/cm^2.

This Hanle signal is, indeed, collision-induced. The peak intensity of the resonance increases rapidly with increasing gas pressure, as shown in Figure 6. The observed narrowing of the linewidth with increasing buffer gas pressure is shown in Figure 7. Dicke narrowing is seen to occur at least above $p_{Ar} \sim 10$ atm. The pressure dependence of the four-wave mixing Hanle resonance is in agreement with theory. As seen in Figures 6 and 7, the intensity is proportional to p_{Ar}^2 in a region where the linewidth stays constant (in this region Dicke narrowing is observed). In the region where the linewidth changes proportionally to p_{Ar}^{-1} the peak intensity is proportional to p_{Ar}^3.

The collision-induced Hanle effect allows one to study the spin-exchange collisions in the ground state of the alkaline atoms since their spin-flip processes determine the linewidth of the collision-induced resonance. Other effects which can be investigated are dynamic Stark effects or

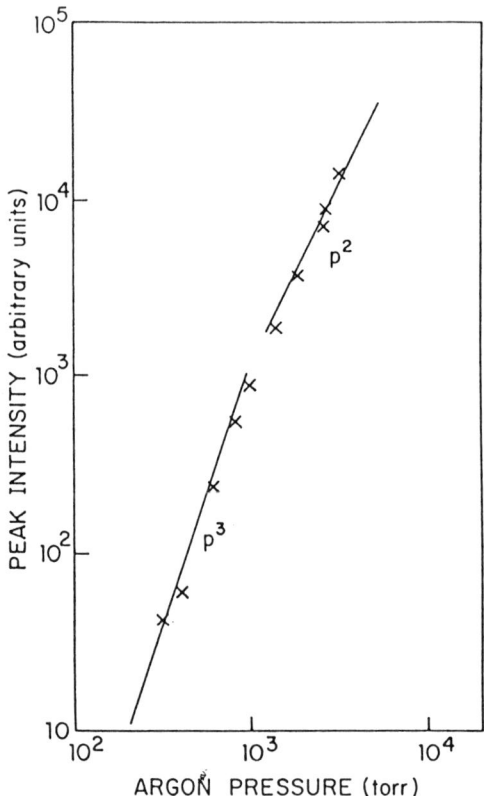

FIGURE 6. Peak intensity of collision-induced Hanle resonance versus argon buffer gas pressure (from Ref. 91). Compare with Figure 7 with respect to the linewidth change.

saturation effects of the absorptive and dispersive parts of the four-wave mixing signal.

The discussion to this point has been entirely based on Zeeman coherences, which can be uniquely identified in the case where an external magnetic field removes all degeneracy between the states. It was shown that the collision-enhanced signal is proportional to the sum of the populations in the pairs of Zeeman levels and occurs in the absence of any optical pumping between the nondegenerate Zeeman sublevels. In fact, rather weak optical pumping effects occur if the polarization of the E_1 and E_2 beams are chosen parallel to each other. The resulting population gratings between nondegenerate states leads to a resonance at $\omega_1 = \omega_2$ even in a strong external magnetic field.[92] The foregoing remarks notwithstanding, there exists a close connection between the collision-enhanced Hanle-type coherence in zero magnetic field and Kastler-type optical pumping (see also Ref. 91). In the strictly degenerate case the coherence due to nonvanishing off-diagonal elements $\sigma_{mm'}$ in a density matrix of Zeeman sublevels can always be transformed to population differences of diagonal elements by rotation of the coordinate frame. The x axis parallel to the direction of the light beams is chosen as the axis of quantization. Since the phase shift between the two perpendicularly polarized beams, $|E_1|\hat{z} \exp(i\mathbf{k}_1 \cdot \mathbf{r} - i\omega t) +$ c.c. and $|E_2|\hat{y} \exp(i\mathbf{k}_2 \cdot \mathbf{r} - i\omega t) + $ c.c., varies as $(\mathbf{k}_1 - \mathbf{k}_2) \cdot \mathbf{r}$, a "pump grating" is created. For $|E_1| = |E_2|$, the electric field will have a right circular polarization at some location $\mathbf{r}_0 + \Delta \mathbf{r}$ with $(\mathbf{k}_1 - \mathbf{k}_2) \cdot \mathbf{r} = \pi(2n+1)$. Collision-assisted transitions to the $3\,^2P_{1/2}$ states will take place. The rapid collisions and subsequent rapid radiative decay permits no significant population build-up in the excited state. At the location $\mathbf{r}_0$ the Kastler-type pumping will produce an excess in the state $M_x = +\frac{1}{2}$. At the location $\mathbf{r}_0 + \Delta \mathbf{r}$ an excess in the state $M_x = -\frac{1}{2}$ will be produced. An $\hat{x}$-directed polarization grating in the Na vapor is thus created, which is probed by the backward wave and leads to a diffracted phase-conjugate beam. At the locations $(\mathbf{k}_1 - \mathbf{k}_2) \cdot \mathbf{r} = \pi(n + \frac{1}{2})$, the optical field will have a linear polarization at an angle of $\pm 45°$ with the $\hat{y}$ and $\hat{z}$ axes. In these locations no polarization will occur. If nuclear spin is taken into account, the $F = 1$ and $F = 2$ levels

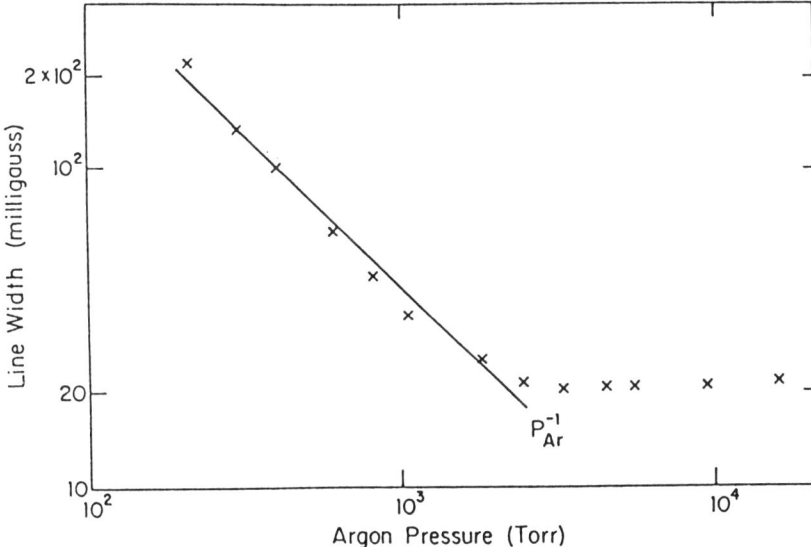

FIGURE 7. Logarithmic plot of the FWHM width of the collision-induced Hanle resonance versus argon buffer gas pressure (see Ref. 91).

might obtain alignment along the 45° angle in the $\hat{y}\hat{z}$ plane. This will not substantially alter the argument of the Hanle signal observed. This discussion shows the close relationship between the collision-enhanced optical Zeeman and hyperfine coherences and the radio-frequency optical pumping schemes.

3.4. Fluctuation-Induced Hanle Resonances

It is not a simple problem to completely account for the effect of laser fluctuations on coherent light scattering (see, for example, Ref. 93); in particular, their influence on coherent four-wave mixing has only recently received attention.

A two-level model of the pressure-induced "extra" resonances in four-wave mixing was proposed by Agarwal and Cooper.[94] Although these resonances require inclusion of three levels, it is, however, possible to project the dynamics onto a two-level space. This model facilitates the description of the effects of laser fluctuations on pressure-induced resonances. The laser fluctuations are treated with phase diffusion models for the incident laser fields, which are assumed to be uncorrelated. The predictions are that the laser bandwidths add to the width of the pressure-induced resonances and that the resonance intensity depends on the laser fluctuations.

Later, Agarwal and Kunasz predicted coherent resonances induced solely by laser fluctuations.[95] Although these resonances occur at $\omega_1 - \omega_2$, they are nonparametric, the emission being at atomic frequencies ω_{eg} and ω'_{eg}. The phase-matching criteria are also different from pressure-induced resonances, and they are destroyed rather than enhanced by collisions. Such resonances confirm the qualitative speculation that the relaxation associated with laser fluctuations can be responsible for cooperative phenomena.

Recently Prior, Schek, and Jortner showed that stochastic phase fluctuations of the pump field are equivalent to dephasing processes such as collisions.[96] Specifically, the laser bandwidth from a phase-diffusion model of the laser field causes exponential relaxation of the off-diagonal elements of the density matrix. Since the phase-diffusion width enters similarly to the collision width, analogous results are obtained; stochastic fluctuation-induced Hanle resonances occur at $\omega_1 - \omega_2$ and have the same emission frequency and phase-matching criterion as pressure-induced resonances. When the fields are correlated, then some time orderings of interactions are preferred and some matrix elements no longer exactly cancel. In this limit of correlated fields, the laser linewidths do not simply add to the resonance width, an observation also confirmed for coherent two-photon absorption.[97]

4. HANLE EFFECT IN STRONG LASER FIELDS

4.1. General Characteristic

The essential difference between the interaction of an atom with weak and strong, coherent electromagnetic fields is that, in the first case, the atom interacts with a photon at most once within the spontaneous lifetime, while in the second case it undergoes coherent interaction with many photons (see Figure 8). (The case of a strong, incoherent field will be discussed later in Section 5.2.)

As seen in Figure 8a for weak light perturbation, the mean time between interactions with successive photons T_p is longer than the atomic lifetime τ. In this case the interactions with subsequent photons are independent and not correlated. For intense light, however, such interactions are correlated. With increasing coherence, the correlation gets stronger and the two-level atom undergoes the well-known Rabi nutations. In this case the optical transition is saturated, i.e., the atom remains about as long in the lower as in the upper state (Figure 8b).

Usually, we are not dealing with an idealized two-level atom but with a real one having at least Zeeman degeneracy in one or both energy levels coupled by the light field. Because of selection rules, each photon couples states differing in their magnetic quantum numbers by $\Delta m = 0, \pm 1$, depending on the polarization. With the weak light source, coherence between sublevels differing by at most $\Delta m = 2$ may thus be created, giving rise to the standard Hanle effect. With the intense light, however, $|\Delta m|$ is no longer limited to 2. In fact, each time an atom undergoes a photon-induced transition, depicted by a straight arrow in Figure 8b, it may also jump to a different magnetic sublevel. As the phase memory of these sublevels is of the order to τ, high-order coherences might be created between levels with $|\Delta m| > 2$. Such higher-order coherences are conveniently described in terms of perturbation theory applied to the solution of the density matrix equations. The perturbative approach has been widely used to describe the Hanle effect with strong light beams.[98-100] Instead of discussing the details,

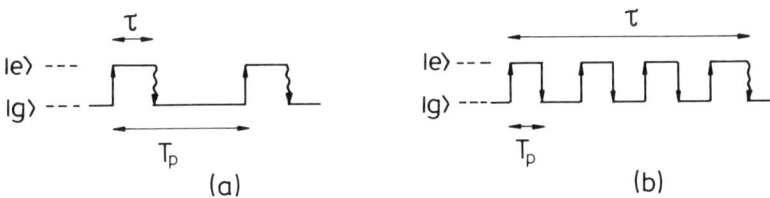

FIGURE 8. Interaction in a weak (a) and strong (b) field. For details see text.

we will present simple diagrams illustrating the essential mechanisms of higher-order couplings in the Hanle effect. In Figure 9 we illustrate coherences induced by linearly polarized light in the multilevel system,

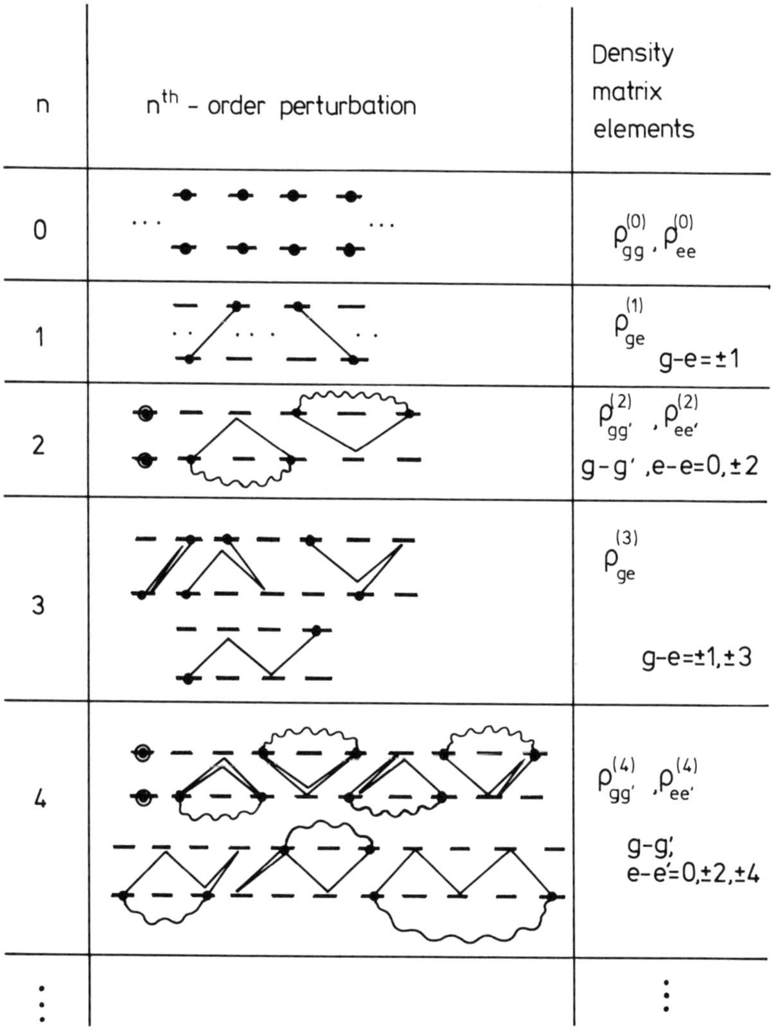

n	n^{th} – order perturbation	Density matrix elements
0		$\rho_{gg}^{(0)}, \rho_{ee}^{(0)}$
1		$\rho_{ge}^{(1)}$ g-e=±1
2		$\rho_{gg'}^{(2)}, \rho_{ee'}^{(2)}$ g-g', e-e=0,±2
3		$\rho_{ge}^{(3)}$ g-e=±1,±3
4		$\rho_{gg'}^{(4)}, \rho_{ee'}^{(4)}$ g-g', e-e'=0,±2,±4
⋮		⋮

FIGURE 9. Diagrams representing various magnetic sublevels of the lower and upper states by horizontal bars, their populations by dots and circles, optical coherences by solid lines, and Zeeman coherences by wavy lines. Multiple lines represent many-photon perturbation of a given transition, sometimes also involving sublevels other than the initial and final ones, which are represented by black dots.

taking into account the Zeeman degeneracy of ground and excited states. The subsequent perturbation orders are displayed from top to bottom.

As seen in Figure 9, the interaction with an intense light beam gives rise to new contributions to the atomic density matrix which are responsible for novel properties of the level-crossing signals. They will be discussed in the following (see also Chapter 4 of this book).

4.1.1. Power Broadening of the Hanle Resonances. Power broadening of Hanle resonances is in principle analogous to the well-known power broadening of optical or radio-frequency resonances of a two-level system (see, for example, Allen and Eberly[101]). As a result of many photon (nonlinear) interactions between the levels coupled by an intense light field (Figure 9) the effective lifetime is shortened and power broadening occurs. Since at least three coupled levels have to be considered for analysis of the Hanle effect, the exact dependence of the resonance width on atomic parameters and the light intensity is far more complex than in the two-level case.[4] The correction for the power broadening of a Hanle signal is essential when lifetimes are determined. The power broadening of the Hanle curves in Ne was investigated in detail by Decomps and Dumont.[102] The importance of proper extrapolation of the experimental linewidth to small laser powers was also pointed out by Cohen-Tannoudji.[4]

4.1.2. Saturation Resonances. It is seen from equations (4) that the level-crossing signals depend directly on possible coherences between sublevel contributions to the detected fluorescence radiation. As a consequence, no intereference contribution resonantly depending on the magnetic field is expected when the state from which the fluorescence is emitted is single ($J = 0$). Nevertheless, the Hanle effect was observed when the state under consideration ($J = 0$) was coupled by strong light to another one where the Zeeman coherence was created[103,104] (see Figure 10). Such an indirect manifestation of the Zeeman coherence induced not in a directly monitored level is known as saturation resonance, and occurs because of mixing of coherences and populations by a strong laser field.

4.1.3. High-Order Coherences. In the weak field limit, i.e., when an atom interacts essentially with a single photon at a time, the light beam can establish coherence only between Zeeman sublevels with $|\Delta m| < 2$. On the other hand, for high light intensity (high perturbation order in the diagrams of Figure 9) several individual linkages obeying selection rules $\Delta m = 0, \pm 1$ couple many magnetic sublevels and establish coherences between sets of sublevels differing in their m-values by more than 2. These higher-order coherences can be attributed to irreducible tensor operators or multipole moments[105-108] of an order higher than 2,[109,110] as can be seen in Figure 9.

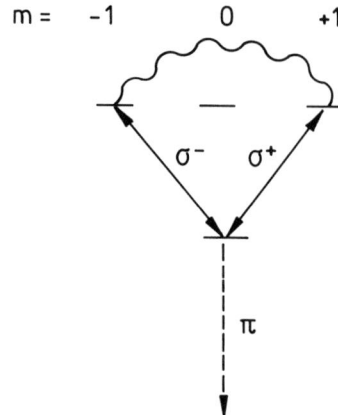

FIGURE 10. Coupling of the Zeeman coherence to a single-level population allows observation of saturation resonances in π-polarized fluorescence from the $J = 0$ level.

These couplings are responsible for generally very complex shapes of the Hanle curves observed with intense laser beams. In addition to the Zeeman coherence effect, there exists another magnetic-field-dependent population effect which also modifies the shape of level- or mode-crossing signals.[104]

With intense excitation, Hanle signals may be observed to be narrower than expected from the natural linewidth of the level from which the fluorescence is emitted. There are two mechanisms responsible for this. On the one hand, this results from convolution of Zeeman and optical coherences. Convolutions of magnetic-field-dependent optical and Zeeman coherences are characterized by lineshapes narrower than those due solely to the Zeeman ones. Therefore, the lineshapes obtained with light intense enough to establish optical coherences are usually narrower than in the case of weak excitation, when only the Zeeman coherences show up.[4,104,111] On the other hand, since the strong field couples the lower-state coherences with the upper-state ones, both will show up in the Hanle signals. If the states have very different relaxation times, e.g., in cases involving ground and excited states, the Hanle curves consist of narrow and broad components. In addition to these two mechanisms, even within a given state the coherences between various pairs of Zeeman sublevels yield different contributions. In particular, the contributions of coherences with different $\Delta m = |m - m'|$ have different widths, first, because of different crossing angles of the appropriate m, m' sublevels in zero field, and second, owing to different power broadenings, which reflect different values of the respective transition probabilities. Different probabilities and Rabi frequencies are also responsible for different ways in which the amplitudes of the contributions with different Δm change with light intensity. Also, upper-state coherences can be transferred to lower states by spontaneous emission,[4,112-114] which can additionally affect the shape of the Hanle signals.

A thorough analysis of couplings between various coherences was presented by Ducloy in terms of nonperturbative formalism and with the broad-line approximation, which is adequate for perturbation by a multimode laser with closely spaced modes.[109] Ducloy also analyzed nonlinear effects in optical pumping of states at the classical limit of large angular momenta.[110] One of the important results of this theory is that the hexadecapole moment (Zeeman coherence with $\Delta m = 4$) is the highest-order multipole that can be observed as Hanle resonance in a fluorescence signal.

4.2. Specific Situations

4.2.1. Fluorescence Detection. First experiments on the Hanle effect with strong light were mostly devoted to studies of nonlinear effects in neon and xenon atoms within gas laser cavities. In this way Fork *et al.*,[115] working with a multimode He-Ne laser, observed so-called mode-crossing resonances occurring in the fluorescence from part of the laser tube in a magnetic field when Zeeman and mode splittings became equal. Their results could only be explained in terms of nonlinear fourth-order theory. Decomps and Dumont[116] applied mode-crossing resonances occurring in fluorescence from an intracavity neon cell to measure the Landé factor. Studying fluorescence from a xenon laser tube in a magnetic field, Tsukakoshi and Shimoda[98] observed fourth-order coupling between alignments (coherences) of the upper and lower levels of a laser transition and determined the lifetimes of these levels from the width of Hanle curves.

Very interesting effects of high-order couplings between upper- and lower-state alignments were observed by Ducloy *et al.* in an intracavity experiment using a neon cell.[117] When the light intensity was strong enough, a contribution of a hexadecapole moment was identified in a complex Hanle curve (Figure 11).

Higher-order couplings between lower- and upper-state coherences were also observed in the intracavity Hanle experiments with a pulsed He-Cd laser by Boyarskii and Kotlikov[118] and with a He-Hg laser by Alipieva and Kotlikov.[119] These observations were analyzed theoretically in a paper by Kotlikov and Kondrateva.[120]

Owing to the high field intensity inside the laser cavity, higher-order nonlinear effects can be studied much more easily and with better signal-to-noise ratio than outside the cavity. However, in the intracavity experiment

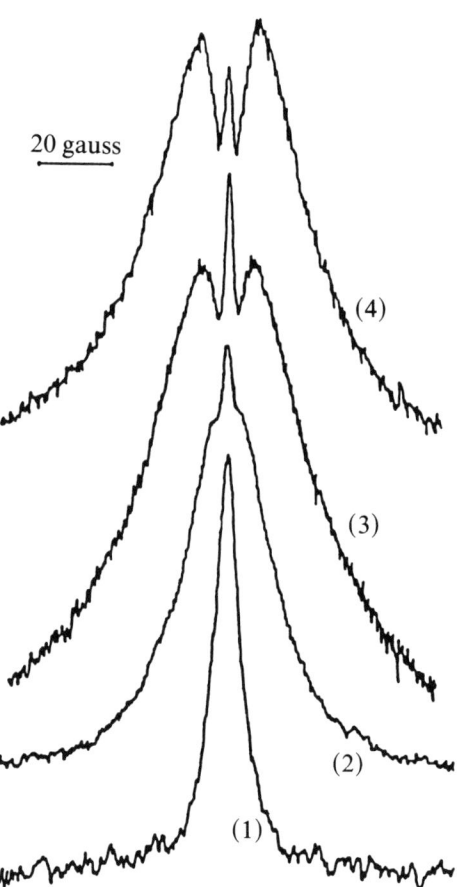

FIGURE 11. $3s_2$ Hanle effect observed on the fluorescence at 7305 Å for laser irradiation at 6328 Å ($3s_2$-$2p_4$). Curves (1), (2), (3), and (4) correspond to laser intensities equal, respectively, to 0.2, 1, 2, and 3.75 mW out of the laser cavity (from Ref. 117).

the laser action itself becomes influenced by the absorbing and dispersive medium under study as well as by the magnetic field. Additional difficulty is due to strong power broadening and to standing-wave and multimode effects in the intracavity measurements. Early intracavity studies were, moreover, limited to several noble gas levels. In spite of these difficulties, thorough investigations of the Paris group yielded results which were also valid in cases studied later. In fact, all of the above-mentioned nonlinear aspects of the Hanle effect in strong fields were studied in these works reviewed elsewhere.[99] Level-crossing studies could be greatly simplified after the advent of tunable dye lasers, which allowed extracavity studies of various elements not necessarily in gas but also in beam form. Theoretical works of Avan and Cohen-Tannoudji[4,111] relating to clean experimental conditions of an atomic beam excited by a single monochromatic running wave clarified and systematized various effects associated with strong light and Hanle effects. Some of their theoretical predictions were verified experimentally.[120-124]

Higher-order coherences were also studied with the Hanle effect in a laser-excited fast ion beam of Ar^+ by Eichhorn et al.[125] By measuring the anisotropy of fluorescence light reemitted by the ion beam after dye laser excitation, the authors detected complex Hanle curves which they attributed to multipole moments of order 2, 4, and 6 and determined lifetimes of two $46p$ levels of Ar^+.

The lower (ground) state coherences are responsible for very interesting effects, which are termed population trapping. In the case where lower-state coherence is established between the Zeeman sublevels which are degenerate in zero magnetic field, population trapping becomes nothing but the ground-state Hanle effect. Such an effect was recently analyzed by McLean et al.[126] for the Ne $2p_3 \to 1s_4$ ($0 \to 1$) transition in an extracavity neon cell with a single-mode dye laser. In a similar way, these authors studied higher-order coherences in the $2p_5 \to 1s_5$ ($J = 1 \to J = 2$) neon transition.[127] In this work a detailed theoretical treatment allowed attribution of specific features of experimental Hanle signals to particular higher-order Zeeman coherences, including the hexadecapole one.

4.2.2. Detection in Absorption or Laser Gain and in Forward Transmission. (a) Stimulated Level and Mode Crossing. When the Zeeman splitting of the sublevels involved in the laser transition equals the frequency separation between the laser modes, coherent mode-crossing effects can be observed in the laser output. Schlossberg and Javan observed such mode crossings in an Xe laser[128] and applied it to the g-value and hyperfine-structure measurements. A similar phenomenon occurring in the laser output is the stimulated level-crossing effect, which manifests itself as a nonlinear change in the induced polarization as appropriate atomic levels overlap within their natural width. Using this effect, Levine et al.[129] precisely determined the hyperfine structure of the $5d$ $(\frac{5}{2})_2^0$ state of Xe.[129] Luntz et al.[130] detected Stark-tuned level-crossing resonances in nonlinear optical absorption of a methane cell in a magnetically tuned 3.39μ He–Ne laser cavity. Stimulated mode-crossing and Hanle effects also influence the polarization and frequency characteristics of radiation emitted by gas lasers in magnetic fields[112,131-135] (see also Ref. 136 and references therein).

A detailed analysis of the stimulated Hanle effect not necessarily confined to laser cavities was presented by Feld et al.,[100] who explored the relationship between the stimulated and spontaneous versions of the level-crossing effect (see also Chapter 4 of this book). Recently, Ståhlberg et al.[137] observed the stimulated Hanle effect in resonance absorption in an extracavity experiment with the 633 nm Ne line, where population and Zeeman coherence effects were clearly resolved.

A very interesting combination of saturated absorption and level-crossing spectroscopy was applied by Colomb et al.[138] to study velocity-changing collisions in an absorption cell placed in a magnetic field.

(b) Forward Scattering. As already mentioned in Section 2.4, forward (or coherent) scattering is the process in which all scattering centers coherently contribute to the scattered intensity, yielding the characteristic N^2 dependence on the atomic (molecular) density N. This offers the possibility of observing effects of interference not only between sublevels of a given atom, but also between sublevels belonging to different atoms. The theory of forward scattering in the context of double resonance and level crossing was elaborated by Corney et al.[35] This type of scattering was studied initially with spectral lamps[35,37,38,41] and later with lasers.[139-143] In the first experiment with a narrow-band CW dye laser, very complex Hanle curves (Figure 12) were observed by Gawlik et al.[140] and attributed to

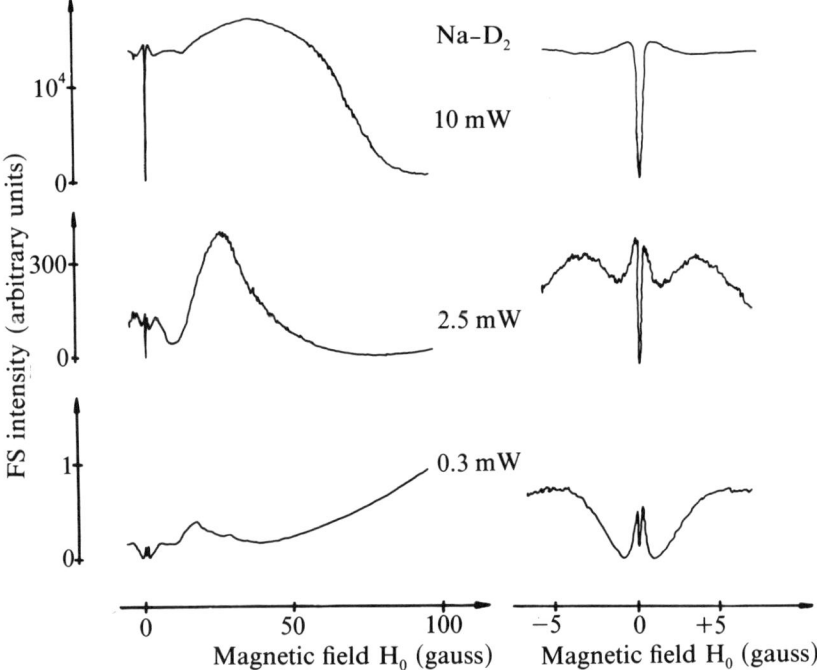

FIGURE 12. Forward-scattering signals from sodium indicating higher-order coherences created in the ground and excited states as a function of the laser power (from Ref. 140).

high-order coherences created in the ground and excited states of sodium. Full interpretation of these signals is very complicated because of the wealth of levels involved and owing to strong saturation, which does not allow perturbative approximation. As seen from recent discussion,[141,142] these results still deserve convincing explanation. Despite open questions concerning interpretation, the results of Gawlik et al.[140] illustrate essential features of the Hanle effect in strong fields very well and in particular the mutual mixing of all states coupled by the light, which leads to high-order coherences. In another experiment with forward scattering of a powerful pulsed dye laser light[139] strong narrowing of the forward-scattering signals was observed as a function of the laser power and ascribed to hexadecapole coherence. Later, more detailed experimental and theoretical analysis[143] showed, however, that these signals were due to excited-state coherence detected under conditions of decoupling between I and J by a strong broad-band laser field. Except for the laser source, the experiments described in Refs. 139 and 143 were nearly identical with that of Ref. 140, yet their results were very much different. In particular, while with monochromatic radiation the forward-scattering signals exhibit very pronounced contribution of the ground-state Hanle effect for both D_1 and D_2 sodium lines, with the pulsed broad-band laser the Hanle effect shows up only in the $3\,^2P_{3/2}$ state, i.e., no Hanle contribution is seen in the case of NaD_1 scattering. These observations were explained by breaking of the hyperfine interaction due to the short coherence time and high power of the laser beam. Because of this decoupling the F quantum number ceases to be a good one and the interaction with the laser field can be described in a reduced J, m_J representation.[144] As there is no possibility of inducing any Zeeman coherence in the $J = \frac{1}{2}$ state with σ linearly polarized light, no ground-state coherence shows up in the experiment with sodium (whose ground state is $3\,^2S_{1/2}$). Similarly, no excited state coherence can be seen for the D_1 ($3\,^2S_{1/2}$–$3\,^2P_{1/2}$) transition and only the excited state coherences can be observed for the D_2 ($3\,^2S_{1/2}$–$3\,^2P_{3/2}$) line. As compared with other detection methods, forward scattering offers very high sensitivity owing to the N^2 signal dependence and an excellent signal-to-noise ratio due to the nearly background-free detection with crossed polarizers. This technique uses the same light beam in order to generate and probe the coherence. Another method was applied by Brand et al.,[145] who used different pump and probe beams for their studies of the Hanle effect in forward scattering in samarium vapor.

Forward-scattering signals observed with strong light beams are, in fact, manifestations of nonlinear Faraday and Voigt effects, depending on whether the magnetic field is oriented longitudinally or transversally to the light beam. From this point of view forward scattering was recently analyzed for a simple $J = 1$–$J = 0$ transition by Drake et al.[146]

It was shown[147] that forward scattering is closely related to the technique of polarization spectroscopy[148] with crossed polarizers. The difference between these methods lies in the way in which the symmetry of the σ^+, σ^- components of the scattered beam is perturbed. In the case of forward scattering it is an external magnetic field which breaks the symmetry, while in polarization spectroscopy it is another, differently polarized light beam, which additionally allows velocity selection, i.e., elimination of Doppler broadening. Should both perturbations act simultaneously, then complex line shapes can be observed which arise from interference between Hanle effect contributions and those due to velocity-selective optical pumping.[147]

(c) *Nonoptical Detection.* As pointed out above, laser-induced couplings between populations and coherences are responsible for transfer of the resonant behavior of the Zeeman coherences to the population of magnetic sublevels, which is known as saturation resonance. Such a coupling of alignments in ground $^2S_{1/2}$ and excited $^2P_{1/2}$ states of sodium with the population of the sublevels of the ground state allowed Picqué to observe the Hanle signal arising from alignments of the two states in a nonoptical way, using a Rabi-type atomic-beam apparatus.[149]

A more practical way of nonoptical detection of level-crossing signals seems to be the much less sophisticated optogalvanic technique described above (Section 3.2).

4.2.3. Optical Hanle Effect.

The essence of the Hanle effect is the generation of coherent superposition of degenerate Zeeman sublevels of an atom by a light beam. As the Zeeman sublevels are split by a magnetic field, the degeneracy is lifted and the coherence decreases, which gives rise to a characteristic magnetic-field dependence of the scattered light intensity or the polarization distribution. The role of the magnetic field is only to lift the degeneracy and to change the coherent mixture between given sublevels. Obviously, it is possible to remove the degeneracy by any other perturbation which acts differently on various m-sublevels. Such perturbation could be either a constant electric field, in which case it is the Stark rather than the Zeeman effect which removes the degeneracy (see Refs. 9 and 10), or an alternating electric field produced by an appropriately polarized light beam, in which case it is the AC Stark or light shift effect[150,151] which removes the degeneracy.

While the Stark splitting has already been used in Hanle effect studies with classical light sources (see Refs. 8-11) it was only possible after the development of strong, tunable lasers to perform an experiment where the light shift owing to a strong off-resonant light field removes the degeneracy. The "optical Hanle effect" was suggested by Kaftandjian *et al.*[152,153] and was experimentally observed by Delsart *et al.*[154] and Kaftandjian *et al.*[155]

with a Ba-atomic beam. Barium is a 0-1 system for which theoretical interpretation is particularly simple. In this experiment the coherence between Zeeman sublevels was created by a linearly polarized resonant, weak laser beam. The role of the magnetic field was played by another strong, nonresonant, circularly polarized laser beam which shifted only one of the Zeeman sublevels. The shift was proportional to the light intensity. When the latter was changed, it was possible to obtain fluorescence signals similar to those of a standard Hanle experiment. The theory of the optical Hanle effect[155] was extended to arbitrary field strengths and bandwidths by Anantha Lakshmi and Agarwal.[156] Furthermore, a study of its spectral features was performed by Ananthalakshmi and Agarwal[157] and the case with Doppler broadening was treated by Kaftandjian et al.[158]

4.2.4. Hanle Effect with Two-Photon Excitation. The strong light intensities available with laser sources make it possible to study the Hanle effect also with two-photon excitation. This could be very attractive from the spectroscopic point of view, since with two-photon excitation higher excited levels of the same parity as the ground state can be reached and the difficulties associated with the detection of scattered light and resonance radiation trapping are eliminated. The two-photon Hanle effect was analyzed theoretically by Kolwas,[159] who distinguished various coherent and population contributions when analyzing the system $J_g = 1 - J_i = 0$ and $J_e = 1$ perturbed by two resonant light beams, where J_g, J_i, and J_e are the angular momenta of the ground, intermediate, and excited states, respectively. A system $J_g = 0 - J_i = 1 - J_e = 0$, complementary to that analyzed by Kolwas,[159] was discussed by Diebold[160] as a two-photon interference and nonlinear analog of the Hanle effect. Experimentally, the two-photon Hanle effect was studied by Jastrzebski and Kolwas,[161] who managed to distinguish the effects of the two-photon coherent transitions discussed in Ref. 159 from the step-by-step ones. Another group of experimental studies of the two-photon Hanle effect was concerned with creating alignment of highly excited states either via two-photon transitions with a single laser[162-165] or via two-step transitions with two lasers.[166] In these experiments, however, no specific two-photon character of such excitation was manifested and they were basically not very different from experiments with classical light sources (see, for example, Ref. 1). In the work of Hunter and Peck,[165] in addition to two-photon excitation, collisional coherence transfer was employed to permit measurements on the ^1F state of Ca, which differs from the ground state ^1S by three units of angular momentum.

A very interesting combination of level-crossing and two-step laser excitation techniques was recently worked out by McLean et al.[167] and applied to determine A/g_j (A being the hyperfine coupling constant) values for four highly excited even-parity and two odd-parity levels of atomic yttrium generated by cathodic sputtering. In this work, laser beam coupling of coherences and populations was exploited to detect level-crossing signals from the upper $e\,^4P_{3/2}$ and intermediate $z\,^4D_{3/2}$ states via the fluorescence from the upper state. This experiment could be regarded as a nonlinear extension of the "multiplicative" stepwise level-crossing effect studied by Kibble and Pancharatnam.[168]

4.2.5. Ramsey Structures. In precision spectroscopy in the radio-frequency and optical regions, the two-field Ramsey method[34] is useful in cases where the resolution is limited by the interaction time of the atoms with the field. An interaction in two subsequent fields can also produce significant narrowing and modification of the Hanle curves. Bertolucci et al.[169] performed an experiment where such narrowing was demonstrated with Hanle signals using a Ca metastable atomic beam. This experiment is closely related to other observations of the Ramsey-like structures with Raman transitions, such as the work of Thomas et al.[170] on ground-state hyperfine coherences and Mlynek et al.[171] on ground-state Zeeman coherences, and also to laser optical pumping studies where polarization rather than alignment was induced by a circularly polarized light beam.[172,173] As follows from Ref. 173, it is possible to observe Ramsey-like structures also in gas cells and not only in beams.

Though ultranarrow resonances were obtained in these experiments, it is important to realize that they offer increased spectral resolution only for the Raman transition in the radio-frequency region and not for the optical one. In particular, the Hanle–Ramsey resonances in zero magnetic field (zero Raman frequency) cannot increase the spectral resolution. With their ultranarrow width, they can nevertheless be very useful for relaxation studies.

4.3. Hanle Effect and Nonlinear Optics

Generally, the Hanle effect is interpreted in terms of single-atom interference. This is absolutely correct and sufficient for describing fluorescence signals, in which case interatomic correlations are usually negligible. Another situation occurs, however, in the case where bulk absorptive or dispersive properties become important, e.g., in the case of forward scattering (Section 4.2.2). The light interaction in such cases is described in terms of bulk polarizability or susceptibility of the scattering medium. The relation of the polarizability tensor with Zeeman coherences has already been pointed out for weak light perturbations by Series[174] and later for laser light by Gawlik.[175]

Closely related to the Hanle effect are also such nonlinear optical effects as magnetically induced second-harmonic generation,[176,177] some four-wave mixing and phase-conjugation phenomena,[178-184] as well as the collision- and fluctuation-induced resonances discussed above in Sections 3.3 and 3.4, respectively. There are also interesting Zeeman coherence effects related to nonlinear optical processes in solids which are beyond the scope of the present chapter (see Chapter 6 of this book).

5. HANLE EFFECT IN QUANTUM OPTICS

5.1. Dressed-Atom Model

The "dressed-atom" model was developed by Cohen-Tannoudji and Haroche[4,185-188] to interpret experiments performed with atoms interacting with radio-frequency fields. In this fully quantum-mechanical approach the field is described in terms of photons and the system under study is the total "atom + field" system which is called the dressed atom.

This model proved useful not only for interpreting well-known phenomena but also for predicting some new physical effects, e.g., modification of the magnetic properties of the atomic system in strong radio-frequency fields.[186,189]

All predictions of the dressed-atom model applied to radio-frequency interaction could in principle be reconstructed in the framework of semiclassical theories. In fact, even the modification of the Landé factor has been interpreted in terms of semiclassical theory,[190] which was experimentally verified by Chapman.[191] Nevertheless, quantization of the radio-frequency field greatly simplifies interpretation since it leads to a time-independent Hamiltonian for the whole isolated atom + field system which is much easier to deal with than the time-dependent Hamiltonian of the classical theories. In particular, higher-order effects such as multiquantum transitions, Bloch-Siegert shift, or Autler-Townes splitting naturally appear in the dressed-atom energy diagrams.

With the development of laser spectroscopy, the dressed-atom model has been extended to study nonlinear phenomena induced by optical photons from intense lasers (the "optically dressed atom").

When eigenenergies of the dressed atom are studied, its energy levels are found to exhibit various crossings and anticrossings. Each of these is responsible for interference between scattering amplitudes if the dressed atom is probed by, for example, light scattering. Such interferences give rise to resonances in the scattered light in the same way as in the standard level-crossing effect. (Detailed investigations on the dressed-atom model in radio-frequency fields are described and referred to by Cohen-Tannoudji[4,188].)

Another interesting experiment interpreted in terms of the dressed-atom model was the level-crossing study of the RF parametric resonance in the $3\,^3P_{3/2}$ state of Na.[192]

The application of the dressed-atom picture to optical interaction is discussed extensively in Ref. 4. It allows a convenient interpretation of the signals in terms of the level anticrossings of the dressed atom. In particular, it also explains quantum optics phenomena such as resonance fluorescence of strongly driven systems and Autler–Townes splitting.

5.2. Hanle Effect with Fluctuating Fields

While the two limiting cases of the Hanle effect excited either with white light or monochromatic excitation are well understood, it is very interesting to study also an intermediate case with strong, yet nonmonochromatic excitation.

The first theoretical work devoted to the Hanle effect with fluctuating field was the paper by Avan and Cohen-Tannoudji.[193] It was found there that level-crossing resonances are quite sensitive to fast fluctuations of the laser beam and could provide information on the higher-order correlation functions of the light field. Recently, Ryan and Bergeman[194] applied a more comprehensive statistical theory[195] to the case of the Hanle effect with a phase-fluctuating field. Their results extend the predictions of Avan and Cohen-Tannoudji to include a wide range of field intensities and fluctuation parameters. These predictions were only recently verified experimentally by Arnett et al.[196] with a very elaborate apparatus. It allowed precise tailoring of the laser band shape and width in such a way that the conditions of the phase diffusion model could be met precisely. The experimental results of Arnett et al. are in good agreement with the predictions of Avan and Cohen-Tannoudji[193] and Ryan and Bergeman.[194]

Other studies of the Hanle effect with nonmonochromatic excitation were performed with forward-scattering detection by Gawlik and Zachorowski.[143] They demonstrated that Hanle signals from sodium atoms obtained when the laser linewidth and/or Rabi frequency was of the order of a hyperfine structure of sodium D lines were very much different from those recorded with nearly monochromatic excitation.[140] This discrepancy is due to the "hyperfine uncoupling" discussed in Refs. 143 and 144. As shown by Zakrzewski and Dohnalik,[197] forward scattering is very sensitive to various properties of the laser field and may therefore also be applied successfully to their study.

Another interesting contribution to studies of the Hanle effect with fluctuating fields is the theory of modulated Hanle signals in partly coherent fields by Saxena and Agarwal,[198] which uses the phase-diffusion model for the laser light and is valid for arbitrary intensities.

5.3. Squeezing in the Hanle Effect

In nonlinear radiation-matter interaction it is possible to generate squeezed light displaying exotic quantum statistics.[199] In some of its aspects this light exhibits less quantum noise than laser light. Consequently, besides serving as a probe to explore new physical phenomena, squeezed light may have applications in sensitive and precision measurement instrumentation.

The physical phenomena which can be used to generate squeezed states include degenerate parametric amplification and four-wave mixing.[200] As shown by Reid et al.,[201] atomic coherence, e.g., between Zeeman sublevels, may lead to nonlinearity appropriate for generation of squeezed states, too. It was also shown that the quantum mechanical features of radiation produced by the Hanle effect can show squeezing characteristics.[202] These phenomena will be discussed in the following.

The situation considered in the calculations is the standard $J = 0$ to $J = 1$ transition. The atoms are excited by a linearly polarized light beam propagating in the magnetic field direction. The fluorescent light is observed perpendicularly to the magnetic field direction. It could be shown that the normally ordered variance of the scattered radiation (when detected phase-sensitively) shows considerable reduction depending on the external parameters such as the magnetic field strength, laser intensity, and polarization.

When the scattered radiation field is detected in the forward direction, i.e., in the same direction as the pump field, the counting distribution of the radiation provides a direct measure of the squeezing characteristics of the radiation produced by the Hanle effect; in this case the problem presented by phase-sensitive detection by using a local oscillator is avoided.

We would like to close this chapter by mentioning that the reduction of fluctuations of fluorescence radiation mentioned here is related to the correlated spontaneous emission and the Hanle laser described in another chapter of this book. It should also be mentioned that fluorescence radiation alone also exhibits squeezing[200] without Zeeman coherences being effective. The actual detection of the reduced noise is in this case, however, far more difficult than in the case of the Hanle effect.

ACKNOWLEDGMENT. One of the authors (W.G.) is grateful to the Alexander von Humboldt Foundation for awarding a grant.

REFERENCES

1. W. HAPPER, *Rev. Mod. Phys.* **44**, 169 (1972); W. HAPPER AND R. GUPTA, in *Progress in Atomic Spectroscopy*, Part A, edited by W. Hanle and H. Kleinpoppen (Plenum Press, New York, 1978), p. 391.
2. V. G. POKAZAN'EV AND G. V. SKROTSKII, *Usp. Fiz. Nauk* **107**, 623 (1972) [*Sov. Phys. Usp.* (*Engl. Transl.*) **15**, 452 (1973)].

3. M. P. CHAIKA, *Interference of Degenerate Atomic States* (Izdatelstvo Leningradskovo Universiteta, Leningrad, 1975), in Russian.
4. C. COHEN-TANNOUDJI, in *Frontiers in Laser Spectroscopy*, Vol. 1, edited by R. Balian, S. Haroche, and S. Liberman, Proc. Les Houches Summer School on Theoretical Physics, Session XXVII, 1975 (North-Holland, Amsterdam, 1977), p. 3.
5. E. HANDRICH, H. KRETZEN, W. LANGE, A. STEUDEL, R. WALLENSTEIN, AND H. WALTHER, in *Proceedings of the International Conference on Optical Pumping and Atomic Line Shape OPaLS, Warsaw, 1968*, edited by T. Skaliński (Państwowe Wydawnictwo Naukowe, PWN, Warszawa, 1969), p. 417.
6. J. P. BARRAT, *J. Phys. Radium* **20**, 541, 633 (1959).
7. J. KAPELEWSKI AND K. ROSIŃSKI, *Acta Phys. Pol.* **28**, 177 (1965).
8. W. HANLE, *Z. Phys.* **35**, 346 (1926).
9. A. KHADJAVI, W. HAPPER, AND A. LURIO, *Phys. Rev. Lett.* **17**, 463 (1966).
10. N. D. BHASKAR AND A. LURIO, *Phys. Rev. A* **10**, 1685 (1974).
11. A. M. BONCH-BRUEVICH AND V. A. KHODOVOI, *Usp. Fiz. Nauk* **93**, 7i (1967) [*Sov. Phys. Usp. (Engl. Transl.)* **10**, 637 (1967)]; K. J. KOLLATH AND M. C. STANDAGE, in *Progress in Atomic Spectrsocopy*, Part B, edited by W. Hanle and H. Kleinpoppen (Plenum Press, New York, 1979), p. 955; G. von OPPEN, *Comments At. Mol. Phys.* **15**, 87 (1984).
12. M. ELBEL, in *Progress in Atomic Spectroscopy*, Part B, edited by W. Hanle and H. Kleinpoppen (Plenum Press, New York, 1979), p. 1299.
13. H. J. ANDRÄ, in *Progress in Atomic Spectroscopy*, Part B, edited by W. Hanle and H. Kleinpoppen (Plenum Press, New York, 1978), p. 829.
14. M. A. BOUCHIAT, J. GUENA, AND L. POTTIER, *J. Phys. (Paris)* **46**, 1897 (1985).
15. M. A. BOUCHIAT, J. GUENA, AND L. POTTIER, *Opt. Commun.* **37**, 265 (1981).
16. M. A. BOUCHIAT, J. GUENA, AND L. POTTIER, *J. Phys. Lett. (Paris)*, **45**, L523 (1984).
17. M. A. BOUCHIAT, J. GUENA, AND L. POTTIER, *Opt. Commun.* **51**, 243 (1984).
18. W. WINIARCZYK, *Acta Phys. Polon.* **A52**, 157 (1977).
19. V. N. GRIGORIEVA, P. R. KARAVASILEV, AND G. C. TODOROV, *Bulg. J. Phys.* **11**, 273 (1984).
20. M. BAUMANN, *Z. Naturforsch., A* **24**, 1049 (1969).
21. G. COPLEY, B. P. KIBBLE, AND G. W. SERIES, *J. Phys. B* **1**, 724 (1968).
22. I. J. MA, G. ZU PUTLITZ, AND G. SCHÜTTE, *Z. Phys.* **208**, 276 (1968).
23. I. J. MA, J. MERTENS, G. ZU PUTLITZ, AND G. SCHÜTTE, *Z. Phys.* **208**, 352 (1968).
24. R. C. HILBORN AND R. L. DE ZAFRA, *J. Opt. Soc. Am.* **62**, 1492 (1972).
25. P. SCHENCK, R. C. HILBORN, AND H. METCALF, *Phys. Rev. Lett.* **31**, 189 (1973).
26. J. S. DEECH, P. HANNAFORD, AND G. W. SERIES, *J. Phys. B* **7**, 1131 (1974).
27. H. FIGGER AND H. WALTHER, *Z. Phys.* **267**, 1 (1974).
28. M. KRAIŃSKA-MISZCZAK, *Bull. Acad. Pol. Sci., Ser. Sci. Math. Astron. Phys.* **15**, 595 (1967).
29. W. E. BAYLIS, *Phys. Lett.* **26A**, 414 (1968).
30. M. KRAIŃSKA-MISZCZAK, *Acta Phys. Pol.* **35**, 745 (1969).
31. M. BAUER, M. BAUMANN, AND H. LIENING, *Phys. Lett.* **60A**, 101 (1977).
32. H. LIENING, *Z. Phys. A* **320**, 363 (1985).
33. W. HARTMANN AND A. OED, *J. Phys. B* **12**, 31 (1979).
34. N. R. RAMSEY, *Molecular Beams* (Oxford University Press, London, 1956), p. 124.
35. A. CORNEY, B. P. KIBBLE, AND G. W. SERIES, *Proc. R. Soc. London, Ser. A* **293**, 70 (1966).
36. A. FLUSSBERG, T. MOSSBERG, AND S. R. HARTMANN, in *Coherence and Quantum Optics IV*, edited by L. Mandel and E. Wolf (Plenum Press, New York, 1978), p. 695.
37. A. V. DURRANT AND B. LANDHEER, *J. Phys. B* **4**, 1200 (1971).
38. I. KRÓLAS AND W. WINIARCZYK, *Acta Phys. Pol.* **A41**, 785 (1972).
39. D. GAWLIK, W. GAWLIK, AND H. KUCAŁ, *Acta Phys. Pol.* **A43**, 627 (1973).
40. A. V. DURRANT, *J. Phys. B* **5**, 133 (1972).

41. R. Hackett and G. W. Series, *Opt. Commun.* **2**, 93 (1970).
42. D. A. Church and T. Hadeishi, *Phys. Rev. A* **8**, 1864 (1973).
43. G. Stanzel, *Phys. Lett.* **47A**, 283 (1974).
44. W. Siegmund and A. Scharmann, *Z. Phys. A* **276**, 19 (1976).
45. W. Gawlik, J. Kowalski, R. Neumann, H. B. Wiegeman, and K. Winkler, *J. Phys. B* **12**, 3873 (1979).
46. S. Giraud-Cotton, V. P. Kaftandjian, and L. Klein, *Phys. Rev. A* **32**, 2211 (1985).
47. M. Ito, S. Murayama, K. Kayama, and M. Yamamoto, *Spectrochim. Acta* **32B**, 347 (1977); H. Debus, W. Hanle, A. Scharmann, and P. Wirz, *Spectrochim. Acta* **36B**, 1015 (1981).
48. G. W. Series, *Proc. Phys. Soc.* **91**, 432 (1967); G. Stanzel, *Phys. Lett.* **41A**, 335 (1972).
49. J. Dupont-Roc, S. Haroche, and C. Cohen-Tannoudji, *Phys. Lett.* **28A**, 638 (1969).
50. C. Cohen-Tannoudji, J. Dupont-Roc, S. Haroche, and F. Laloë, *Phys. Rev. Lett.* **22**, 758 (1969).
51. J. Dupont-Roc, *Rev. Phys. Appl.* [*Suppl. J. Phys.* (*Paris*)] **5**, 853 (1970).
52. C. Cohen-Tannoudji, J. Dupont-Roc, S. Haroche, and F. Laloë, *Rev. Phys. Appl.* [*Suppl. J. Phys.* (*Paris*)] **5**, 95, 102 (1972).
53. E. B. Aleksandrov, A. M. Bonch-Bruevich, and V. A. Khodovoi, *Opt. Spektrosk.* **23**, 282 (1967 [*Opt. Spectrosc.* **23**, 151 (1967)].
54. L. N. Novikov, V. G. Pokazaniev, and G. V. Skrotskii, *Usp. Fiz. Nauk* **101**, 273 (1970) [*Sov. Phys. Usp.* (*Engl. Transl*) **13**, 384 (1970)].
55. A. L. Kotkin, V. V. Majorshin, and R. M. Umarkhodzhaev, *Radiotekh. Elektron.* **25**, 1038 (1980) [*Radio Eng. Electron. Phys.* (*Engl. Transl.*) **25**, 104 (1980)].
56. J. Dupont-Roc, *J. Phys.* (*Paris*) **32**, 135 (1971).
57. S. A. Kazantsev and A. V. Subbotenko, *J. Phys. D* **20**, 741 (1987).
58. W. E. Lamb, *Phys. Rev.* **105**, 559 (1957).
59. J. C. Pebay-Peyroula, in *Physics of the One- and Two-electron Atoms*, edited by F. Bopp and H. Kleinpoppen (North-Holland, Amsterdam, 1969).
60. A. Corney, *Atomic and Laser Spectroscopy* (Clarendon Press, Oxford, 1974).
61. A. Faure, O. Nedelec, and J. C. Pebay-Peyroula, *C.R. Acad. Sci. Paris* **256**, 5088 (1963).
62. C. B. Carrington and A. Corney, *J. Phys. B* **4**, 849 (1971).
63. M. B. Gorny, B. G. Matisov, and S. A. Kazantsev, *Z. Phys. A* **322**, 25 (1985).
64. L. Frasiński, *Acta Phys. Pol.* **A60**, 867 (1981).
65. D. Z. Zhechev and I. I. Koleva, *Phys. Scr.* **34**, 221 (1986).
66. M. Chenevier, J. Dufayard, and J. C. Pebay-Peyroula, *Phys. Lett.* **25A**, 283 (1967).
67. M. Lombardi and J. C. Pebay-Peyroula, *C.R. Acad. Sci. Paris* **261**, 1485 (1965).
68. C. G. Carrington, *J. Phys. B* **5**, 1572 (1972).
69. D. W. Fahey, W. F. Parks, and L. D. Schearer, *J. Phys. B* **12**, 4619 (1979).
70. E. E. Gibbs and P. Hannaford, *J. Phys. B* **9**, L225 (1976).
71. D. Zhechev, *Opt. Spektrosk.* **49**, 465 (1980) [*Opt. Spectrosc.* (Engl. Transl.) **49**, 253 (1980)].
72. M. P. Chaika, *Opt. Spektrosk.* **30**, 822 (1971) [*Opt. Spectrosc.* (*Engl. Transl.*) **30**, 443 (1971)]; see also N. I. Kaliteevski and M. Tschaika, in *Atomic Physics* 4, edited by G. zu Putlitz, E. W. Weber, and A. Winnacker (Plenum Press, New York, 1975), p. 19.
73. P. D. Foote and F. L. Mohler, *Phys. Rev.* **26**, 195 (1925).
74. R. B. Green, R. A. Keller, G. G. Luther, P. K. Schenck, and J. C. Travis, *Appl. Phys. Lett.* **29**, 727 (1976).
75. J. E. H. Goldsmith, A. I. Ferguson, J. E. Lawler, and A. L. Schawlow, *Opt. Lett.* **4**, 230 (1979).
76. N. Beverini and M. Inguscio, *Nuovo Cimento Lett.* **20**, 10 (1980).

77. P. HANNAFORD AND G. W. SERIES, *J. Phys. B* **14**, L661 (1981).
78. P. HANNAFORD AND G. W. SERIES, *Phys. Rev. Lett.* **48** 1326 (1982).
79. P. HANNAFORD AND G. W. SERIES, *Opt. Commun.* **41**, 427 (1982).
80. J. E. LAWLER, *Phys. Rev. A* **22**, 1025 (1980).
81. N. BLOEMBERGEN, *Nonlinear Optics* (Benjamin, New York, 1965), p. 29.
82. M. W. DOWNER, A. R. BOGDAN, AND N. BLOEMBERGEN, in *Laser Spectroscopy V*, edited by A. R. W. McKeller, T. Oka, and B. Stoicheff (Springer, Heidelberg, 1981), p. 157.
83. N. BLOEMBERGEN, H. LOTEM, AND R. T. LYNCH JR., *Indian J. Pure Appl. Phys.* **16**, 151 (1978).
84. Y. PRIOR, A. R. BOGDAN, M. DAGENAIS, AND N. BLOEMBERGEN, *Phys. Rev. Lett.* **46**, 111 (1981).
85. G. GRYNBERG, *J. Phys. B* **14**, 2089 (1981).
86. G. GRYNBERG, *Opt. Commun.* **38**, 439 (1981).
87. R. SCHOLZ, J. MLYNEK, A. GIERULSKI, AND W. LANGE, *Appl. Phys.* **B28**, 191 (1982).
88. R. SCHOLZ, J. MLYNEK, AND W. LANGE, *Phys. Lett.* **51**, 1761 (1983).
89. N. BLOEMBERGEN, M. W. DOWNER, AND L. J. ROTHBERG, in *Atomic Physics 8*, edited by I. Lindgren, A. Rosen, and S. Svanberg (Plenum Press, New York, 1983), p. 71.
90. L. J. ROTHBERG AND N. BLOEMBERGEN, *Phys. Rev. A* **30**, 820 (1984).
91. N. BLOEMBERGEN, Y. H. ZOU, AND L. J. ROTHBERG, *Phys. Rev. Lett.* **54**, 186 (1985); N. BLOEMBERGEN, *Ann. Phys. Fr.* **10**, 681 (1985).
92. L. J. ROTHBERG AND N. BLOEMBERGEN, *Phys. Rev. A* **30**, 2327 (1984).
93. G. S. AGARWAL AND S. SINGH, *Phys. Rev. A* **25**, 3195 (1982).
94. G. S. AGARWAL AND J. COOPER, *Phys. Rev. A* **26**, 2761 (1982).
95. G. S. AGARWAL AND C. V. KUNASZ, *Phys. Rev. A* **27**, 996 (1983).
96. Y. PRIOR, I. SCHEK, AND J. JORTNER, *Phys. Rev. A* **31**, 3775 (1985).
97. D. S. ELLIOTT, M. W. HAMILTON, K. ARNETT, AND S. J. SMITH, *Phys. Rev. Lett.* **53**, 439 (1985).
98. M. TSUKAKOSHI AND K. SHIMODA, *J. Phys. Soc. Jpn.* **264**, 146 (1969).
99. B. DECOMPS, M. DUMONT, AND M. DUCLOY, in *Laser Spectroscopy of Atoms and Molecules*, edited by H. Walther, Topics in Applied Physics, Vol. 2 (Springer, Berlin, 1976), p. 284.
100. M. S. FELD, A. SANCHEZ, A. JAVAN, AND B. J. FELDMAN, in *Methodes de spectroscopie sans largeur Doppler de niveaux excites de systemes moleculaires simples*, Colloques Internationaux du C.N.R.S., No. 217 (1974), p. 87.
101. L. ALLEN AND J. H. EBERLY, *Optical Resonance and Two-level Atoms* (Wiley, London, 1975).
102. B. DECOMPS AND M. DUMONT, *C.R. Acad. Sci. Paris* **262B**, 1004 (1966).
103. M. DUCLOY, *Opt. Commun.* **3**, 205 (1971).
104. M. DUMONT, *Phys. Rev. Lett.* **28**, 1357 (1972); *J. Phys. (Paris)* **33**, 971 (1972).
105. U. FANO, *Rev. Mod. Phys.* **29**, 74 (1957).
106. M. I. DYAKONOV, *Zh. Eksp. Teor. Fiz.* **47**, 2213 (1964) [*Sov. Phys. JETP (Engl. Transl.)* **20** 1484 (1965)].
107. M. DUMONT AND B. DECOMPS, *J. Phys. (Paris)* **29**, 181 (1968).
108. A. OMONT, *Prog. Quantum Electron.* **5**, 69 (1977).
109. M. DUCLOY, *Phys. Rev. A* **8**, 1844 (1973); **9**, 1319 (1974).
110. M. DUCLOY, *J. Phys. B* **9**, 357; 699 (1976).
111. P. AVAN AND C. COHEN-TANNOUDJI, *J. Phys. Lett. (Paris)*, **36**, L85 (1975).
112. M. I. DYAKONOV AND V. I. PEREL, *Opt. Spektrosk.* **20**, 472 (1966) [*Opt. Spectrosc. (Engl. Transl.)* **20**, 257 (1966)].
113. M. DUCLOY AND M. DUMONT, *J. Phys. (Paris)* **31**, 419 (1970).
114. M. DUCLOY, E. GIACOBINO-FOURNIER, AND B. DECOMPS, *J. Phys. (Paris)* **31**, 533 (1970).

115. R. L. FORK, L. E. HARGROVE, AND M. A. POLLACK, *Phys. Rev. Lett.* **12**, 705 (1964).
116. B. DECOMPS AND M. DUMONT, *C.R. Acad. Sci. Paris* **262B**, 1695 (1966).
117. M. DUCLOY, M. P. GORZA, AND B. DECOMPS, *Opt. Commun.* **8**, 21 (1973).
118. K. K. BOYARSKII AND E. N. KOTLIKOV, *Kvant. Elektron.* **2**, 23 (1975) [*Sov. J. Quantum Electron.* (*Engl. Transl.*) **5**, 10 (1975)].
119. E. A. ALIPIEVA AND E. N. KOTLIKOV, *Opt. Spektrosk.* **45**, 1026 (1978) [*Opt. Spectrosc.* (*Engl. Transl.*) **45**, 833 (1978)].
120. E. N. KOTLIKOV AND V. A. KONDRATEVA, *Opt. Spektrosk.* **48**, 667 (1979) [*Opt. Spectrosc.* (*Engl. Transl.*) **48**, 367 (1980)].
121. W. RASMUSSEN, R. SCHIEDER, AND H. WALTHER, *Opt. Commun.* **12**, 315 (1974).
122. H. BRAND, W. LANGE, J. LUTHER, B. NOTTBECK, AND H. W. SCHRÖDER, *Opt. Commun.* **13**, 286 (1975).
123. J. L. PICQUÉ, *J. Phys. B* **11**, L59 (1978).
124. J. D. CRESSER, J. HÄGER, G. LEUCHS, M. RATEIKE, AND H. WALTHER, in *Dissipative Systems in Quantum Optics, Resonance Fluorescence, Optical Bistability, Superfluorescence*, edited by R. Bonifacio (Springer, Berlin, 1982), p. 22.
125. A. EICHHORN, M. ELBEL, W. KAMKE, R. QUAD, AND H. J. SEIFNER, *Physica* **24C**, 282 (1984).
126. R. J. MCLEAN, R. J. BALLAGH, AND D. M. WARRINGTON, *J. Phys. B* **18**, 2371 (1985).
127. R. J. MCLEAN, R. J. BALLAGH, AND D. M. WARRINGTON, *J. Phys. B* **18**, 3477 (1986).
128. H. R. SCHLOSSBERG AND A. JAVAN, *Phys. Rev.* **150**, 267 (1966); *Phys. Rev. Lett.* **17**, 1242 (1966).
129. J. S. LEVINE, P. A. BONCZYK, AND A. JAVAN, *Phys. Rev. Lett.* **22**, 267 (1969).
130. A. C. LUNTZ, R. G. BREWER, K. L. FORSTER, AND J. D. SWALEN, *Phys. Rev. Lett.* **23**, 951 (1969).
131. W. CULSHAW AND J. KANNELAND, *Phys. Rev.* **133**, A691 (1964); **136**, A1209 (1964); **141**, 228 (1966).
132. M. I. DYAKONOV, *Zh. Eksp. Teor. Fiz.* **49**, 1169 (1965) [*Sov. Phys. JETP* (*Engl. Transl.*) **22**, 812 (1966)]; M. I. DYAKONOV AND V. I. PEREL, *Zh. Eksp. Teor. Fiz.* **50**, 448 (1966) [*Sov. Phys. JETP* (*Engl. Transl.*) **23**, 298 (1966)]; M. I. DYAKONOV AND S. A. FRIDRIKHOV, *Usp. Fiz. Nauk* **90**, 565 (1966) [*Sov. Phys. Usp.* (*Engl. Transl.*) **9**, 837 (1967)].
133. W. VAN HAERINGEN, *Phys. Rev.* **158**, 256 (1967).
134. M. SARGENT III, W. E. LAMB JR., AND R. L. FORK, *Phys. Rev.* **164**, 436, 450 (1967).
135. W. J. TOMLINSON AND R. L. FORK, *Phys. Rev.* **164**, 466 (1967).
136. A. P. VOITOVICH, *J. Sov. Laser Research* (*Engl. Transl.*) **8**, 551 (1987).
137. B. STÅHLBERG, M. LINDBERG, AND P. JUNGNER, *J. Phys. B* **18**, 627 (1985).
138. I. COLOMB, M. GORLICKI, AND M. DUMONT, *Opt. Commun.* **21**, 289 (1977).
139. W. GAWLIK, J. KOWALSKI, R. NEUMANN, AND F. TRÄGER, *Phys. Lett.* **48A**, 283 (1974).
140. W. GAWLIK, J. KOWALSKI, R. NEUMANN, AND F. TRÄGER, *Opt. Commun.* **12**, 400 (1974).
141. S. GIRAUD-COTTON, V. P. KAFTANDJIAN, AND L. KLEIN, *Phys. Lett.* **88A**, 453 (1982).
142. W. GAWLIK, *Phys. Lett.* **89A**, 278 (1982).
143. W. GAWLIK AND J. ZACHOROWSKI, *J. Phys. B* **20**, 5939 (1987).
144. A. EKERT AND W. GAWLIK, *Phys. Lett.* **121A**, 175 (1987).
145. H. BRAND, K. H. DRAKE, W. LANGE, AND J. MLYNEK, *Phys. Lett.* **75A**, 345 (1980).
146. K. H. DRAKE, W. LANGE, AND J. MLYNEK, *Opt. Commun.* **66**, 315 (1988).
147. W. GAWLIK AND G. W. SERIES, in *Laser Spectroscopy IV*, edited by H. Walther and K. W. Rothe (Springer, Berlin, 1979), p. 210.
148. C. WIEMAN AND T. W. HÄNSCH, *Phys. Rev. Lett.* **36**, 1170 (1976).
149. J. L. PICQUÉ, *J. Phys. B* **11**, L105 (1978).
150. W. HAPPER, *Prog. Quantum Electron.* **1**, 51 (1971).
151. C. COHEN-TANNOUDJI AND J. DUPONT-ROC, *Phys. Rev. A* **5**, 968 (1972).

152. V. P. Kaftandjian and L. Klein, *Phys. Lett.* **62A**, 317 (1977).
153. V. P. Kaftandjian, L. Klein, and W. Hanle, *Phys. Lett.* **65A**, 188 (1978).
154. C. Delsart, J. C. Keller, and V. P. Kaftandjian, *Opt. Commun.* **32**, 406 (1980).
155. V. P. Kaftandjian, C. Delsart, and J. C. Keller, *Phys. Rev. A* **23**, 1363 (1981).
156. P. Anantha Lakshmi and G. S. Agarwal, *Phys. Rev. A* **23**, 2553 (1981).
157. P. Ananthalakshmi and G. S. Agarwal, *Phys. Rev. A* **25**, 3379 (1982).
158. V. P. Kaftandjian, Y. Botzanowski, and B. Talin, *Phys. Rev. A* **28**, 1173 (1983).
159. M. Kolwas, *J. Phys. B* **10**, 583 (1977).
160. G. J. Diebold, *Phys. Rev. A* **32**, 2739 (1985).
161. W. Jastrzebski and M. Kolwas, *J. Phys. B* **17**, L855 (1984).
162. F. Biraben, K. Beroff, G. Grynberg, and E. Giacobino, *J. Phys. (Paris)* **40**, 519 (1979).
163. L. Hunter, E. Commins, and L. Roesch, *Phys. Rev. A* **25**, 885 (1982).
164. L. R. Hunter, G. M. Watson, D. S. Weiss, and A. G. Zajonc, *Phys. Rev. A* **31**, 2268 (1985).
165. L. R. Hunter and S. K. Peck, *Phys. Rev. A* **33**, 4452 (1986).
166. A. G. Zajonc, *Phys. Rev. A* **25**, 2830 (1982).
167. R. J. McLean, P. Hannaford, H. A. Bachor, P. J. Manson, and R. J. Sandeman, *Z. Phys. D* **8**, 135 (1988).
168. B. P. Kibble and S. Pancharatnam, *Proc. Phys. Soc. (London)* **86**, 1351 (1965).
169. G. Bertolucci, N. Beverini, M. Galli, M. Inguscio, and F. Strumia, *Opt. Lett.* **11**, 351 (1986).
170. J. E. Thomas, P. R. Hemmer, S. Ezekiel, C. C. Leiby, R. H. Picard, and C. R. Wills, *Phys. Rev. Lett.* **48**, 867 (1982).
171. J. Mlynek, R. Grimm, E. Buhr, and V. Jordan, *Appl. Phys.* **B45**, 77 (1988).
172. R. Schieder and H. Walther, *Z. Phys.* **270**, 55 (1974).
173. S. Nakayama, G. W. Series, and W. Gawlik, *Opt. Commun.* **34**, 389 (1980).
174. G. W. Series, *Proc. Phys. Soc.* **88**, 995 (1966).
175. W. Gawlik, *J. Phys. B* **10**, 2561 (1977).
176. A. Flussberg, T. Mossberg, and S. R. Hartmann, in *Coherence and Quantum Optics, IV*, edited by L. Mandel and E. Wolf (Plenum Press, New York, 1978), p. 695.
177. H. Uchiki, H. Nakatsuka, and M. Matsuoka, *Opt. Commun.* **30**, 345 (1979).
178. R. Saxena and G. S. Agarwal, *Phys. Rev. A* **31**, 877 (1985).
179. S. N. Jabr, L. K. Lam, and R. W. Hellwarth, *Phys. Rev. A* **24**, 3264 (1981).
180. J. F. Lam, D. G. Steel, R. A. McFarlane, and R. C. Lind, *Appl. Phys. Lett.* **38**, 977 (1981).
181. J. F. Lam and R. L. Abrams, *Phys. Rev. A* **26**, 1539 (1982).
182. G. P. Agrawal, *Phys. Rev. A* **28**, 2286 (1983).
183. D. Bloch and M. Ducloy, *J. Phys. B* **14**, L471 (1981).
184. C. Schmidt-Iglesias, G. Orriols, and F. Pi, *Appl. Phys.* **B47**, 27 (1988).
185. C. Cohen-Tannoudji and S. Haroche, *J. Phys. (Paris)* **30**, 125 (1969).
186. C. Cohen-Tannoudji and S. Haroche, *J. Phys. (Paris)* **30**, 153 (1969).
187. S. Haroche, *Ann. Phys. (Paris)* **6**, 189, 327 (1971).
188. C. Cohen-Tannoudji, in *Fundamental and Applied Laser Physics*, Proc. Esfahan Symp., edited by M. Feld, A. Javan, and N. Kurnit (Wiley, New York, 1973), p. 791.
189. C. Cohen-Tannoudji and S. Haroche, *C.R. Acad. Sci. Paris* **262**, 268 (1966).
190. D. T. Pegg and G. W. Series, *J. Phys. B* **3**, 33 (1970).
191. G. D. Chapman, *J. Phys. B* **3**, L26 (1970).
192. T. Dohnalik, H. Kucał, and T. Lubowiecka, *Acta Phys. Pol.* **A49**, 81 (1976).
193. P. Avan and C. Cohen-Tannoudji, *J. Phys. B* **10**, 171 (1977).
194. R. Ryan and T. Bergeman, private communication.
195. S. N. Dixit, P. Zoller, and P. Lambropoulos, *Phys. Rev. A* **21**, 1289 (1980).

196. K. Arnett, S. J. Smith, R. E. Ryan, T. Bergeman, H. Metcalf, M. W. Hamilton, and J. R. Brandenberger, *Phys. Rev. A* **41**, 2580 (1990).
197. J. Zakrzewski and T. Dohnalik, *J. Phys. B* **16**, 2119 (1983).
198. R. Saxena and G. S. Agarwal, *Phys. Rev. A* **25**, 2123 (1982).
199. D. F. Walls, *Nature* **306**, 141 (1983).
200. R. Loudon and P. L. Knight, *J. Mod. Opt.* **34**, 709 (1987).
201. M. D. Reid, D. F. Walls, and B. J. Dalton, *Phys. Rev. Lett.* **55**, 1288 (1985).
202. P. Anantha Lakshmi and G. S. Agarwal, *Phys. Rev. A* **32**, 1643 (1985).

CHAPTER 3

THE HANLE EFFECT AND LEVEL-CROSSING SPECTROSCOPY ON MOLECULES

H. G. WEBER

1. INTRODUCTION

The Hanle effect and level-crossing spectroscopy were applied to molecules rather late. A theoretical study on molecular level-crossing spectroscopy by Zare[1] in 1966 was soon followed by experiments, as shown by a first report[2] on applications of this technique to molecules in 1969. These first experiments demonstrated clearly why it was difficult to apply the technique of the Hanle effect and level-crossing spectroscopy to molecules. There were no spectral lamps, as for atoms, to prepare molecules into selected states, because any electronic state of a molecule exhibits a large number of vibrational and rotational levels. With the advent of lasers, expecially of tunable continuous lasers, the Hanle effect and level-crossing spectroscopy became more attractive to the investigation of molecules. The new light source enabled the preparation of a sufficient number of molecules into selected states, and made possible systematic investigations of molecules in several states. Such systematic studies were performed on diatomic molecules especially by Lehmann and co-workers. The review articles of these authors[3-6] serve also best as an introduction to the application of the Hanle effect and level-crossing spectroscopy to diatomic molecules. However, when these articles were written, there was only little known on the application of this experimental technique to polyatomic molecules. This situation has changed, especially owing to recent experiments[7] on nitrogen dioxide.

H. G. WEBER • Heinrich-Hertz-Institut and Optisches Institut der Technischen Universität Berlin, D-1000 Berlin 10, Federal Republic of Germany.

The Hanle effect experiments on nitrogen dioxide started with the surprising result, that the lifetime of the collision-free molecule evaluated from the width of the Hanle effect signal was by a factor 10 to 100 shorter than the lifetime obtained from time-resolved fluorescence spectroscopy. This finding, which has no analog in experiments on atoms and diatomic molecules, stimulated numerous investigations on nitrogen dioxide. The results often contradict expectations based on experimental and theoretical work with atoms and diatomic molecules. Up to now, NO_2 is the only polyatomic molecule which has been studied by the Hanle effect and level-crossing spectroscopy sufficiently in detail to allow a comparison with theory.

In this chapter we intend to describe the present (1988) state of application of the Hanle effect and level-crossing spectroscopy to molecules. We exclude from our discussion some topics which appear equally well also in experiments on atoms and which are discussed in other chapters of this book. Moreover, as the present chapter reviews for the first time applications of the Hanle effect and level-crossing spectroscopy also to polyatomic molecules, we consider especially the experiments on nitrogen dioxide in detail. This presentation is naturally biased by the author's own work.

2. MOLECULAR LEVEL-CROSSING SIGNAL

In the Hanle effect and level-crossing experiments, which we consider in this chapter, the degree of polarization of the fluorescence light

$$P = \frac{I_\| - I_\perp}{I_\| + I_\perp} \tag{1}$$

or sometimes quantities related to P, are measured. Here, $I_\|$ and $I_\perp$ are the fluorescence intensities with linear polarization respectively parallel and perpendicular to the linear polarization of the incident light. We consider only optical excitation in the following. The Hanle effect, for instance, appears as a change in the degree of polarization P when a magnetic field B directed perpendicularly to the linear polarization of the incident light is swept through $B = 0$. We do not consider the Hanle effect or level-crossing experiments in which the transmission of a light beam through a sample of molecules is measured versus an electric or magnetic field.

The theory of the Hanle effect and level-crossing spectroscopy starts conveniently with the Breit–Franken formula.[8] Let us consider an ensemble of molecules in its ground state $|g\rangle$, which is excited by light absorption into a coherent superposition of excited states $|e_k\rangle$ and $|e_j\rangle$, having energies W_k and W_j, respectively. The fluorescence decay into final states $|f\rangle$ is detected. We assume $|e_k\rangle = |\alpha J m_k\rangle$ and $e_j = |\alpha J m_j\rangle$, and correspondingly for the states $|g\rangle$ and $|f\rangle$, where the quantum numbers J and m characterize, respectively, the total angular momentum of the molecule and the projection of the total angular momentum on the space-fixed quantization axis; α stands for all other quantum numbers. The rate at which light with polarization $\hat{a}$, i.e., with polarization vector $\hat{a}$, is absorbed and fluorescence light with polarization $\hat{b}$ is detected, is given by the Breit–Franken formula [equation (40) in Chapter 1], which can be written in the form

$$I(\hat{a}, \hat{b}) = c \sum_{m_k m_j} \frac{F_{m_k m_j} G_{m_j m_k}}{1 - i\tau(W_k - W_j)/\hbar} \tag{2}$$

Here c is a parameter proportional to the intensity of the incident light and depending on geometrical factors, τ the lifetime of the states $|e_k\rangle$ and $|e_j\rangle$ of energy W_k and W_j, and $\hbar$ the Planck constant; $F_{m_k m_j}$ is the excitation matrix and $G_{m_j m_k}$ the detection matrix with

$$F_{m_k m_j} = \sum_g \langle \alpha J m_k | \hat{a}\hat{D} | g \rangle \langle g | \hat{a}^* \hat{D} | \alpha J m_j \rangle$$

and

$$G_{m_j m_k} = \sum_f \langle \alpha J m_j | \hat{b}\hat{D} | f \rangle \langle f | \hat{b}^* \hat{D} | \alpha J m_k \rangle \tag{3}$$

where $\hat{D}$ represents the electric-dipole transition operator. The excitation matrix describes population and alignment of the excited-state levels prepared by light absorption while the detection matrix describes in which way the excited-state population and alignment affect the polarization and angular distribution of the fluorescence light.

Equation (2) is valid for all molecules if $|e_k\rangle$, $|e_j\rangle$, etc., are eigenstates of the molecule. With the assumption $|e_k\rangle = |\alpha J m_k\rangle$ and $|e_j\rangle = |\alpha J m_j\rangle$, equation (2) describes zero-field level-crossing experiments, which reduce to the usual Hanle effect if the energy difference $W_k - W_j$ is proportional to an external magnetic field. The corresponding equaton for high-field level-crossing signals has the same form as equation (2) but the involved states $|e_k\rangle$, $|e_j\rangle$, etc., must, in general, be defined appropriately.

Calculations of the excitation and detection matrices and also of the degree of polarization P were performed[1,9] for molecules whose rotational motion is represented by the motion of a symmetric top. In a symmetric-top molecule the electric-dipole transition moment is rigidly attached to the molecular frame and has a well-defined component along the total angular momentum of the molecule. The calculations do not differ from the corresponding calculations for atoms, especially if the irreducible tensor formalism is applied. The same degree of polarization P as for atoms is obtained if an upper state with total angular momentum quantum number J is selectively excited, and the fluorescence decay to a lower state with total angular momentum quantum number J' is selectively detected. In general, the optically excited molecule has different decay channels with the total angular momentum quantum numbers of the final state being $J' = J$ (Q branch), $J' = J + 1$ (P branch), and $J' = J - 1$ (R branch). The degree of polarization of the Q branch always has a sign opposite to the degree of polarization of the P and R branches. If all three branches of emission are detected simultaneously, the degree of polarization depends on the relative intensities of the Q, P, and R branches. These relative intensities are known for a rigid symmetric-top molecule[10] and enable a calculation of the degree of polarization. The theory applies well to diatomic molecules.

Polyatomic molecules are often asymmetric-top molecules. This means that there is no molecular axis which carries out a simple precession movement about the space-fixed total angular momentum, as does the symmetry axis of a symmetric top. However, with respect to the molecular fixed frame the total angular momentum still follows periodic trajectories, such that the component of the electric-dipole transition moment along the total angular momentum exhibits in general a constant average value. The Hanle-effect signal will be described by the same formula as for atoms. However, the signal strength is difficult to evaluate because the relative intensities of the P, Q, and R branch transitions have to be calculated numerically.[11]

Rigid symmetric or rigid asymmetric rotors are sometimes not a good approximation for the rotational motion, because centrifugal and Coriolis forces strongly couple the vibrational and rotational motion. Therefore, even in symmetric-top molecules the projection of the total angular momentum on the molecular frame may not always be a constant of motion, because energy is exchanged between rotation and vibration. This is expected[12] to cause a lower degree of polarization of the fluorescence light than for a rigid molecule, because it results in a breakdown of the selection rules in the fluorescence decay process, for instance. However, this uncertainty affects only the signal strength, i.e., the nominator in equation (2) but not the denominator, which determines the width of the Hanle-effect signal and also the position of a high-field level-crossing signal. The essential condition for the application of the above-described level-crossing theory is that the excited states $|e_k\rangle$, $|e_j\rangle$ are eigenstates of the molecule.

The theory of the Hanle effect and level-crossing spectroscopy as described above was developed for classical light sources. At strong light intensity, as provided for instance by a narrow-band laser, one observes a broadening of the width and a change in the shape of the Hanle-effect signal.[13,14] Strong light intensity may cause a mixing between the upper and the lower state of a laser-induced transition. Consequently, the Hanle-effect signal contains information on the upper and the lower state. This was used to perform studies in the ground state of some diatomic molecules.[15-17] Laser saturation effects, as the effects associated with intense laser light are often termed, are a source of systematic errors in Hanle-effect and level-crossing experiments. Consequently it is necessary in all experiments to vary the light intensity in order to eliminate these systematic errors.

3. COMPARISON WITH QUANTUM BEAT EXPERIMENTS

A main advantage of the Hanle effect and level-crossing spectroscopy is the simple experimental setup, with no need for fast photodetectors and fast electronics. A serious disadvantage is that, in general, a second independent experiment is needed for the interpretation of the results of the Hanle effect and level-crossing spectroscopy. For instance, from Hanle-effect experiments one obtains the product $g\tau$ of the Landé g-factor and the lifetime of the excited state. Consequently, one of the two quantities has to be measured in a separate experiment.

Instead of the Hanle effect and level-crossing spectroscopy one can choose their transient counterpart, quantum beat spectroscopy, for the study of molecular excited states. Quantum beat spectroscopy requires short light pulses with a high repetition rate and a fast photodetector and transient digitizer to monitor the fluorescence decay. Quantum beat experiments on molecules were reported with quantum interfrences between Zeeman sublevels,[18-23] between hyperfine structure levels,[24] and between rotational levels.[25] The references give only a few examples. The excitation and detection geometry in these experiments is the same as in Hanle-effect and level-crossing experiments. Especially, polarization-sensitive detection is required.

However, there are also other manifestations of quantum beat effects in the fluorescence of polyatomic molecules which need no polarization-sensitive detection, and which have no counterpart in atomic spectroscopy. A sinusoidal temporal modulation in the dispersed fluorescence detection was attributed to intramolecular energy relaxation processes such as intersystem crossing,[26] internal conversion,[27] and intramolecular vibrational energy redistribution.[28] Again, we have listed only a selection of references. These polarization-independent modulations of the fluorescence light result from a cyclic transfer of energy between a prepared initial state (optically active state) and other isoenergetic states (optically inactive states). On the other hand, the modulation of the polarization of the fluorescence light in the aforementioned quantum beat experiments is a manifestation of the temporal evolution of the orientation of the molecules. One has to expect that, in general, the two physically different quantum beat phenomena, those associated with the molecular orientation and those associated with intramolecular energy transfer, will mix in the measured signal unless care is taken to separate them.

For the time being, there are many more quantum beat experiments than Hanle-effect or level-crossing experiments reported on molecules, especially on polyatomic molecules. There are only few steady-state experiments on energy transfer processes in polyatomic molecules compared to the many transient experiments. An example is given by investigations on singlet–triplet interaction in Glyoxal by level-anticrossing spectroscopy.[29] If one compares quantum beat spectroscopy with the Hanle effect and level-crossing spectroscopy, it seems that at present, as far as molecules are concerned, several advantages are in favor of quantum beat spectroscopy. Tunable pulsed lasers are available over a much wider range of wavelengths than tunable continuous wave lasers, especially in the UV. Moreover, quantum beat experiments are easier to interpret than their steady-state counterpart without additional independent measurements.

However, as the experiments on NO_2 reported in Section 9 reveal, Hanle-effect experiments on NO_2 give information about this molecule which, at present, is not available from quantum beat experiments. This information seems to open up a new field of application for the Hanle effect and level-crossing spectroscopy, which will probably be most relevant in understanding polyatomic systems.

4. EXCITATION OF MOLECULES

The preparation of molecules into selected excited states is essential for Hanle-effect and level-crossing experiments. A classical light source for selective excitation of atoms is the spectral lamp. There are some examples of Hanle-effect experiments on molecules with use of "molecular spectral lamps," for instance in experiments on OH and OD,[30,31] NS,[32] NP,[33] and CO.[34] However, in general, "molecular spectral lamps" are not useful because of the many rotational and vibrational levels associated with each electronic state. The lack of appropriate spectral lamps led scientists to use white light sources, such as high-pressure Xenon lamps, in conjunction with a monochromator, in Hanle-effect experiments on CS_2[35] and SO_2.[36] Also, electronic beam excitation[37,38] or radio-frequency discharge[39] was used to produce aligned H_2 in excited states or to prepare the intermediate state in two-step excitation of molecular hydrogen.[40] A radio-frequency discharge was also used in experiments on N_2 and O_2.[41] More often molecules were excited with "atomic lines" having an accidental coincidence with a molecular absorption line. Examples are the Hanle-effect experiments on NO,[42-45] CS,[46,47] S_2,[48] OH and OD,[49,50] and high-field level-crossing experiments on OD.[51]

The advent of lasers changed this unsatisfactory experimental situation. Experiments were performed with the fixed frequency lines of argon and krypton ion lasers, for instance, on Na_2,[52] I_2,[53,54] Se_2,[55,56] and on Br_2.[57] Also, pulsed nitrogen lasers were used in Hanle experiments on CS_2.[58,59] But only the advent of tunable lasers made the Hanle effect and level-crossing spectroscopy fully applicable to molecules. Tunable lasers, in fact, allow systematic studies on excited states not restricted to accidental coincidences with frequency-fixed light sources.

The spectra of gas-phase molecules at room temperature are often congested and preclude therefore state-selective excitation. Even continuous single-mode lasers simultaneously excite, in general, many rotational levels within the Doppler width of about 500 to 1000 MHz, except for some simple and light molecules. One reason is the high density of rotational and vibrational levels in the excited state. A second reason is the high number of populated vibrational and rotational levels in the ground state. Except for some diatomic molecules it is therefore necessary to simplify the spectra by use of the molecular beam technique. This technique enables a strong reduction of the Doppler width to less than 50 MHz, for instance. Moreover, especially in supersonic molecular beams, the molecules may be cooled to a rotational temperature of less than 100 K, which strongly reduces the number of populated vibrational and rotational levels in the ground state.[60] However, very often molecules must first be generated by a chemical reaction. In this case it is very difficult and often impossible to use molecular beams.

Figure 1 shows an excitation spectrum of NO_2 under molecular beam conditions. The residual Doppler width is about 50 MHz. One takes excitation spectra by monitoring the total fluorescence while a laser is scanned through the region of interest. If one tunes the laser light to one of these absorption lines in order to perform a Hanle-effect experiment, one needs to know how many states are excited and what these states are. An often used method to identify the "lines" in an absorption or excitation spectrum is the technique of wavelength-resolved laser-induced fluorescence spectroscopy (LIF). The laser light is tuned to the desired line. Then the wavelength-resolved emission spectrum is recorded. The spectrum is labeled with the final-state levels, i.e., ground-state levels, of the fluorescence transition that the prepared level exhibits. From prior knowledge of the ground-state level structure, i.e., energy splittings of the ground state, the excited state may be assigned, at least partially, from the fluorescence pattern it produces. Examples of such work are spectroscopic investigations on NO_2.[61] Nitrogen dioxide has an unpaired electron. Each rotational level splits up into two fine structure levels. From the LIF technique we know that most lines in the excitation spectrum in Figure 1 represent the excitation of a single fine structure level. Moreover, we know also the assignment of the

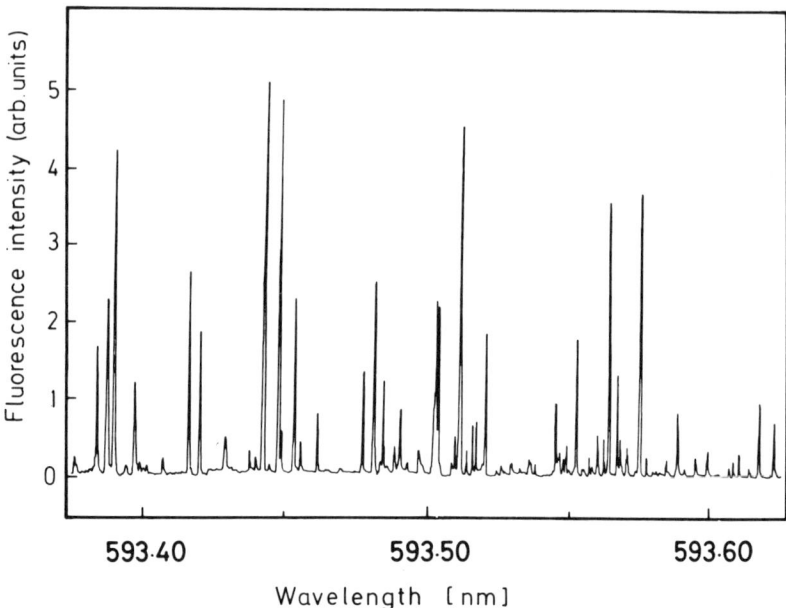

FIGURE 1. Excitation spectrum of NO_2 under molecular beam conditions. The strong lines in this spectrum were identified. They correspond to a laser-induced transition between a well-defined upper and lower fine structure level of the molecule.

angular momentum quantum numbers of the upper and lower states of the laser-induced transitions on most lines in this excitation spectrum. In the predominant isotopic form $^{14}N^{16}O_2$ each of the two fine structure levels is further split into three hyperfine structure levels. The hyperfine structure is not resolved in Figure 1. The hyperfine structure splitting of the ground and excited state is between 5 and 50 MHz.[62,63]

In the special case of Figure 1 the assignment of the LIF technique was reexamined by optical radio-frequency double-resonance measurements, with the laser light tuned to the desired line in this spectrum. The hyperfine structure appears as resolved in the optical radio-frequency double-resonance spectrum. Measurements of the Landé g-factors enabled an independent assignment of the angular momentum quantum numbers, which confirmed in most cases the assignment of the LIF technique.[64] Thus optical radio-frequency double-resonance spectroscopy gives not only the Landé g-factors, which are required to evaluate the lifetime from the Hanle-effect signal, but it supports also the identification of the excited states.

The lifetime of excited NO_2 is longer than 30 μs. This enables the application of special techniques to identify laser-induced transitions. With the use of two separate light beams crossing the molecular beam at a given distance, magnetic resonances simultaneously in the ground and excited state have been observed with a width which is solely determined by the time of flight between both light beams.[65] The experimental results were Landé g-factors simultaneously in the ground and excited state. This enabled an assignment of both the upper and the lower state of the laser-induced transition.

5. LIFETIME INVESTIGATIONS

The Hanle effect is a classical method for measuring lifetimes of excited states. The lifetime τ is related to the full width at the half maximum ΔB of the Hanle signal ($\Delta m = 2$ coherence) by $\tau = \hbar/\mu_B|g|\Delta B$, where μ_B is the Bohr magneton and g the Landé g-factor of the considered state. In static gas experiments the lifetime τ is expected to depend on the pressure according to

$$\frac{1}{\tau} = \frac{1}{\tau_R} + k_2 n \tag{4}$$

where n is the number denisty of the gas and the rate constant k_2 encompasses all collisional processes that remove alignment or population from the excited molecule. The lifetime τ_R of the collision-free molecule is obtained by extrapolating the data to zero pressure.

There are several sources of systematic errors which affect lifetime measurements by the Hanle effect.[66] In general these error sources are the same as in experiments on atoms. There are only few differences. For instance, multiple scattering of fluorescence light is expected to constitute no problem in molecular gases, because the branching ratio of the fluorescence light back to an absorbing level is generally small. Nevertheless, this effect was observed on CS because of favorable Frank–Condon factors.[47] An effect which is of special importance for molecular states having large rotational angular momenta is magnetic broadening of the absorption line. Although the applied magnetic field only has to be large enough to cause an energy splitting of the order of the natural linewidth, the total spread in energy of all magnetic sublevels can be very large compared to the natural linewidth. This may cause a deformation of the Hanle-effect signal in experiments with narrow-band laser excitation.[52]

In its application to lifetime measurements, the Hanle effect competes with other techniques, mostly with time-resolved fluorescence spectroscopy.

For the time being, the transient technique is almost exclusively used for lifetime measurements on molecules for the same reasons that quantum beat spectroscopy is mostly used now. There were very few Hanle-effect experiments performed after 1977 and the review article by Broyer et al.[5] discusses nearly all known Hanle-effect experiments on diatomic molecules. As regards polyatomic molecules, the main interest is presently to investigate whether the results of Hanle-effect experiments agree with the results of time-resolved fluorescence spectroscopy. The experiments on NO_2 reported below reveal that, for instance, equation (4) is not valid. In the following we report first about Hanle-effect experiments on diatomic molecules and subsequently about Hanle-effect experiments on polyatomic molecules.

In many Hanle-effect experiments, broad-band excitation was applied and often many rotational levels were excited simultaneously. The Hanle-effect signals were analyzed by making assumptions about the contribution of the different excited levels and their Landé g-factors, and estimated lifetimes were derived. Examples are experiments on NS,[32] NP,[33] and CO.[34,67] More often Hanle-effect experiments were not supplemented by measurements of the corresponding Landé g-factors, so that only the product $g\tau$ was obtained. As calculations of the Landé g-factors are uncertain, the results of these experiments may not always be fully satisfying. Examples are experiments on Na_2,[52] S_2,[48,68] N_2,[69] K_2,[70] and CS.[47]

In the following investigations the Landé g-factors were obtained by a separate independent experiment. On H_2 both Hanle-effect and radio-frequency magnetic-resonance experiments (Brossel-Bitter experiments) were performed[38] in several rotational and vibrational levels of the $(1s\,3p)$ $^3\Pi_u$ state excited by electron impact. Lifetimes, Landé g-factors, and disalignment rate constants k_2 were all found to be independent of the vibrational quantum number. Lifetimes, disalignment rate constants, hyperfine structure constants, and Landé g-factors were also reported[71] for a number of rotational levels of the $3d$ electronic complex and the nearby doubly excited $K^1\Sigma_g^+$ state of H_2. Hanle-effect and optical radio-frequency double-resonance techniques were used[31,30,49,50] to measure the lifetimes and g-factors of several rotational levels of the $v' = 0$, $A^2\Sigma^+$ state of OH and OD. On NO, Hanle-effect experiments were performed[42-45] in two rotational levels of the $v' = 1$, $A^2\Sigma^+$ state. Together with an optical radio-frequency double-resonance experiment,[45] lifetimes and disalignment rate constants were obtained.

Very detailed and systematic investigations have been performed on Se_2 and I_2. The $B^3\Sigma_u^-$ state of Se_2 was studied by the Hanle effect,[55,56] light modulation,[56] and Zeeman quantum beat spectroscopy.[72,73] The experimental results show that the Landé g-factors of close-lying rotational levels vary strongly by up to a factor 3, and that the lifetime varies between 60 ns and 90 ns. The rapid variation of the quantities in close-lying rotational levels was attributed to a local perturbation in the molecule. The experimental results are in good agreement with theory. I_2 was also investigated in detail by the Hanle effect and other experimental techniques.[54,74-76] A large number of vibrational and rotational levels of the $B^3\Pi_{0^+u}$ state was studied by the various lines of an argon or krypton ion laser. Landé g-factors were measured by light-modulation spectroscopy. While conducting Hanle-effect experiments it was found that there was significant circular polarization in the fluorescence emitted in the direction of the magnetic field. This observation led to a detailed study of the effect of predissociation in the B state of I_2.[76] Coriolis, hyperfine structure, Zeeman interaction, and interference terms between these interactions cause predissociation. The circular polarization in the fluorescence emitted in the direction of the magnetic field was attributed to an interference between natural predissociation (Coriolis interaction) and magnetic predissociation (Zeeman interaction). The experimental results are well described by theory.

Next we describe investigations on polyatomic molecules. Under collision-free conditions Hanle-effect measurements are expected to give the same lifetime as radiative decay measurements. Experiments on atoms and diatomic molecules confirmed this always. However, the first Hanle-effect experiments on NO_2 gave surprising results. An experiment performed in the yellow-red region near 593 nm of the absorption spectrum revealed lifetimes which are about a factor 10 shorter than those obtained by time-resolved fluorescence decay measurements.[77] These experiments were performed on static gas samples and corrections were made for the effect of collisions by extrapolating the measurements to zero pressure. A collisional origin of the effect was excluded. This was even more clearly shown in subsequent Hanle-effect experiments in the green and blue region of the absorption spectrum.[78] These experiments were performed in molecular beams and revealed lifetimes which are about a factor 100 smaller than those obtained by time-resolved fluorescence spectroscopy under the same experimental conditions. On the other hand, in subsequent Hanle-effect experiments[64,79-84] good agreement was obtained with time-resolved fluorescence decay measurements.[85] The controversial results of Hanle-effect experiments on NO_2 will be discussed later in Section 9.1.

No other polyatomic molecule was investigated by Hanle-effect experiments as thoroughly as NO_2. In experiments on CS_2 a broad spectral region near 320 nm was excited with the aid of a white light source, and an effective $g\tau$ value representing an average value with contributions of many states was obtained.[35] Subsequent Hanle-effect experiments on CS_2 used a nitrogen laser.[58,59] A vibrational level was excited near 337 nm. Under the same experimental conditions the time-resolved fluorescence decay of CS_2 is not a single exponential and may be represented by two characteristic decay times. The short-lived component of about 3 μs was associated with the Hanle-effect signal. A Hanle-effect signal was also observed in SO_2 under broad-band excitation.[36] Hanle-effect experiments in NH_2 revealed a lifetime of about 0.35 μs in several rotational levels of the (0, 10, 0) vibrational level of the $^2A_1$ electronic state, when the Landé g-factor was evaluated assuming Hund's coupling case b.[86] Lifetime measurements by time-resolved fluorescence spectroscopy in the (0, 9, 0) vibrational level of the same electronic state gave a value of about 10 μs. Hanle-effect experiments in BO_2 led to lifetime values in agreement with results from radiative decay measurements if the Landé g-factor was evaluated assuming Hund's coupling case a.[87,88] BO_2 is a linear molecule in the ground and excited state, contrary to all other investigated polyatomic molecules.

6. LANDÉ g-FACTORS

If the lifetime τ of an excited state is known, the Landé g-factor may be derived from the measured product $g\tau$ of a Hanle-effect experiment. In most Hanle-effect experiments on molecules, however, several hyperfine structure levels are excited simultaneously. This may result in an ambiguity of the derived g-factor. Therefore, these g-factor determinations are very rare. Consequently Hanle-effect experiments were mostly measurements of lifetimes, and Landé g-factors had to be determined separately. Landé g-factors may be measured by optical radio-frequency double resonance. An example is given by experiments on NO_2.[64,80,81] This technique is difficult to apply in molecular states having short lifetimes and small Landé g-factors, e.g., in molecular states with large rotational angular momentum. In this case light-modulation spectroscopy may be used. The laser light is modulated at a fixed frequency ω and the modulation of the fluorescence light is detected at the same frequency ω. Excitation and detecton must be coherent. Resonances appear in the fluorescence light when the Zeeman splitting equals the frequency ω. An example is given by experiments on I_2.[54,89,75] Another technique to measure Landé g-factors is Zeeman quantum beat spectroscopy.[18,21,23,73]

In few cases Landé g-factors and lifetimes may both be obtained separately in Hanle-effect experiments as was demonstrated in experiments on I_2.[53,54] The total nulcear spin of $^{127}I_2$ has either the even or odd integer values between 0 and 5. This results in 15 or 21 hyperfine structure components for each rotational level. In the considered experiments all hyperfine structure components of a given rotational level were excited simultaneously. The Hanle-effect signal was therefore a superposition of 15 or 21 Lorentzian curves. For large rotational Landé g-factor g_J and large rotational angular momentum, the Landé g-factors of all hyperfine structure components are about equal to g_J. In this case all 15 or 21 Lorentzian curves have approximately the same width. The result is a pure Lorentzian curve with a width characteristic of $g_J\tau$, enabling no separate determination of g_J and of the lifetime τ. However, if g_J and the rotational angular momentum are small, the Landé g-factors of the hyperfine structure components differ strongly. The resulting superposition of Hanle-effect curves is non-Lorentzian. In this case it was possible to derive the Landé g-factor g_J and the lifetime τ separately from the shape of the Hanle curve.

As reported earlier in Section 5, very often Landé g-factors were calculated from an assumed angular momentum coupling scheme. In the following we give some examples which indicate that Landé g-factors in molecules may change strongly even for small perturbations. It is therefore not reliable to use calculated g-values in the evaluation of the Hanle effect and level-crossing experiments. The first example is the strong variation of the g-values in the B state of I_2 close to the dissociation limit.[54,89] The rotational Landé g-factor g_J and the diamagnetic correction σ_E to the nuclear g-value were found to increase strongly as the excited-state levels approach the dissociation limit. The value of g_J changes by more than a factor 10, and the diamagnetic correction ("chemical shift") which is generally less than 10^{-3} for atoms or molecules in the ground state assumes values of about $|\sigma_E| = 17$. Also, the hyperfine structure shows a similar change near the dissociation limit. All these properties of molecular states near the dissociation limit of a diatomic molecule could well be described by theory.[90] A second example of a strong variation in the Landé g-factor is given by the experiments on Se_2 [72,73] as reported in the previous section. Here a local perturbation causes a rapid change in the Landé g-factors of close-lying rotational levels.

Landé g-factors were also measured in NO_2. Because the hyperfine structure splitting in the $\tilde{A}^2 B_2$ state of NO_2 is small compared to the separation of the two fine structure components, one generally assumes that the rotational angular momentum $\hat{N}$ couples with the electron spin $\hat{S}$ to $\hat{J} = \hat{N} + \hat{S}$ and the nuclear spin $\hat{I}$, which has the quantum number $I = 1$ in the predominant isotopic form $^{14}N^{16}O_2$, couples with $\hat{J}$ to the total angular momentum $\hat{F} = \hat{J} + \hat{I}$. Measurements of the Landé g-factor by optical radio-frequency double resonance[80,81,64] and by Zeeman quantum beat spectroscopy[18] revealed that the coupling scheme $\hat{F} = \hat{J} + \hat{I}$ is well behaved, while the coupling scheme $\hat{N} + \hat{S} = \hat{J}$ is valid for some levels but not for others. The experimental values differ irregularly from the expected values. Also, the hyperfine structure splitting exhibits the same irregular behavior and both experimental results are strongly correlated.[91] The experimental results were explained by the assumption that N is not a good quantum number because spin–orbit interaction couples rotational levels with different N.

7. ELECTRIC-FIELD LEVEL CROSSING

Zero-electric-field level-crossing experiments are mostly investigations of electric-dipole moments of excited molecular states. Although this quantity is of much interest in molecular physics, there are only few experiments.[92-94] Electric field experiments are often more difficult to analyze that the corresponding magnetic field experiments, because the Stark effect in molecules is mostly a second-order effect only.[10] To point out the difficulties we describe in this section a zero-electric-field level-crossing experiment on NO_2 [94] in detail.

Figure 2 gives a schematic view of the experimental apparatus in the NO_2 experiment. Light from a continuous single-mode dye laser propagates along the y axis. The laser light, which is linearly polarized parallel to the x axis, excites NO_2 molecules, which propagate in a beam along the x axis. The laser-induced molecular fluorescence is viewed by two photomultipliers, which detect fluorescence light emitted in opposite directions along the z axis with linear polarization parallel to the x and y axis, respectively. The difference of the two fluorescence intensities is recorded versus an electric field E, which is directed parallel to the z axis. The electric field E is generated by two plain stainless steel meshes separated by a distance of about 2 cm and between with a voltage up to 10 kV is applied. Both meshes represent two conducting "plates" oriented parallel to the xy plane of the coordinate system, so that the molecular beam and the light beam cross between these plates. The fluorescence light is viewed by both photodetectors through the meshes. A magnetic field can also be applied parallel to the electric field, i.e., parallel to the z axis, with use of a pair of Helmholz coils.

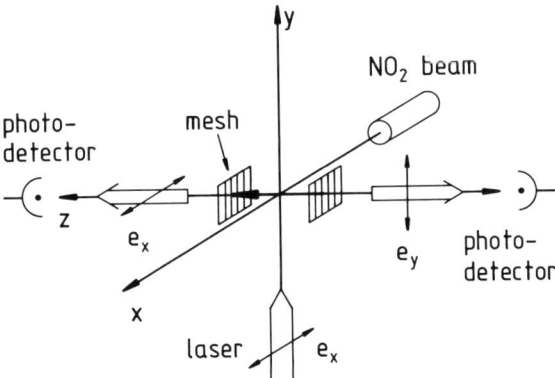

FIGURE 2. Schematic diagram of a zero-electric-field level-crossing experiment. The electric field is produced by two plain stainless steel meshes separated by a distance of about 2 cm and between which a voltage is applied. e_x, e_y, and e_z represent linear polarizations of the exciting or detected light, respectively.

With the electric field off, for instance without the two meshes, the arrangement corresponds to a Hanle-effect experiment. With the magnetic field $B = 0$ the arrangement was used in zero-electric-field level-crossing experiments.

As in Section 2 we consider the crossing of upper state levels $|\alpha F m_k\rangle$ and $|\alpha F m_j\rangle$. However, we now label the quantum number of the total angular momentum F, and not J, because we will explicitly consider the presence of hyperfine structure. The second-order Stark effect gives the following dependence of the energy W_k of the state $|\alpha F m_k\rangle$ on the electric field E:

$$W_k = W_0 + A_{\alpha F} E^2 + B_\alpha C_{\alpha F} m_k^2 E^2 \qquad (5)$$

In the level-crossing signal, equation (2), the difference $W_k - W_j = B_\alpha C_{\alpha F} E^2 (m_k^2 - m_j^2)$ enters with $m_k - m_j = \pm 2$ in the denominator; $C_{\alpha F}$ is a known quantity[94] depending on the angular momentum quantum numbers of the excited state only, while the Stark constant B_α is the quantity to be measured. The difference $W_k - W_j$ is also dependent on the quantum number m. This is different in the corresponding zero-magnetic-field level-crossing signal. As long as the linear Zeeman effect is valid, for instance, as long as no hyperfine structure decoupling is observed, the energy W_k depends linearly on the magnetic quantum number m. The difference $W_k - W_j$ is therefore independent of the magnetic quantum number in Hanle-effect experiments.

The appearance of the magnetic quantum number m in the energy difference $W_k - W_j$ in an electric-field level-crossing experiment complicates the evaluation of the zero-electric-field level-crossing signal considerably. Figure 3 shows a measurement of the degree of polarization P versus the electric field E. The laser light excites all three hyperfine structure components of one fine structure level of NO_2. The resulting resonance signal is, however, not a superposition of three resonance curves as in the corresponding Zeeman effect experiment but a superposition of many more curves depending on the quantum numbers F and m. It was nevertheless possible to obtain approximate values for the Stark constant B_α by an appropriate fitting procedure. In the next section we describe an experiment which gives exact values for B_α.

Zero-electric-field level-crossing spectroscopy was used to measure the Stark constant B_α of two vibrational levels of the $A^1\Sigma^+$ state of BaO[92] and of three vibrational levels of the $A^1\Sigma^+$ state of LiH.[93] In both experiments, low rotational levels were excited in order to alleviate the problem associated with the dependence of the energy difference $W_k - W_j$ on the magnetic quantum number m. An evaluation of the electric dipole moment from the measured values of B_α was possible with the knowledge of the lifetime and the rotational constant of the excited states. Experiments on

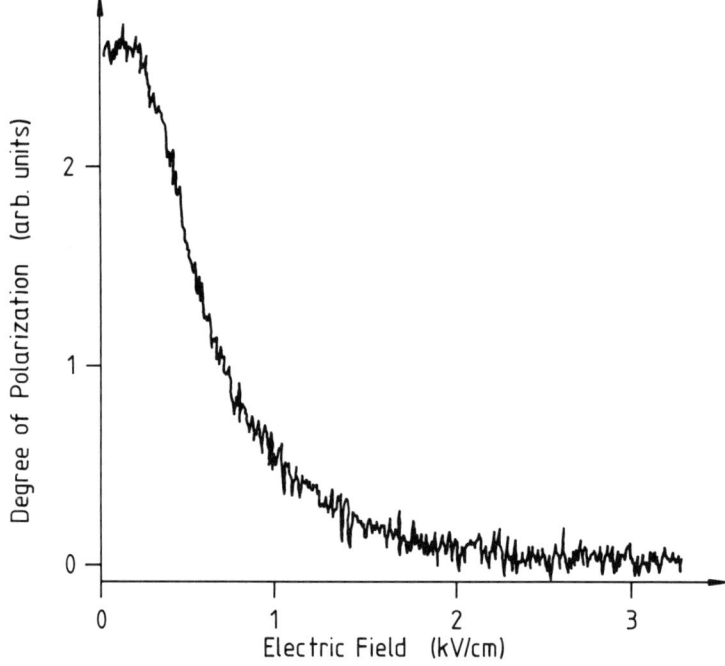

FIGURE 3. A zero-electric-field level-crossing signal on NO_2. The measured quantity is the difference between the fluorescence intensities with polarization e_x and e_y as depicted in Figure 2.

CS[46] yielded, in addition to the Stark constant, also the lambda doubling in the $A^1\Pi$ state. If the excited state of the molecule has no permanent electric dipole moment, the molecule may interact with the electric field via its polarizability.[10] In a zero-electric-field level-crossing experiment[95] on I_2 the anisotropic polarizability of I_2 in the $B^3\Sigma_{0^+_u}$ state was determined.

8. STARK–ZEEMAN RECROSSING AND HIGH-FIELD LEVEL CROSSING

We will show that some of the problems which appear in zero-electric-field level-crossing experiments are eliminated in Stark–Zeeman recrossing experiments. We discuss experiments[94] on NO_2 which were performed in the arrangement depicted in Figure 2. A description of this arrangement was given earlier in Section 7. With both the electric and magnetic fields being applied, the energy W_k of the state $|\alpha F m_k\rangle$ is given by

$$W_k = W_0 + A_{\alpha F}E^2 + B_\alpha C_{\alpha F}m_k^2 E^2 - g_F\mu_B m_k B \qquad (6)$$

where μ_B is the Bohr magneton, g_F the Landé g-factor, and all other quantities already defined in Section 7. Level-crossing signals are observed as a resonant change in the degree of polarization P, if the levels $|\alpha F, m+1\rangle$ and $|\alpha F, m-1\rangle$ have the same energy with $m = \pm\frac{1}{2}, \pm\frac{3}{2}, \ldots, \pm(F-1)$. These level crossings are expected at the magnetic fields $B = 2mB_1$ where $B_1 = -B_\alpha C_{\alpha F} E^2/\mu_B g_F$. Figure 4 shows Stark–Zeeman recrossing signals. The electric field is kept constant with $E = 2660$ V/cm and the magnetic field B is scanned over the crossings $B = \pm B_1$ and $B = \pm B_2 = \pm 3B_1$. As the three hyperfine structure components F_+, F_0, F_- of a single fine structure level with $J = \frac{7}{2}$ were excited simultaneously, one observes level-crossing signals associated with each hyperfine structure component. From the position of the level-crossing signals the Stark constant B_α was derived, because the Landé g-factors had been measured before by optical radio-frequency double-resonance experiments.

The Stark constants B_α were measured in NO_2 on many absorption lines in the excitation spectrum shown in Figure 1. The value of B_α varies strongly between close-lying levels. Differences up to a factor 10 were obtained even for fine structure components belonging to one rotational level. The evaluation of excited-state electric-dipole moments from the measured Stark constants B_α requires a knowledge of all nearby levels that

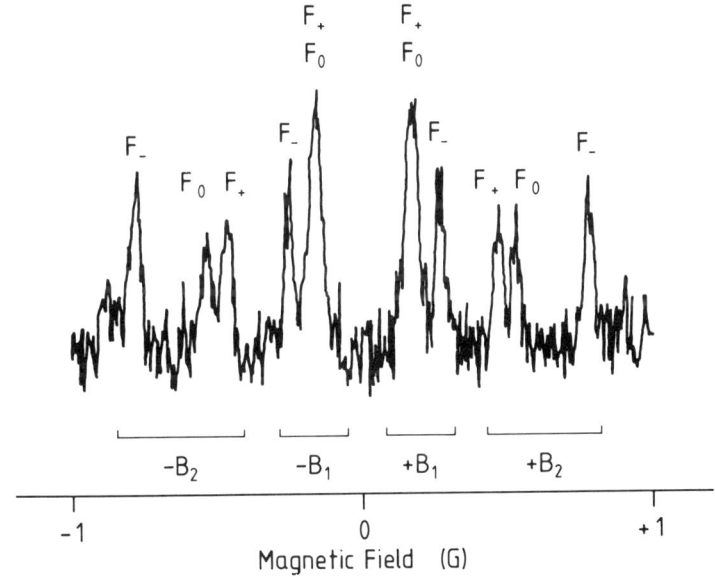

FIGURE 4. Level-crossing signals in a Stark–Zeeman recrossing experiment. The experimental arrangement depicted in Figure 2 was used with a magnetic field applied parallel to the constant electric field $E = 2660$ V/cm. F_+, F_0, and F_- designate the three hyperfine structure components of the selectively excited fine structure level.

can be connected to the level under study by the electric-dipole operator. As this knowledge is incomplete for NO_2 only estimated values for the electric-dipole moments were derived.[20] Moreover, the strong variation in B_α for close-lying levels could not be explained.

Stark–Zeeman recrossing experiments were also performed on the $A^1\Pi$ state of the molecule CS[46] and on the $A^2\Sigma^+$ state of the molecule OD.[51] In combination with high-field level-crossing or zero-field level-crossing experiments, electric-dipole moments and hyperfine constants or lambda doubling were determined.

In a zero-field level-crossing experiment all magnetic sublevels cross at the same applied field, namely, the zero field. Consequently all crossings add up to one signal. This is different in the Stark–Zeeman recrossing experiment and generally in all high-field level-crossing experiments. The signal amplitudes in high-field level-crossing experiments are therefore about a factor $2F + 1$ weaker than in the corresponding zero-field level-crossing experiments. Moreover, contrary to zero-field level crossings, in high-field level crossings the magnetic or electric fields at which the crossings appear have to be searched for. Because of this search problem and the expected weak signal strength, it is therefore not surprising that only very few high-field level-crossing experiments were reported.

High-magnetic-field level-crossing experiments were reported[37,96] for the molecule H_2. Fine structure constants were determined in several levels with low rotational quantum number of the $1s\,3p\,^3\Pi_u$ and the $1s\,4p\,^3\Pi_u$ state of parahydrogen, when the Zeeman sublevels of adjacent rotational levels cross versus the applied magnetic field. Preliminary fine structure parameters determined before in optical radio-frequency double-resonance experiments were used to largely eliminate the search problem. High-magnetic-field level-crossing experiments were also performed[51,97] on the $A^2\Sigma^+$ state of the molecule OD. Hyperfine structure constants were derived from the crossing of Zeeman sublevels of adjacent hyperfine structure components of one fine structure level.

9. HANLE EFFECT ON NO_2

9.1. The Influence of Detection Geometry

As already reported, the first Hanle-effect experiments[78] on NO_2 under collision-free conditions resulted in lifetimes which were nearly a factor 100 shorter than the lifetimes from time-resolved fluorescence spectroscopy. On the other hand, in subsequent Hanle-effect experiments[64,79-81] good agreement with results from radiative decay measurements was obtained. This section deals with this discrepancy of experimental results on NO_2.

Figure 5 is a schematic diagram of the experimental configuration of the Hanle-effect experiments on NO_2 described in the following. Light from a continuous single-mode dye laser or argon ion laser propagates along the y axis. The laser light is tuned, for instance, to one of the absorption lines of the NO_2 excitation spectrum shown in Figure 1. The molecules propagate in a beam along the x axis. In one kind of experiment the laser-induced molecular fluorescence is monitored by one photomultiplier, and in another kind of experiment one detects the fluorescence light emitted in opposite directions along the z axis by means of two photomultipliers and records the difference signal. In all experiments the signal is recorded versus the magnetic field B. The magnetic field is directed along either the z or the x axis. The polarization of the light beam and the orientation of the polarizers in the fluorescence detection path can be chosen appropriately. To specify these polarizations the polarization vectors e_x, e_y, and e_z are introduced and represent linear polarizations along the x, y, and z axis, respectively.

Figure 6 depicts four magnetic resonance signals which were obtained in the following arrangements:

Arrangement $[\sigma \uparrow \pi \downarrow]$: The magnetic field B is directed along the x axis. The laser light has linear polarization e_z, representing σ-polarized excitation. Fluorescence light I_x with linear polarization e_x is detected, representing π-polarized fluorescence detection.

Arrangement $[\sigma \uparrow \sigma \downarrow]$: The magnetic field B and the polarization of the laser light are the same as in arrangement $[\sigma \uparrow \pi \downarrow]$. Fluorescence light

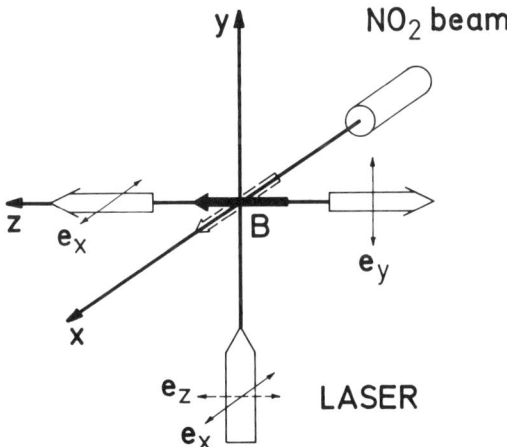

FIGURE 5. Schematic diagram used to describe several arrangements for magnetic resonance experiments in which the magnetic field B is swept through $B = 0$. Compare Figure 6. e_x, e_y, and e_z represent linear polarizations of the exciting or detected light, respectively.

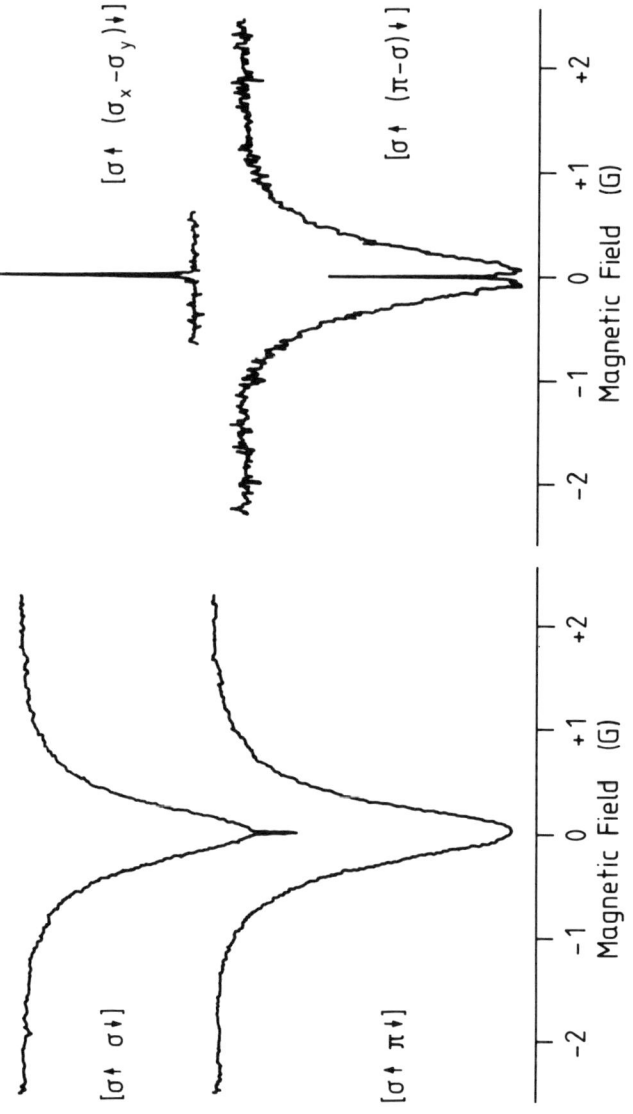

FIGURE 6. Magnetic resonance signals obtained with different experimental arrangements according to Figure 5.

I_y with linear polarization e_y is detected, representing σ-polarized fluorescence detection.

Arrangement $[\sigma \uparrow (\pi - \sigma) \downarrow]$: The magnetic field B and the polarization of the laser light are the same as in arrangement $[\sigma \uparrow \pi \downarrow]$. Fluorescence light I_x and I_y with linear polarization e_x and e_y, respectively, is detected with use of both photomultipliers and the difference $I_x - I_y$ is recorded.

Arrangement $[\sigma \uparrow (\sigma_x - \sigma_y) \downarrow]$: The magnetic field B is directed along the z axis. The laser light has linear polarization e_x, representing σ-polarized excitation. Fluorescence light $I_x - I_y$ is detected.

The magnetic resonance signals in Figure 6 are, in general, a superposition of a narrow and a broad resonance, both with Lorentzian shape. Both resonances disappear if linear polarization of the laser light parallel to the magnetic field, i.e., π-polarized excitation, is used. The narrow resonance has a width characteristic of the lifetime $\tau_R \approx 40 \, \mu s$, and the broad resonance has a width characteristic of the time τ_u with

$$\frac{1}{\tau_u} = \frac{1}{\tau_0} + \frac{2.1 \pm 0.2}{T_L}$$

$$\tau_0 = (3.2 \pm 0.6) \, \mu s \tag{7}$$

In these equations T_L is the transit time of the molecules through the light field, which is defined by $T_L = d/u$ where u is the average velocity of the molecules along the x axis and d is the diameter of the laser beam as determined by an aperature in the light path. The width of the narrow resonance is independent of the transit time T_L. Both widths are also independent of the light intensity as measurements with several light intensities between $0.2 \, mW/mm^2$ and $20 \, mW/mm^2$ showed.

The narrow resonance represents a change in the polarization of the fluorescence light. It therefore has the properties of the expected Hanle-effect signal and is named the Hanle-effect signal in the following. On the other hand, the broad resonance represents predominantly a change in the intensity of the fluorescence light. This change amounts to 20% of the total fluorescence intensity. The broad resonance is called σ-resonance in the following. Both resonances were observed on all investigated absorption lines near 593 nm and 514 nm.[82] It was shown that the σ-resonance is different from the "line-crossing" resonance,[98] which represents also a change in the fluorescence intensity versus an applied magnetic field if σ-polarized excitation is used. The "line-crossing" resonance is a well-understood laser saturation effect.

The measurements represented in Figure 6 explain the controversial results of previous Hanle-effect experiments on NO_2. The experiments,[64,79-81] which resulted in Hanle-effect signals having a width characteristic of the radiative lifetime $\tau_R \approx 40\ \mu s$, used the arrangement $[\sigma \uparrow (\sigma_x - \sigma_y) \downarrow]$. In this arrangement only the narrow resonance is observed. (This statement is valid for the experimental conditions used most often.[64,79-81] See, however, the more detailed discussion of the resonance in Section 9.2.) On the other hand, the Hanle-effect experiments,[78] which resulted in lifetimes of about 1 μs, used essentially the arrangement $[\sigma \uparrow (\pi - \sigma) \downarrow]$. As Figure 6 shows, in this arrangement a superposition of the broad σ-resonance and the Hanle-effect signal appears. Inhomogeneities of the magnetic field affect primarily only the Hanle-effect signal and cause a strong decrease in this resonance. A rapid magnetic-field scan and a long time constant in the electronic signal processing have a similar effect. It is therefore likely that the narrow resonance was overlooked. As the width of the σ-resonance is primarily determined by the transit time $T_L \leq 2\ \mu s$ in this experiment,[78] the lifetime deduce from the σ-resonance is about $T_L/2 \leq 1\ \mu s$, in agreement with the results.

As described in Section 5, Hanle-effect experiments on NO_2 were also performed on static gas samples and corrections were introduced for the effect of collisions by extrapolating the measurements to zero pressure. Lifetimes much shorter than the radiative lifetime τ_R were also reported in these experiments[77] when the arrangement $[\sigma \uparrow \sigma \downarrow]$ was used. Since the σ-resonance is present also in experiments under static gas conditions, it is likely that the σ-resonance was misinterpreted also in these experiments. However, the σ-resonance is more complicated in static gas experiments because there is no well-defined transit time. Moreover, also the Hanle-effect signal is more complicated in static gas experiments as shown below.

Optical radio-frequency double-resonance experiments[99] on NO_2 yield resonances which correspond to the Hanle-effect signal and to the σ-resonance. The counterpart to the Hanle-effect signal appears as three separate resonances each having a width corresponding to the time $\tau_R \approx 40\ \mu s$. There are three resonances in these experiments, because the laser light excites all three hyperfine structure components of a fine structure level simultaneously. The hyperfine structure is resolved in the optical radio-frequency double-resonance experiment, but not in the Hanle-effect experiment. The narrow resonance in Figure 6 is therefore a superposition of three independent resonances with approximately the same width because the three associated Landé g-factors differ, in general, by not very much, and because the hyperfine structure splitting is large compared to the spectral width of the laser light. The counterpart to the σ-resonance in the optical radio-frequency double-resonance experiment appears as six separate resonances, three associated with the Landé g-factors of the upper state hyperfine structure levels and three associated with the Landé g-factor of the lower state hyperfine structure levels. Each resonance has the same properties as the σ-resonance, especially a width characteristic of the time τ_u as given by equations (7). An average of all six Landé g-factors was used to evaluate the time τ_u from the width of the σ-resonance.

9.2. Details of the Hanle-Effect Signal

A closer examination of the Hanle-effect signal in the arrangement $[\sigma\uparrow(\sigma_x - \sigma_y)\downarrow]$, i.e., of the narrow resonance in Figure 6, reveals[100] that this signal is not always Lorentzian shaped with a width corresponding to the radiative lifetime $\tau_R \approx 40$ μs. The appearance of this signal depends on the following experimental parameters: the intensity I of the exciting laser light, the transit time T_L of the molecules through the light field, and the spectral window of fluorescence detection. This window is defined as follows. Fluorescence light with wavelength λ is detected if $\lambda_x \leq \lambda \leq \lambda_0$, where λ_x is set by cutoff filters in the fluorescence detection pass. These filters transmit light with $\lambda \geq \lambda_x$ and suppress light with $\lambda < \lambda_x$ by a factor 10^4. Originally the cutoff filters were introduced to suppress laser stray light. This is possible because the fluorescence emission of NO_2 is strongly red-shifted with respect to the excitation wavelength. With the laser light tuned to some of the absorption lines of the NO_2 excitation spectrum shown in Figure 1, for instance, cutoff filters with $\lambda_x = 630$ nm, 715 nm, and 780 nm were used. The upper wavelength limit is $\lambda_0 \approx 850$ nm, determined by the sensitivity of the photomultipliers.

Figure 7 shows two measurements of Hanle-effect signals in the arrangement $[\sigma\uparrow(\sigma_x - \sigma_y)\downarrow]$ that differ only in the intensity I of the laser light. In these measurements the cutoff wavelength is $\lambda_x = 715$ nm and the diameter of the light beam is $d = 5$ mm, resulting in a transit time $T_L = 8.2$ μs of the molecules through the light field. The dots superimposed on both measurements represent the sum of two Lorentzian curves, which are fitted to the measurements. The two Lorentzian curves have amplitudes with the same sign for low light intensity and with opposite signs for high light intensity. In both measurements the narrow Lorentzian curve has a width corresponding to the time $\tau_R \approx 40$ μs, and the broad Lorentzian curve has a width corresponding to the time $\tau_0 \approx 3$ μs, if one assumes the same relation between lifetime and width (the same g-factor) for both components. Both components appear as dispersion-shaped signals if the polarizers in the fluorescence detection path are rotated by $\pi/4$ about the z axis. The widths of both components show no significant dependence on the light intensity I and on the transit time T_L. The broad component is definitively different from the σ-resonance reported above in Section 9.1. In the following we name the narrow component the narrow Hanle-effect signal and the broad component the broad Hanle-effect signal.

The broad Hanle-effect signal was not seen in the experiments reported in Section 9.1. The important quantity here is the cutoff wavelength λ_x in

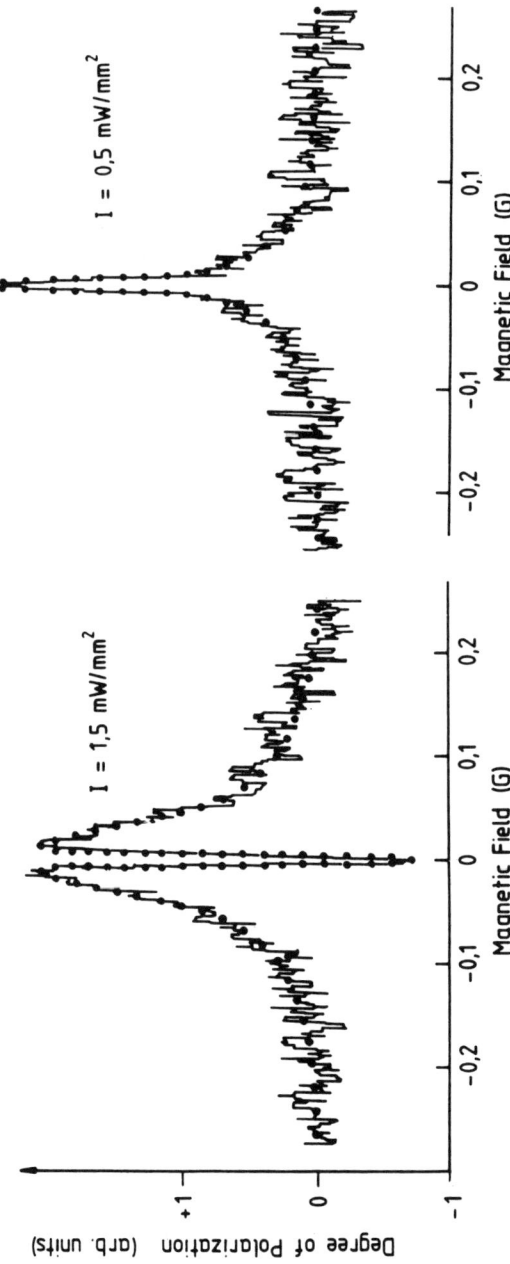

FIGURE 7. Hanle-effect signal in the experimental arrangement with wavelength-selective detection for two light intensities I. The signal consists of the narrow and broad Hanle-effect signals. Dots represent the superposition of two Lorentzian curves which are fitted to the experimental results.

the fluorescence detection paths. We compare the signal amplitudes of the narrow Hanle-effect signal S_n and of the broad Hanle-effect signal S_b, when both assume their maximum values as functions of light intensity and transit time T_L (the inverted narrow Hanle-effect signal, see below). The experiments show that S_b is independent of λ_x while S_n decreases strongly with increasing λ_x, so that $S_b/S_n \leq 0.1$ for $\lambda_x = 630$ nm, $S_b/S_n \approx 0.5$ for $\lambda_x = 715$ nm, and $S_b/S_n \geq 1$ for $\lambda_x = 780$ nm. In the experiments reported in Section 9.1 the cutoff wavelength $\lambda_x = 630$ nm was used. The broad Hanle-effect signal was therefore not seen. We note that the σ-resonance, normalized to the total fluorescence intensity, is independent of λ_x.

The narrow Hanle-effect signal exhibits an unusual property, which was called the inversion effect.[101] Figure 7 shows that the narrow Hanle-effect signal changes with increasing light intensity from an upward directed signal to a downward directed signal. Similar behavior is also observed versus the transit time T_L. Figure 8 shows measurements of the amplitudes of the narrow and broad Hanle-effect signals both normalized to the total fluorescence intensity [see equation (1)] versus the transit time T_L. The experimental conditions are: intensity of the laser light 1 mW/mm^2 and

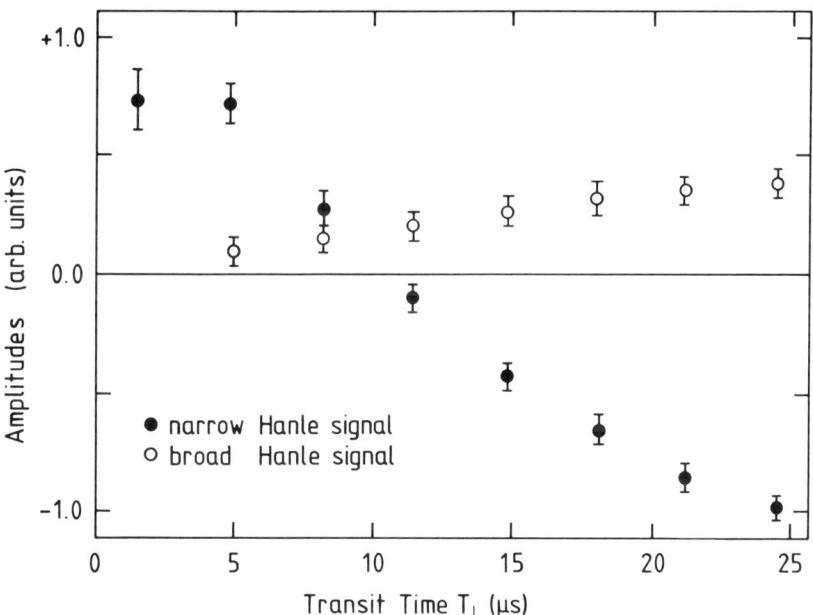

FIGURE 8. Signal amplitudes of the narrow and broad Hanle-effect signals on NO_2 versus the transit time T_L of the molecules through the laser light beam. The change in the sign of the amplitude of the narrow Hanle-effect signal demonstrates the inversion effect.

cutoff wavelength $\lambda_x = 715$ nm. There is an inversion for the narrow Hanle-effect signal but not for the broad Hanle-effect signal. Figure 9 demonstrates the inversion of the narrow Hanle-effect signal in an experiment with cutoff wavelength $\lambda_x = 630$ nm and transit time $T_L \approx 3$ μs versus the intensity of the laser light. Under these experimental conditions the broad Hanle-effect signal is too small to be observable.

The inversion effect does not appear for all absorption lines. With the available light intensity ($I \approx 800$ mW) the inversion effect was observed only on transitions involving low ($N \leq 4$) rotational quantum numbers N. Moreover, the effect is different for P and R branch transitions to the same upper state.

Several investigations were performed in order to check whether the inversion effect is an experimental artifact. It was verified that the inversion effect is not caused by cascading of fluorescence light or collisional depolarization,[101] nor by nonstationary interaction of the molecules with the light beam[102] as was suggested,[103] nor by alignment of the molecules

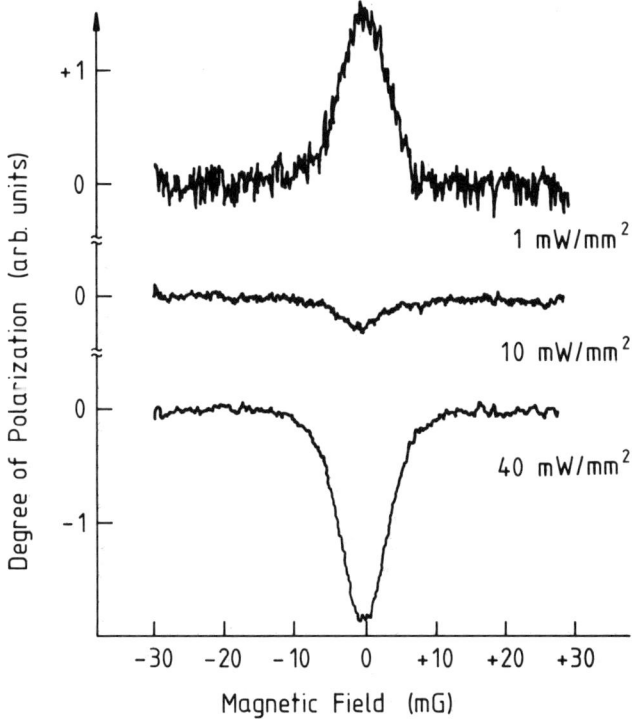

FIGURE 9. The Hanle-effect signal on NO_2 for a transit time $T_L = 3$ μs and for different intensities of the exciting laser light. The change in the sign of the Hanle-effect signal demonstrates the inversion effect.

in the molecular beam,[104] nor by a breakdown of the selection rules at strong light intensities, as for instance $\Delta F = +1$ and $\Delta F = 0$ transitions in the P branch.[105] The inversion effect appears also in the experimental arrangements $[\sigma\uparrow, \sigma\downarrow]$ and $[\sigma\uparrow(\pi-\sigma)\downarrow]$, and in optical radio-frequency double-resonance experiments as an inversion of the resonances for all three hyperfine structure components.

9.3. Collisions

This section deals with Hanle-effect experiments on static gas samples[83] of pure NO_2 with use of the experimental arrangement $[\sigma\uparrow(\sigma_x - \sigma_y)\downarrow]$ defined in Section 9.1. Supplementary to the Hanle-effect measurements, optical radio-frequency double-resonance experiments were always performed in order to select absorption lines which enable excitation of a single fine structure level also under static gas conditions.

Figure 10 shows a Hanle-effect signal obtained at the NO_2 gas pressure of 1 μbar and at an intensity of the laser light 15 mW/mm². The experimental results, represented by the dots, are well fitted by the solid line, which is the sum of two Lorentzian-shaped curves. The narrower curve represents the narrow Hanle-effect signal, and the broader curve represents the broad Hanle-effect signal. Both signals have the same properties as their counterparts in the molecular beam experiments. The narrow Hanle-effect signal displays the inversion effect versus the light intensity, which is absent for the broad Hanle-effect signal. The widths of both signals are independent of the light intensity and exhibit, for pressures larger than 1 μbar, a dependence on NO_2 pressure or NO_2 density n with the characteristic decay rate $1/\tau = 1/\tau_s + k_s n$ for the narrow Hanle-effect signal and $1/\tau' = 1/\tau_b + k_b n$ for the broad Hanle-effect signal.

Usually one expects that the linear extrapolation to zero pressure is in agreement with the molecular beam results, i.e., according to equation (4) $\tau_s = \tau_R$ for the narrow Hanle-effect signal and $\tau_b = \tau_0$ for the broad Hanle-effect signal. However, the experiments always revealed $\tau_s < \tau_R$ and $\tau_b < \tau_0$. This result was investigated in detail on the narrow Hanle-effect signal. Figure 11 shows the experimental dependence of the width of the narrow Hanle-effect signal on the NO_2 pressure. For pressures below 1 μbar, the pressure dependence of the width of the static gas signal becomes nonlinear and tends to the width of the molecular beam signal, represented by the black dot in Figure 11. A detailed analysis revealed[83] that neither wall collisions, nor magnetic field inhomogeneities, nor an inadequate fitting procedure of the two signal components to the experimental results cause this nonlinear pressure dependence.

The results in Figure 11 were obtained with laser excitation near 514 nm. Measurements with laser excitation near 593 nm resulted in $\tau_s \approx \tau_R/2$. The experimental results in this spectral region can be compared to the results of quantum beat experiments[19] on the same absorption lines, which show a decay rate in excellent agreement with the results of the narrow Hanle-effect signal described here. Especially they give also $\tau_s \approx \tau_R/2$. In the quantum beat experiments the discrepancy between τ_R and τ_s was attributed to magnetic field inhomogeneities, however, without any detailed analysis. Moreover, these experiments were not performed in the pressure range below 1 μbar. On the other hand, magnetic field inhomogeneities could be excluded in the Hanle-effect experiments. Also, in the quantum beat experiments only a signal corresponding to the narrow Hanle-effect signal was seen, because of the short interaction time T_L of the molecules with the light field.

The observation of the broad Hanle-effect signal in the molecular beam experiment requires spectral selectivity in the fluorescence detection, because only for red-shifted fluorescence detection do both components have comparable signal amplitudes. Static gas experiments do not need this

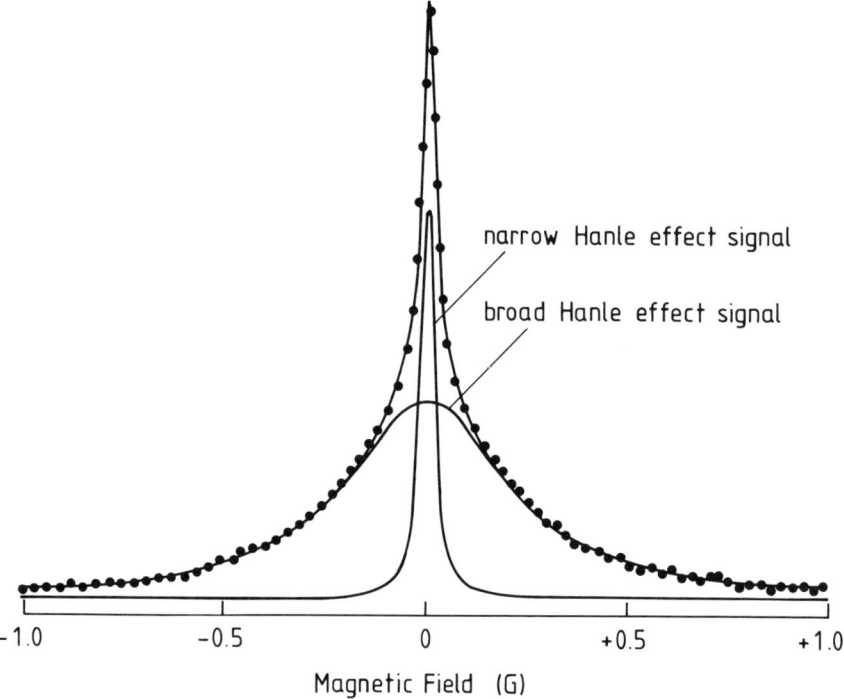

FIGURE 10. The Hanle-effect signal on NO_2 under static gas conditions. The experimental results represented by the dots are fitted by the solid line. The solid line is the sum of two Lorentzian-shaped curves, which represent the narrow and broad Hanle-effect signals.

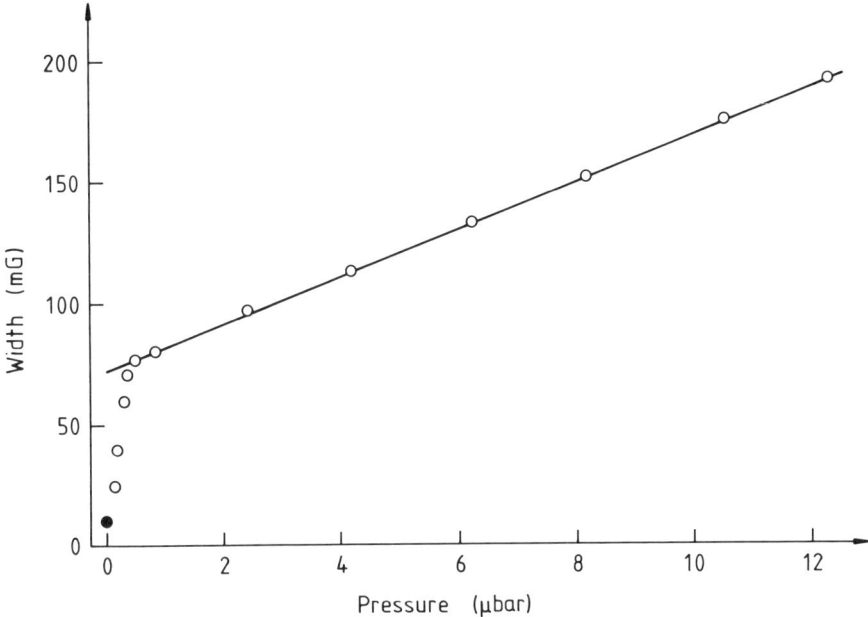

FIGURE 11. Width of the narrow Hanle-effect signal versus the NO_2 pressure in the static gas experiment. The black dot at zero pressure represents the experimental result of the molecular beam experiment.

strongly red-shifted fluorescence detection. The signal amplitude of the broad Hanle-effect signal is not affected significantly by the NO_2 pressure, but the signal amplitude of the narrow Hanle-effect signal is nearly by a factor 5 smaller than the amplitude of the corresponding component in the molecular beam experiment already at pressures less than 1 μbar. Thus, the NO_2 pressure seems to have the same effect on the amplitude of the narrow Hanle-effect signal as the red-shifted fluorescence detection in the molecular beam experiment.

9.4. Discussion of Hanle-Effect Experiments on NO_2

The conditions in the Hanle-effect experiments on NO_2, described in the previous sections, are to our knowledge analogous to those of experiments on atoms and diatomic molecules, namely, preparation of a molecule into a well-defined fine structure level and detection of the radiative decay of this state. The presence of hyperfine structure causes no problems, because the hyperfine structure splitting is large compared to the spectral width of the laser light. Each molecule is excited within a spectral width of about 0.3 MHz[65] into a single isolated hyperfine structure level, i.e., into an expected eigenstate of the molecule. Consequently, we expect a Hanle-effect signal, which is a superposition of three independent resonances each exhibiting about the same width characteristic of the radiative lifetime τ_R. The narrow Hanle-effect signal is in agreement with this prediction. There are also other measurements which verify the expected experimental situation: measurements of Landé g-factors, Stark constants, hyperfine structure splittings, and radiative decay times. However, the σ-resonance, the broad Hanle-effect signal, the inversion effect, and the nonlinear pressure dependence depicted in Figure 11 do not fit into this picture.

Every effort was made to verify that these unexpected results are no experimental artifact. For instance, as already mentioned in the previous sections, we see no possibility to explain these results by known effects associated with laser excitation,[13,14] as, for instance, optical pumping with lasers or mixing of upper and lower states of a laser-induced transition. It seems that these known laser saturation effects are covered by a new phenomenon in NO_2.

One feature of this new phenomenon is the observation of two characteristic times $\tau_R \approx 40$ μs and $\tau_0 \approx 3$ μs associated with each hyperfine structure level of this molecule. Quantity τ_R was measured in time-resolved fluorescence decay measurements as a single exponential radiative decay time[85] and in Hanle-effect and optical radio-frequency double-resonance experiments. The value of τ_0 was obtained from the width of the σ-resonance and of corresponding magnetic resonances in optical radio-frequency double-resonance experiments,[99] from the width of the broad Hanle-effect signal and from the width of resonances in laser scanning experiments.[65] The time τ_0 has also been known for a long time from investigations of the integrated absorption coefficient.[106,107] The experiments reported in the previous sections showed that it is not possible to associate τ_0 with the lifetime of the lower state of the laser-induced transition, with two photon transitions, with hyperfine structure coherence, or with properties of the exciting light source. It is inevitable to attribute both τ_R and τ_0 to a single isolated hyperfine structure level. Consequently, we must conclude that there is a substructure underlying the expected eigenstates.

The following model was suggested for this substructure.[108] Underlying an expected eigenstate $|e\rangle$ is a substructure represented by the two "states" $|e_1\rangle$ and $|e_2\rangle$. An optical transition populates only $|e_1\rangle$. Being in $|e_1\rangle$ the molecule evolves irreversibly from $|e_1\rangle$ to $|e_2\rangle$ in time τ_0. Both $|e_1\rangle$ and $|e_2\rangle$ decay also with the radiative decay time τ_R. The only quantity to discriminate between $|e_1\rangle$ and $|e_2\rangle$ is the degree of polarization of fluorescence light.

The considered model describes the molecular beam experiments reported in Sections 9.1 and 9.2. The broad Hanle-effect signal is attributed to fluorescence decay of $|e_1\rangle$, and the narrow Hanle-effect signal to fluorescence decay of $|e_2\rangle$. As $|e_1\rangle$ and $|e_2\rangle$ have the same radiative lifetime τ_R and presumably also approximately the same spectral distribution of the fluorescence light, it is necessary to associate with $|e_1\rangle$ and $|e_2\rangle$ different degrees of polarization of the fluorescence light. If, for instance, the degree of polarization of fluorescence light is zero for $|e_2\rangle$ but not for $|e_1\rangle$, one observes only the broad Hanle-effect signal having a width characteristic of τ_0 and not of the radiative lifetime τ_R. The model under consideration describes also the inversion effect associated with the narrow Hanle-effect signal[108] and it describes the σ-resonance.[82] The latter was explicitly shown for its counterpart in the optical radio-frequency double-resonance experiment.[99,108] Moreover, the dependence of the signal strength S_n of the narrow Hanle-effect signal on the cutoff wavelength λ_x in the fluorescence detection path can be explained[100] in agreement with this model.

In order to describe also the effect of collisions in the static gas experiments, it is necessary to add some refinements to the model.[109] Especially is it necessary to represnet $|e_2\rangle$ by a larger manyfold of "states." To our knowledge the considered model is in agreement with the whole body of experimental results on NO_2.

This model is not in agreement with the standard quantum mechanical description of molecules. The usual molecular Hamiltonian gives $|e\rangle$ as stationary states (expected eigenstates), but not $|e_1\rangle$ and $|e_2\rangle$. Therefore, this Hamiltonian does not provide the full description of the experimental results. The different degrees of polarization of fluorescence light associated with $|e_1\rangle$ and $|e_2\rangle$ may be attributed to different shapes of the molecule in these "states." Changes of shape, such as equal or unequal bond lengths of NO_2, may affect the transition probability of P-, Q- and R-branch radiative transitions, without affecting the overall decay rate strongly. Presumably, this makes the substructure observable. Moreover, this also restricts the observations of the considered substructure to those systems where changes in shape (arrangement of nuclei) causes changes in symmetry. This excludes atoms and diatomic molecules.

Presumably the environment of the molecule is the origin of molecular shape[110] and consequently also the origin of the substructure examined here. This explains satisfactorily the postulated irreversible evolution associated with the substructure. On the other hand, the experimental results (except those related to the nonlinear pressure dependence) were not seen to show any dependence on the environment despite all efforts to realize isolated NO_2 molecules. This suggests the interpretation that isolation of NO_2 is presumably not possible. In other words, the assumed irreversible evolution of the molecule from $|e_1\rangle$ to $|e_2\rangle$ will presumably take place even in intergalactic space.

If the interpretation of the Hanle-effect experiments on NO_2, as given here, is correct, it will certainly have a strong impact on our understanding not only of polyatomic systems, but eventually in a more general sense of the transition from microscopic to macroscopic objects. However, to fully accept this interpretation, it is most desirable to have more detailed investigations on polyatomic molecules, for instance, on SO_2, CS_2, or NH_2.

10. CONCLUSION

The application of the Hanle effect and level-crossing spectroscopy to molecules yields lifetimes, Landé g-factors, electric-dipole moments, polarizabilities of molecules, energy splittings as hyperfine structure or lambda doubling, collision cross sections etc. This kind of information, which is much richer than the information obtainable from atoms, due to the much richer structure of molecules, is expected in principle. However, as the experiments on NO_2 revealed, there is also information available from these experiments that is unexpected, because it is not in agreement with the experience of the Hanle effect and level-crossing spectroscopy on atoms and because it is not in agreement with the present understanding of the structure and dynamics of polyatomic systems. The investigation and interpretation of this kind of information is still a challange.

Except for some changes in Section 9, this chapter was written in 1987/1988. Some relevant publications have been published since and are appended with their titles as Refs. 111 to 115.

ACKNOWLEDGMENTS. The author is indebted to Dr F. Bylicki and Dr G. Miksch for most exciting cooperation in exploring NO_2 and to Prof. W. Hanle for his continuous interest in these investigations. In the preparation of the manuscript the author was supported by his wife Gudrun Weber.

REFERENCES

1. R. N. ZARE, *J. Chem. Phys.* **45**, 4510-4518 (1966).
2. R. N. ZARE, in *Methodes de Spectroscopie sans Larguer Doppler de Niveaux Excités de Systémes Moléculaires Simples*, Aussois, France, 23-26 May, 1973 (Natl Centre Sci. Rec., Paris, 1974), pp. 29-40.
3. J. C. LEHMANN, in *Frontiers in Laser Spectroscopy*, Les Houches, Session XXVII, edited by R. Balian, S. Haroche, and S. Liberman (North-Holland, Amsterdam, 1977), Vol. 2, pp. 476-527.
4. J. C. LEHMANN, *Rep. Prog. Phys.* **41**, 1609-1663 (1978).
5. M. BROYER, G. GOUÉDARD, J. C. LEHMANN, AND J. VIGUÉ, in *Advances in Atomic and Molecular Physics*, edited by D. R. Bates and B. Bederson (Academic Press, New York, 1976), Vol. 12, pp. 165-213.
6. N. BILLY, M. BROYER, B. GIRARD, G. GOUÉDARD, J. C. LEHMANN, AND J. VIGUÉ, *Ann. Phys.* (*Paris*) **10**, 1101-1116 (1985).

7. H. G. WEBER AND F. BYLICKI, *Acta Phys. Pol.* **A69**, 699-722 (1986).
8. P. FRANKEN, *Phys. Rev.* **121**, 508-512 (1961).
9. G. GOUÉDARD AND J. C. LEHMANN, *J. Phys.* (*Paris*) **34**, 693-699 (1973).
10. C. H. TOWNES AND A. L. SCHAWLOW, *Microwave Spectroscopy* (Dover, New York, 1975).
11. G. W. LOGE AND C. S. PARMENTER, *J. Chem. Phys.* **74**, 29-35 (1981).
12. G. M. NATHANSON AND G. M. CLELLAND, *J. Chem. Phys.* **85**, 4311-4321 (1986); **81**, 629-642 (1984).
13. C. COHEN-TANNOUDJI, in *Atomic Physics* 4, edited by G. zu Putlitz, E. W. Weber, and A. Winnacker (Plenum Press, New York, 1975), pp. 589-614.
14. B. DECOMPS, M. DUMONT, AND M. DUCLOY, in *Laser Spectroscopy of Atoms and Molecules*, edited by H. Walther (Springer-Verlag, Berlin, 1976), pp. 283-347.
15. M. P. AUZINSH, M. YA. TAMANIS, and R. S. FERBER, *Opt. Spectrosc.* **59**, 828-829 (1985); **63**, 582-588 (1987).
16. R. S. FERBER, O. A. SHMIT, AND M. YA. TAMANIS, *Chem. Phys. Lett.* **61**, 441-444 (1979).
17. M. YA. TAMINIS, R. S. FERBER, AND O. A. SHMIT, *Opt. Spectrosc.* **53**, 449-450 (1982).
18. P. J. BRUCAT AND R. N. ZARE, *J. Chem. Phys.* **78**, 100-111 (1983).
19. P. J. BRUCAT AND R. N. ZARE, *J. Chem. Phys.* **81**, 2562-2570 (1984).
20. P. J. BRUCAT AND R. N. ZARE, *Mol. Phys.* **55**, 277-285 (1985).
21. N. OCHI, H. WATANABE, S. TSUCHIYA, AND S. KODA, *Chem. Phys.* **113**, 271-285 (1987).
22. G. W. LOGE, J. J. TIEE, AND F. B. WAMPLER, *J. Chem. Phys.* **84**, 3624-3629 (1986).
23. H. WATANABE, S. TSUCHIYA, AND S. KODA, *J. Chem. Phys.* **82**, 5310-5317 (1985).
24. M. DUBS, P. SCHMIDT, AND J. R. HUBER, *J. Chem. Phys.* **85**, 6335-6339 (1986).
25. J. S. BASKIN, P. M. FELKER, AND A. H. ZEWAIL, *J. Chem. Phys.* **86**, 2483-2499 (1987).
26. J. KOMMANDEUR, W. A. MAJEWSKI, W. L. MEERTS, AND D. W. PRATT, *Ann. Rev. Phys. Che.* **38**, 433-462 (1987).
27. M. JVANCO, J. HAGER, W. SHARFIN, AND S. C. WALLACE, *J. Chem. Phys.* **78**, 6531-6540 (1983).
28. P. M. FELKER AND A. H. ZEWAIL, *J. Chem. Phys.* **82**, 2961-3002 (1985).
29. E. PEBAY-PEYROULA, R. JOST, M. LOMBARDI, AND J. P. PIQUE, *Chem. Phys.* **106**, 243-257 (1986).
30. R. L. DE ZAFRA, A. MARSHALL, AND H. METCALF, *Phys. Rev. A* **3**, 1557-1567 (1971).
31. A. MARSHALL, R. L. DE ZAFRA, AND H. METCALF, *Phys. Rev. Lett.* **22**, 445-449 (1969).
32. S. J. SILVERS AND CHI-LIAN CHIU, *J. Chem. Phys.* **61**, 1475-1479 (1974).
33. M. B. MOELLER, M. R. MCKEEVER, AND S. J. SILVERS, *Chem. Phys. Lett.* **31**, 398-400 (1975).
34. W. C. WELLS AND R. C. ISLER, *Phys. Rev. Lett.* **24**, 705-708 (1970).
35. J. W. MILLS AND R. N. ZARE, *Chem. Phys. Lett.* **5**, 37-41 (1970).
36. H. M. POLAND AND R. N. ZARE, *Bull. Am. Phys. Soc.* **15**, 347 (1970).
37. P. BALTAYAN AND O. NEDELEC, *Phys. Lett.* **37A**, 31-32 (1971).
38. M. A. MARECHAL, R. JOST, AND M. LOMBARDI, *Phys. Rev. A* **5**, 732-740 (1972).
39. J. VAN DER LINDE AND F. W. DALBY, *Can. J. Phys.* **50**, 287-297 (1972).
40. M. GLASS-MAUJEAU, *J. Phys.* (*Paris*), *Lett.* **38**, 427-428 (1977).
41. J. DUFAYARD, M. LOMBARDI, AND O. NEDELEC, *C.R. Hebd. Seances. Acad. Sci.* (France) *B* **276**, No. 12, 471-474 (1973).
42. G. GOUÉDARD AND J. C. LEHMANN, *C.R. Hebd. Seances Acad. Sci., Ser. B* **270**, No. 26, 1664-1667 (1970).
43. G. GOUÉDARD, *Ann. Phys.* (*Paris*) **7**, 159-198 (1972).
44. K. R. GERMAN, R. N. ZARE, AND D. R. CROSLEY, *J. Chem. Phys.* **54**, 4039-4044 (1971).
45. E. M. WEINSTOCK, R. N. ZARE, AND L. A. MELTON, *J. Chem. Phys.* **56**, 3456-3462 (1972).
46. S. J. SILVERS, T. H. BERGEMAN, AND W. KLEMPERER, *J. Chem. Phys.* **52**, 4385-4399 (1970).
47. S. J. SILVERS AND CHI-LIAN CHIU, *J. Chem. Phys.* **65**, 5663-5667 (1972).

48. K. A. MEYER AND D. R. CROSLEY, J. Chem. Phys. **59**, 1933-1941 (1973).
49. K. R. GERMAN AND R. N. ZARE, Phys. Rev. **186**, 9-13 (1969).
50. K. R. GERMAN, T. H. BERGEMAN, E. M. WEINSTOCK, AND R. N. ZARE, J. Chem. Phys. **58**, 4304-4317 (1973).
51. E. M. WEINSTOCK AND R. N. ZARE, J. Chem. Phys. **58**, 4319-4326 (1973).
52. M. MCCLINTOCK, W. DEMTRÖDER, AND R. N. ZARE, J. Chem. Phys. **51**, 5509-5521 (1969).
53. M. BROYER AND J. C. LEHMAN, Phys. Lett. **A40**, 43-44 (1972).
54. M. BROYER, J. C. LEHMANN, AND J. VIGUÉ, J. Phys. (Paris) **36**, 235-241 (1975).
55. F. W. DALBY, J. VIGUÉ, AND J. C. LEHMANN, Can. J. Phys. **53**, 140-144 (1975).
56. G. GOUÉDARD AND J. C. LEHAMNN, C.R. Hebd. Seances Acad. Sci., Ser. B **280**, No. 15, 471-474 (1975).
57. G. DE VLIEGER AND H. EISENDRATH, J. Phys. B **10**, L463-L467 (1977).
58. S. J. SILVERS AND M. R. MCKEEVER, Chem. Phys. **27**, 27-31 (1978).
59. P. MINGUZZI AND F. BERGHI, Nuovo Cimento B **64**, 94-99 (1981).
60. R. E. SMALLEY, L. WHARTON, AND D. H. LEVY, J. Chem. Phys. **63**, 4977-4989 (1975).
61. C. G. STEVENS AND R. N. ZARE, J. Mol. Spectrosc. **56**, 167-187 (1975).
62. G. R. BIRD, J. C. BAIRD, A. W. JACHE, J. A. HODGESON, R. F. CURL, A. C. KUNKLE, J. W. BRANDSFORD, J. RASTRUP-ANDERSON, AND J. ROSENTHAL, J. Chem. Phys. **40**, 3378-3390 (1964).
63. H. J. VEDDER, M. SCHWARZ, H. J. FOTH, AND W. DEMTRÖDER, J. Mol. Spectrosc. **97**, 92-116 (1983).
64. F. BYLICKI AND H. G. WEBER, J. Chem. Phys. **78**, 2899-2909 (1983).
65. H. G. WEBER, Z. Phys. D **1**, 403-420 (1986).
66. R. E. IMHOF AND F. H. READ, Rep. Prog. Phys. **40**, 1-104 (1977).
67. R. L. BURNHAM, R. G. ISLER, AND W. C. WELLS, Phys. Rev. A **6**, 1327-1340 (1972).
68. T. A. CAUGHEY AND D. R. CROSLEY, Chem. Phys. **20**, 467-475 (1977).
69. E. N. KOTLIKOU, Opt. Spectrosc. **41**, 434-436 (1976).
70. S. JSAKSEN AND P. S. RAMANUJAM, J. Chem. Phys. **69**, 2259-2260 (1978).
71. C. W. T. CHIEN, F. W. DALBY, AND J. VAN DER LINDE, Can. J. Phys. **56**, 827-837 (1978).
72. G. GOUÉDARD AND J. C. LEHMANN, J. Phys. Lett. **38**, L85-L86 (1977).
73. G. GOUÉDARD AND J. C. LEHMANN, Faraday Discuss. Chem. Soc. **71**, 143-149 (1981).
74. M. BROYER, J. VIGUÉ, AND J. C. LEHMANN, C.R. Hebd. Seances Acad. Sci., Ser. B **273**, No. 7, 289-292 (1971).
75. M. BROYER AND J. VIGUÉ, in Methods de Spectroscopie sans Largeur Doppler de Niveaux Excités de systémes Moléculaires Simples, Aussois, France, 23-26 May, 1973 (Natl Centre Sci. Rec., Paris, 1974), pp. 185-194.
76. J. VIGUÉ, M. BROYER, AND J. C. LEHMANN, J. Phys. (Paris) **42**, 937-978 (1981).
77. H. FIGGER, D. L. MONTS, AND R. N. ZARE, J. Mol. Spectrosc. **68**, 388-398 (1977).
78. I. R. BONILLA AND W. DEMTRÖDER, Chem. Phys. Lett. **53**, 223-227 (1978).
79. F. BYLICKI AND H. G. WEBER, Chem. Phys. Lett. **79**, 355-359 (1981).
80. F. BYLICKI AND H. G. WEBER, Chem. Phys. Lett. **79**, 517-520 (1981).
81. F. BYLICKI, H. G. WEBER, H. ZSCHEEG, AND M. ARNOLD, J. Chem. Phys. **80**, 1791-1795 (1984).
82. H. G. WEBER AND F. BYLICKI, Chem. Phys. **116**, 133-140 (1987).
83. H. G. WEBER AND F. BYLICKI, Z. Phys. D **8**, 279-288 (1988).
84. F. BYLICKI, F. KÖNIG, AND H. G. WEBER, Chem. Phys. Lett. **88**, 142-144 (1982).
85. G. PERSCH, H. J. VEDDER, AND W. DEMTRÖDER, Chem. Phys. **105**, 471-479 (1986).
86. M. KROLL, J. Chem. Phys. **63**, 1803-1809 (1975).
87. S. MCINTOSH, R. A. BEAUDET, AND D. A. DOWS, Chem. Phys. Lett. **78**, 270-272 (1981).
88. D. COULTER, C. Y. R. WU, AND D. A. DOWS, Chem. Phys. Lett. **60**, 51-54 (1978).

89. G. Gouédard, M. Broyer, J. Vigué, and J. C. Lehmann, *Chem. Phys. Lett.* **43**, 118-121 (1976).
90. M. Broyer, J. Vigué, and J. C. Lehmann, *J. Phys. (Paris)* **39**, 591-609 (1978).
91. F. Bylicki, H. G. Weber, G. Persch, and W. Demtröder, *J. Chem. Phys.* **88**, 3532-3538 (1988).
92. G. Dohnt, A. Hese, A. Renn, and H. S. Schweda, *Chem. Phys.* **42**, 183-190 (1979).
93. P. J. Dagdigian, *J. Chem. Phys.* **73**, 2049-2051 (1980).
94. F. Bylicki and H. G. Weber, *Chem. Phys.* **70**, 299-305 (1982).
95. F. W. Dalby, M. Broyer, and J. C. Lehmann, in *Méthodes de Spectroscopie sans largeur Doppler de Niveaux Exités de Systémes Moléculaires Simples*, Aussois, France, 23-26 May, 1973 (Natl Centre Sci. Rec., Paris, 1974), pp. 227-229.
96. P. Baltayan, *Phys. Lett. A* **42**, 435-436 (1973).
97. K. R. German, *J. Chem. Phys.* **64**, 4192-4196 (1976).
98. M. Broyer, F. W. Dalby, J. Vigué, and J. C. Lehmann, *Can. J. Phys.* **51**, 226-228 (1973).
99. H. G. Weber and G. Miksch, *Phys. Rev. A* **31**, 1477-1487 (1985).
100. H. G. Weber, *Phys. Lett. A* **129**, 355-358 (1988).
101. H. G. Weber, F. Bylicki, and G. Miksch, *Phys. Rev. A* **30**, 270-279 (1984).
102. H. G. Weber, *Phys. Rev. A* **35**, 2747-2749 (1987).
103. W. Gawlik, *Phys. Rev. A* **35**, 2744-2746 (1987).
104. H. G. Weber, *Z. Phys. D* **6**, 73-80 (1987).
105. F. Bylicki (to appear).
106. A. E. Douglas, *J. Chem. Phys.* **45**, 1007-1015 (1966).
107. V. M. Donnelly and F. Kaufmann, *J. Chem. Phys.* **69**, 1456-1460 (1978).
108. H. G. Weber, *Phys. Rev. A* **31**, 1488-1493 (1985).
109. H. G. Weber (to appear).
110. R. G. Woolley, *Adv. in Phys.* **25**, No. 1, 27-52 (1976).
111. M. P. Auzinsh, Polarization of laser-excited fluorescence of diatomic molecules and the magnetic field effect, *Opt. Spectrosc.* **63**, 721-725 (1987).
112. M. P. Auzinsh, Hanle effect in the ground electronic state of dimers with allowance for the finite value of angular momentum, *Opt. Spectros.* **65**, 153-156 (1988).
113. M. P. Auzinsh and R. S. Ferber, Optical orientation and alignment of high-lying vibrational-rotational levels of diatomic molecules under their fluorescence population, *Opt. Spectros.* **66**, 158-163 (1989).
114. I. P. Klintsare, A. V. Stolyarov, M. Ya. Tamanis, R. S. Ferber, and Ya. A. Kharya, Anomalous behavior of Lande factors of the Te_2 (BO_u^+) molecule of the intensities of the BO_u^+-$X1_g^+$ transition, *Opt. Spectrosc.* **66**, 595-697 (1989).
115. G. W. Loge and J. J. Tiee, Lifetime of the $A^2\Sigma^+$, $v' = 0$ level of HS measured using the Hanle effect, *J. Chem. Phys.* **89**, 7167-7171 (1988).

CHAPTER 4

THE NONLINEAR HANLE EFFECT AND ITS APPLICATIONS TO LASER PHYSICS

Giovanni Moruzzi, Franco Strumia, and
Nicolò Beverini

1. THE NONLINEAR HANLE EFFECT AND ITS EXPERIMENTAL OBSERVATION

We say that a Hanle effect occurs whenever the following two conditions are simultaneously satisfied: (1) an external field removes the degeneracy of one or more energy levels of a quantum system, and (2) this removal originates a change in some spectroscopic property of the sample (other than the Zeeman frequency shifts). The transition between the zero-field (degenerate) and the high-field (nondegenerate) case occurs in a field region whose extension depends on the homogeneous width of the Zeeman levels and on the gyromagnetic factors of the involved states. In a classical Hanle effect the angular distribution of the polarized components of the fluorescence radiation changes as a function of the field intensity while the external field is swept through this transition region. Different behaviors are observed according to the experimental geometry. We have seen in the preceding chapters that the classical Hanle effect can be described in terms of absorption and spontaneous emission, disregarding stimulated emission. The presence of stimulated emission becomes important when the radiation intensity is high enough to introduce visible changes in the level populations, and thus a visible decrease in the total absorption coefficient. In these conditions a degeneracy removal in the upper and/or lower state of a transition gives origin to a nonlinear (saturation) effect which, while still

---

Giovanni Moruzzi, Franco Strumia, and Nicolò Beverini • Dipartimento di Fisica dell'Università di Pisa, 56126 Pisa, Italy.

affecting the angular distribution of the polarized components of the fluorescence radiation, affects also the global absorption of the sample. This new phenomenon is known as the *nonlinear Hanle effect* (NLHE), and is of particular relevance in laser physics.

The high brightness and spectral purity of the laser radiation can be exploited in a large number of sub-Doppler techniques, based on the nonlinear response of saturated transitions. The number of possible experiments can be further increased by applying static or RF electromagnetic fields to the sample. The NLHE, contrary to other saturation phenomena, requires only one laser beam, and does not require the presence of a standing wave. This aspect is very important, because it makes the resolution of the experiment largely independent of the distortions in the wave front of the laser beam. A NLHE signal is observed when the degeneracy in either the lower or the upper level of the saturated transition is removed by an external electric or magnetic field. The signal is thus independent of the laser frequency within the Doppler linewidth. Since, as noted above, the NLHE affects also the total absorption of the sample, it can be detected even in experimental conditions where fluorescence is absent or difficult to observe. It is interesting to note that the NLHE is characterized by the occurrence of an absorption *increase*, while an absorption *decrease* is typical of practically all other saturation phenomena.

A transition from degeneracy to nondegeneracy can give origin to a NLHE only if at least three levels are connected by the laser radiation. Three different classes of experiments are possible. In two of them the energies of at least two of the levels must be field-dependent, so that they are degenerate at a particular value of the external field: (1) the removal of the zero-field degeneracy (stimulated Hanle effect, Figure 1a); and (2)

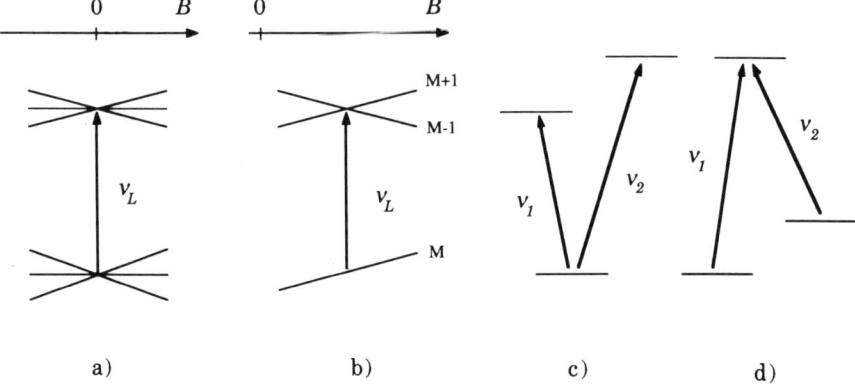

FIGURE 1. Multilevel coherences leading to the nonlinear Hanle effect: (a) zero-field stimulated Hanle effect; (b) stimulated level crossing; (c) and (d) optical–optical double resonance.

high-field crossings or anticrossings of two sublevels (stimulated level crossing, Figure 1b). In the third class of experiments, (3) the degeneracy between two of the levels is induced by a two-frequency laser beam, the difference between the two laser frequencies ω_1 and ω_2 being equal to the separation between the two levels: $\Omega = \omega_1 - \omega_2$ and $\Omega_1 \approx \omega_1 \pm \Delta\omega_D$, $\Omega_2 \approx \omega_2 \pm \Delta\omega_D$ (Figures 1c and 1d). This last method is also known as *optical-optical double resonance* (OODR) spectroscopy and can be considered, in some respects, as an extension of the optical-microwave double-resonance spectroscopy. The NLHE spectroscopy was first proposed by Javan in 1964[1] and clearly described in subsequent work.[2-4]

In order to get further insight, let us consider the situation of Figure 1a in the simple case of a $J = 1 \leftarrow 0$ transition. We shall also assume that the Doppler width $\Delta\omega$ of the transition is much larger than the homogeneous width γ; due to the combined effect of the natural and collisional broadenings of the transition. At zero field the absorption coefficient for a running wave of frequency ω, taking saturation into account (see Section 2 for further details), can be written in the form

$$\alpha(\omega) = \frac{b|\mu|^2}{\sqrt{1+\sigma|\mu|^2}} \exp\{-[(\omega - \omega_0)/\Delta\omega]^2\}$$

$$= \frac{b|\mu|^2}{\sqrt{1+S}} \exp\{-[(\omega - \omega_0)/\Delta\omega]^2\} \qquad (1)$$

where μ is the matrix element of the dipole moment for the given transition, ω_0 its central frequency, and σ a coefficient proportional to the intensity of the saturating beam; b is a quantity which depends on the population densities of the upper and lower levels in the absence of the excitation light beam. The parameter S, which is known as the *saturation parameter*, equals I/I_s, where I is the intensity of the light beam and I_s is the saturation intensity, i.e., the light intensity at which the population difference between the levels is reduced by a factor $\frac{1}{2}$, to be discussed in the next section.

Now we introduce an external field which removes the degeneracy of the upper state, so that the spacing between two consecutive sublevels is larger than γ. For the sake of simplicity, we assume that the light beam is linearly polarized and propagating along the field direction. In these conditions only the $\Delta M = \pm 1$ absorptions are excited, and their intensities are equal. The $\Delta M = \pm 1$ transitions are then coupled to the applied field independently of each other, and the saturated absorption can be written as

$$\alpha^F(\omega) = \frac{b|\mu^+|^2}{\sqrt{1+\sigma|\mu^+|^2}} \exp\{-[(\omega - \omega_+)/\Delta\omega]^2\}$$
$$+ \frac{b|\mu^-|^2}{\sqrt{1+\sigma|\mu^-|^2}} \exp\{-[(\omega - \omega_-)/\Delta\omega]^2\} \qquad (2)$$

In the case of a $J = 1 \leftarrow 0$ transition we have simply $|\mu^+|^2 = |\mu^-|^2 = |\mu|^2/2$, and the relative absorption increase, when the degeneracy is removed, is given by

$$R(S) = \frac{\alpha^F(\omega)}{\alpha(\omega)} = \sqrt{(1+S)/(1+S/2)} \qquad (3)$$

assuming $\Delta\omega$ much larger than $\omega^+ - \omega^-$, so that the exponentials can be neglected. The value of $R(S)$ ranges from 1 to $\sqrt{2}$ for $S \to \infty$. Therefore, in favorable experimental conditions, the NLHE can be large, and easily detectable as a change in the total absorption.

Like the classical Hanle effect, also the NLHE may be detected by observing the fluorescence radiation.[1,5] When this technique is feasible, the anisotropy and polarization of the fluorescence light provide a deeper insight into the experiment, since the evolutions of the diagonal and off-diagonal elements of the density matrix can be detected independently. However, this technique is not practicable for very dense samples and in the infrared spectral region, where transmission experiments are more suitable.

Also, monitoring the absorption of the sample (i.e., the decrease or increase of the transmitted light intensity) can be difficult in some physically interesting conditions. For instance, when the absorbing medium is optically thin, the absorbed power is only a small fraction of the incoming light intensity, and the effect is a small change in this small fraction. This is why the NLHE was first observed either directly on laser active media[1-4,6-11] or on a resonant absorber located inside the laser cavity.[12-14]

A typical experimental apparatus for the detection of the NLHE is shown in Figure 2a. It consists of a single-mode laser and an absorption cell. The absorption increase vs. the applied external field, as expected from equation (2), is shown in Figure 2b. It is the superposition of a Gaussian line, due to the Zeeman scanning of the Doppler profile by the $\Delta M = \pm 1$ transitions, and a Lorentzian (or nearly Lorentzian) central dip which corresponds to the NLHE. The HWHM ΔB of this Lorentzian equals the FWHM of the $J = 1$ level and is given by

$$\text{FWHM} = 2g_J\mu_0 \Delta B$$

where g_J is the Landé factor and μ_0 is Bohr's magneton. On the other hand,

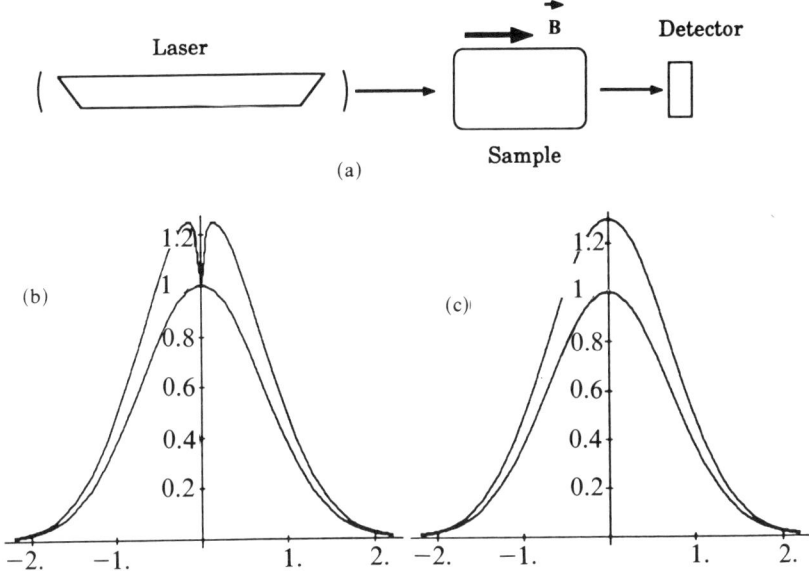

FIGURE 2. (a) Schematic experimental setup for the observation of the NLHE. (b) Expected NLHE absorption increase vs. the external field (**B** or **E**) intensity. (c) Expected NLHE absorption increase vs. the frequency of the light beam.

the fractional absorption increase in the NLHE is expected to be independent of the frequency tuning of the laser within the Doppler profile of the transition, as shown in Figure 2c. This is an important feature of the NLHE because a high resolution can be obtained also when the laser is not frequency-stabilized.

An example of an experimental result obtained with the experimental apparatus of Figure 2a is shown in Figure 3.[14] The absorption of the CO_2 $9P(26)$ laser line by CH_3F vapor is displayed as a function of the applied Stark field. The CO_2 laser is operating simultaneously on two modes. It is possible to observe the zero-field NLHE due to the interaction of the molecules with only one mode (case of Figure 1a) and, at high electric field, the interaction with the two laser modes (case of Figure 1c). The electric dipole moment of the CH_3F molecule has been determined with very high accuracy from the frequency spacing between the two laser modes.

A first general theory of the NLHE and of the related stimulated level-crossing effect was developed by Feld, Sanchez, Javan, and Feldman.[3] This theory provided a qualitative explanation of the physical origin of the effect and yielded a formal expression for the absorption profile at small saturation intensities by means of a perturbative density matrix approach, as had been done in earlier theoretical works.[15–18] At that time, however,

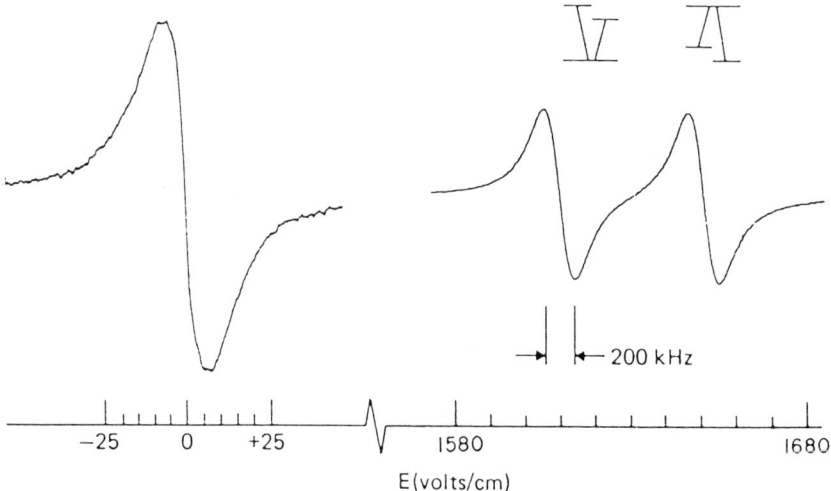

FIGURE 3. Experimental dependence of the NLHE and of the OODR for the $Q(J = 12, K = 2)$ ν_3 line of CH_3F on the applied electric field. The excitation light is provided by two CO_2 lasers, both oscillating on the $9P(20)$ line, with a frequency difference of 39.629 MHz (from Ref. 14).

it was practically impossible to perform an experimental quantitative verification of the size of the effect.

In Ref. 3 equations (1)–(3) were generalized to the case of a coupled three-level system with arbitrary matrix elements μ_1 and μ_2 (Figure 1b). It was demonstrated that when two levels become degenerate, the coupled Schrödinger equation reduces to that of a two-level system with an equivalent matrix element

$$\bar{\mu} = \sqrt{\mu_1^2 + \mu_2^2} \quad (4)$$

The gain as given in equations (1) and (2) can be rewritten in a more general form,

$$\alpha(\omega) = \frac{b(|\mu_1|^2 + |\mu_2|^2)\exp\{-[(\omega - \omega_0)/\Delta\omega]^2\}}{\sqrt{1 + \sigma(|\mu_1|^2 + |\mu_2|^2)}} \quad (5)$$

and

$$\alpha^F(\omega) = \frac{b|\mu_1|^2 \exp\{-[(\omega - \omega_1)/\Delta\omega]^2\}}{\sqrt{1 + \sigma|\mu_1|^2}} + \frac{b|\mu_2|^2 \exp\{-[(\omega - \omega_2)/\Delta\omega]^2\}}{\sqrt{1 + \sigma|\mu_2|^2}} \quad (6)$$

Again, an absorption decrease is expected when coupled transitions become degenerate, also at nonzero Stark or Zeeman fields. The signal is maximum when $|\mu_1| = |\mu_2|$, and decreases with increasing difference between the dipole moments. A further generalization of the above equations to the case of Figure 1c (OODR) in the presence of two independent laser fields was presented by Feld, Parks, and Schlossberg.[2]

After the earlier papers of the sixties, Dumont, Decomps, and Ducloy performed a detailed theoretical and experimental study of the magnetic-field dependence of the fluorescence radiation emitted by an atomic Ne vapor, pumped by a multimode gas laser.[19,20] Dumont's paper provided a theoretical interpretation of the phenomenon. A general theory was developed by Ducloy[21] in terms of a multipole expansion valid for small angular momenta and at the limit of *broad line approximation* (BLA). The NLHE effectively plays a role in this experiment, but one must also take into account that the Zeeman sublevels are affected by a strong optical pumping. The OODR evolved into a well-established spectroscopic technique, particularly useful in the IR spectral region.[22-27,230,231] In particular, subnatural spectral resolution could be obtained by using high spectral purity dye lasers.[28-30] In fact, in the case of OODR the observed linewidth is also a function of the laser linewidth.

New nonconventional detection techniques, like optoacoustic or optogalvanic techniques, were applied to the NLHE detection in the eighties. These techniques, simple and very sensitive at the same time, allow the detection of small changes in the absorption of a vapor sample, and have thus opened new experimental possibilities.[31,234,235] It is noteworthy that these techniques can be applied to media for which the observation of the fluorescence radiation, and hence of the classical Hanle effect or level crossing, is often difficult, if not, as in the case of the infrared molecular optoacoustic spectroscopy, impossible.

The first optoacoustic recording of the NLHE in a molecular vapor was observed by inserting a microphone into an absorption cell external to the CO_2 laser cavity.[32] The signal corresponded to an absorption increase in the CO_2 laser radiation by a CH_3OH or CH_3F molecular vapor when an external electric field was applied to the cell. The optoacoustic signal showed that the absorption coefficient increases significantly (by more than 20%) when the linear Stark effect (the molecule has a large permanent electric dipole moment) removes the degeneracy between the different M components of the transition. The experimental results for CH_3OH were satisfactorily interpreted by a rate-equation approach.[33,34] The first optoacoustic NLHE signal was observed almost simultaneously with the first optoacoustic sub-Doppler saturation Lamb-dip signal.[35] It is worth noting that the NLHE signal was larger than the Lamb-dip signal under similar experimental conditions.

The optogalvanic technique exploits the impedance change due to the absorption of radiation by the atoms present in an electric discharge for detecting this atomic absorption. The NLHE detection by optoacoustic methods suggested a similar possibility for the optogalvanic spectroscopy.[36] The optogalvanic detection of a NLHE in an atomic vapor located in a hollow cathode discharge was reported elsewhere.[37-39] The first quantitative analysis of the NLHE in the typical experimental conditions of optogalvanic spectroscopy, based on the rate-equation approach, and its experimental verification was reported by Beverini et al.[40] This approach to the NLHE will be treated in Section 6. More recently Sokabe and co-workers[41] have developed a similar rate-equation theory in order to interpret their NLHE experiment on the IR absorption of CH_3OH by an optoacoustic technique. In addition, the stimulated level crossing has been observed by means of the optogalvanic technique in a hollow cathode lamp.[42]

The experimental setups for observing the NLHE by the optoacoustic and optogalvanic techniques are sketched in Figure 4. The intensity of the laser radiation is modulated mechanically by a chopper. The signal, i.e., the acoustic signal detected by a microphone located inside the absorption cell (optoacoustic technique), or the variation in the voltage across the ballast resistor in series with the discharge (optogalvanic technique), is fed into a lock-in amplifier phase-locked to the chopper frequency. An electric or magnetic field is applied to the sample and the signal is recorded while sweeping the field across the zero value. The experimental results will be discussed in Section 8.

Of course, the NLHE can give origin not only to an increase in the absorption coefficeint of a saturated transition, but also, in the presence of population inversion, to an increase in the gain coefficient. This can relevantly affect the output power of a laser. The actual size of the effect depends on all the parameters governing the oscillation of the active medium in a laser cavity. In particular, the cavity length will affect the oscillation frequency, while the cavity losses will affect the effective saturation of the transition and thus the gain increase. Intracavity devices, like the Brewster windows or the waveguide, if present, determine the polarization of the EM field, whose direction, relative to the direction of the applied static field, determines the degeneracy removal in the saturated transition. Finally, the output power of a laser depends also on the saturation intensity and on the effective volume of the laser cavity. The power increase induced by the NLHE on a laser transition can thus be very relevant (power increases by factors up to 4 or 5 have been observed) or negligible according to the balance of the roles played by all the constraints mentioned above.

A power increase in the He-Ne laser was observed already in the early sixties,[43,44,110] but the NLHE is small and of little practical interest for

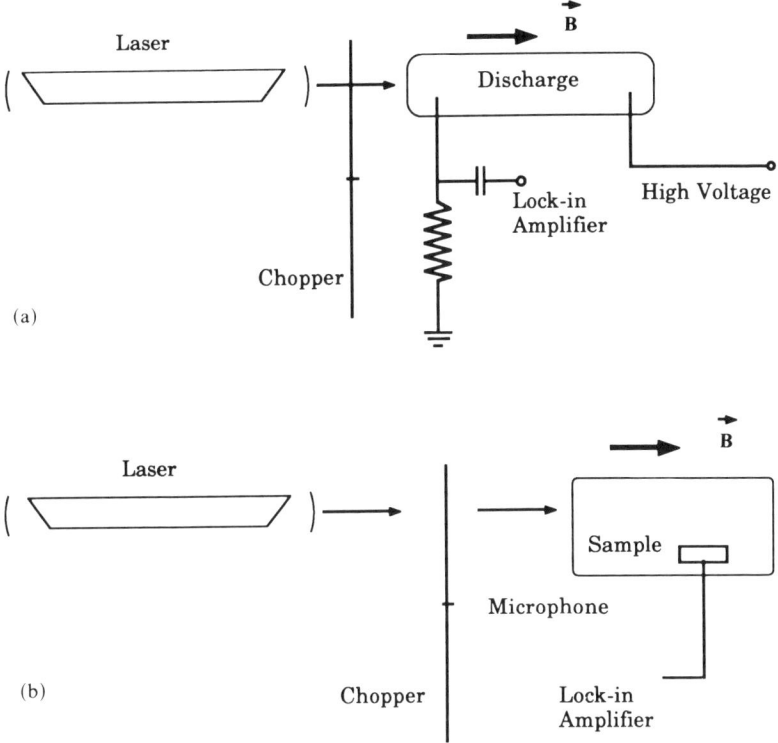

FIGURE 4. Experimental setups for the observation of the NLHE: (a) optogalvanic detection; (b) optoacoustic detection.

He–Ne and similar lasers. For these lasers, however, the NLHE produces interesting frequency effects, like frequency splitting, pulling, modulation, and tuning.[46,47]

The power enhancement induced by an axial magnetic field was soon observed also in the Ar^+ and Kr^+ ion lasers,[45] but at first this effect was wrongly attributed to plasma confinement only.[48,49] The power increase for these lasers is large and of practical importance. Only several years after the discovery of the effect was it recognized that it was a NLHE or, at least, that a large contribution to the power enhancement is due to a NLHE.[50]

In optically pumped lasers the investigation of the NLHE is facilitated by the absence of discharges. Also in this case the power enhancement can be very large because, in favorable experimental conditions, the NLHE can affect both the absorption of the pump transition and the gain of the lasing transition.[32,51–53] An example is shown in Figure 5. Section 9 will be devoted to the NLHE in active laser media.

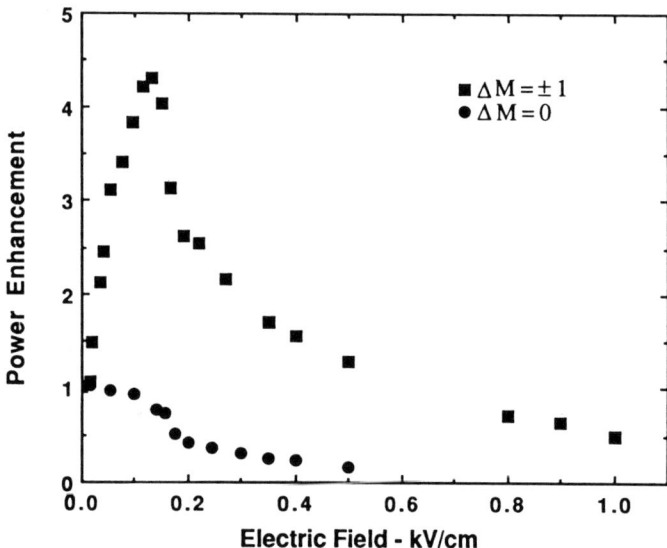

FIGURE 5. NLHE power increase in the emission of an optically pumped far-infrared CH_3OH molecular laser at 70.5 μm induced by an electric field. The power increase is observed when the CO_2 pump laser beam is polarized in order to excite $\Delta M = \pm 1$ transitions. The power decrease at large electric fields corresponds to a Stark effect larger than the Doppler linewidth. When the pump selection rule is $\Delta M = 0$ the power enhancement disappears (from Ref. 53).

2. SATURATION INTENSITY AND SATURATED LINEWIDTH

In order to understand the physical meaning of the nonlinear Hanle effect, we shall first briefly review the concepts of saturation intensity and saturated linewidth. Let us first consider an atomic or molecular vapor interacting with a monochromatic light beam traveling along the z axis of a Cartesian reference frame. We shall assume that the angular frequency ω of the light beam is close to resonance with a nondegenerate atomic (or molecular) transition between a lower level $|i\rangle$ and an upper level $|k\rangle$, with central angular frequency ω_0.

The gain coefficient $\alpha(\omega, z)$ for the light beam is defined by the relation

$$\frac{dI}{I} = \alpha(\omega, z)\, dz \qquad (7)$$

where I is the intensity of the light beam inside the vapor at coordinate z (invariance for x and y translations is assumed), and $I + dI$ is its intensity at coordinate $z + dz$. A semiclassical treatment shows that the small signal gain coefficient $\alpha_0(\omega, z)$, i.e., the gain coefficient when the light intensity is so low that the populations of the upper and lower levels of the transition are not sensibly perturbed, can be written

$$\alpha_0(\omega, z) = \frac{\pi\omega}{\varepsilon_0 c\hbar}(N_k - N_i)|\langle k|\hat{\mathbf{e}} \cdot \mathbf{P}|i\rangle|^2 f(\omega) \tag{8}$$

where N_k and N_i are the population densities at z of the upper and lower state, respectively. If $N_k > N_i$, as for a lasing transition, α_0 is positive; if $N_k < N_i$, as for thermal equilibrium, α_0 is negative, and $-\alpha_0$ is the usual absorption coefficient. As in Chapter 1, we are denoting unit vectors by circumflexes; here $\hat{\mathbf{e}}$ is the unit vector describing the light polarization. Quantity $\mathbf{P}$ is the electric dipole moment operator, and $f(\omega)$ is a normalized function ($\int f d\omega = 1$) describing the shape of the absorption line. In the absence of Doppler broadening $f(\omega)$ would be a Lorentzian of half width (HWHM) γ, where γ is due to the natural broadening:

$$f(\omega) = \frac{1}{\pi}\frac{\gamma}{(\omega - \omega_0)^2 + \gamma^2} \tag{9}$$

For a Doppler-broadened line, with Doppler half-width $\Delta\omega_D \gg \gamma$, we have

$$f(\omega) = \frac{1}{\Delta\omega_D}\frac{\sqrt{\log 2}}{\sqrt{\pi}} \exp\{-\log 2\,[(\omega - \omega_0)/\Delta\omega_D]^2\} \tag{10}$$

In the intermediate cases, i.e., when the orders of magnitude of γ and $\Delta\omega_D$ are comparable, $f(\omega)$ is the convolution of a Lorentzian line with a Gaussian line, known as the Voigt profile.

While the shape, height, and width of the gain profile do not depend on the intensity of the light beam as long as the intensity is low enough not to perturb the population difference ($N_k - N_i$) in a sensible way, changes are observed at higher intensities. As a *Gedankenexperiment*, let us assume that we sweep the resonance region with a strictly monochromatic light beam of variable frequency and fixed intensity, and measure the transmitted light. If the light intensity is high enough, the radiation-induced transitions will tend to equalize the populations of the upper and lower levels (saturation effect), thus decreasing the absolute value of the ($N_k - N_i$) factor appearing in equation (8) and increasing the fraction of transmitted intensity. The (negative) gain profile is correspondingly lowered. The factor by which $|N_k - N_i|$ decreases is not constant since, at constant beam intensity, the saturation effect is stronger for laser frequencies closer to the central frequency of the transition. This corresponds to a broadening of the gain profile.

As so often occurs in quantum mechanics, the best quantitative approach to the problem is provided by the density matrix formalism. For an arbitrary light intensity there is no formula of general validity like equation (8), but each physical situation must be treated separately. In this section we shall consider the simplest case, i.e., a two-level system. This case can be approximated, for instance, by an atomic vapor with a ground atomic state $|i\rangle$ and a highly excited state $|k\rangle$ (highly with respect to kT). We shall assume that our atom has no levels close to the ground state, so that only $|i\rangle$ is populated at thermal equilibrium. We shall exclude the presence of pumping and collisional relaxation mechanisms. We shall assume that the frequency of the light beam is close to resonance with the $|k\rangle \leftarrow |i\rangle$ transition, and that the beam polarization and the selection rules for the transition are such to forbid multiple quantum transitions. A further assumption is that our transition is nondegenerate. All these conditions are verified, for instance, for a transition from a $|\mu = 0\rangle$ ground level to an $|m = +1\rangle$ excited level and σ^+ polarization of the light beam.

The time evolution of the density matrix ρ representing our vapor is

$$\frac{d\rho}{dt} = -\frac{i}{\hbar}[\mathcal{H}_0 + \mathcal{V}, \rho] + \Gamma\rho \tag{11}$$

where $\mathcal{H}_0$ is the unperturbed atomic Hamiltonian, $\mathcal{V}$ the perturbation Hamiltonian describing the interaction of the vapor with the light beam, and Γ is a superoperator (an operator operating on operators) describing the combined effects of relaxation and spontaneous decay (only spontaneous decay, with our assumptions). The interaction Hamiltonian for electric dipole transitions is $\mathcal{V} = -\mathbf{E} \cdot \mathbf{P}$, where $\mathbf{E}$ is the electric field of the wave and $\mathbf{P}$ the atomic (or molecular) electric dipole operator. The electric field at the location z of the atom can be written, for a plane polarized beam,

$$\mathbf{E} = E_0 \hat{\mathbf{x}} \cos(kz - \omega t) = E_0 \hat{\mathbf{x}} \tfrac{1}{2}[e^{i(kz-\omega t)} + e^{-i(kz-\omega t)}]$$

where $\hat{\mathbf{x}}$ is the polarization unit vector and $\mathbf{k} = \hat{\mathbf{z}}(2\pi/\lambda)$ is the wave vector. Since the atomic linear dimensions are much smaller than the wavelength of an optical wave, we can, as usual, neglect the space dependence and write

$$\mathbf{E} = E_0 \hat{\mathbf{x}} \tfrac{1}{2}(e^{-i\omega t} + e^{i\omega t}) \tag{12}$$

By neglecting the space dependence, a σ^+ circularly polarized wave can be written in the form

$$\mathbf{E} = E_0[\hat{\mathbf{x}} \cos(\omega t) + \hat{\mathbf{y}} \sin(\omega t)]$$

or

$$\mathbf{E} = \frac{E_0}{\sqrt{2}} \left[\frac{1}{\sqrt{2}} (\hat{\mathbf{x}} + i\hat{\mathbf{y}}) e^{-i\omega t} + \frac{1}{\sqrt{2}} (\hat{\mathbf{x}} - i\hat{\mathbf{y}}) e^{i\omega t} \right] \quad (13)$$

Thus the intensity of the linearly polarized wave in equation (12) is

$$I = \frac{c\varepsilon_0}{2} E_0^2$$

while the intensity of the circularly polarized wave in equation (13) is

$$I = c\varepsilon_0 E_0^2$$

The factor 2 between the two intensities is due to the different meaning of E_0 in equations (12) and (13). In fact the root-mean-square (rms) electric field of the rotating wave (13) equals E_0, since the modulus of the electric field is constant for a circularly polarized wave, while the rms electric field of the linearly polarized wave (12) is $E_0/\sqrt{2}$. Equations (12) and (13) can be rewritten in the common form

$$\mathbf{E} = \frac{\sqrt{I}}{\sqrt{2c\varepsilon_0}} (\hat{\mathbf{e}} \, e^{-i\omega t} + \hat{\mathbf{e}}^* \, e^{i\omega t}) \quad (14)$$

where $\hat{\mathbf{e}}$ is the appropriate polarization unit vector. This leads to the following form for the matrix elements of the interaction Hamiltonian:

$$\mathcal{V}_{ik} = -\frac{\sqrt{I}}{\sqrt{2c\varepsilon_0}} \langle i|(\hat{\mathbf{e}} \, e^{-i\omega t} + \hat{\mathbf{e}}^* \, e^{i\omega t}) \cdot \mathbf{P}|k\rangle \quad (15)$$

Let us now return to equation (11) and write down the matrices explicitly:

$$\rho = \begin{bmatrix} \rho_{kk} & \rho_{ki} \\ \rho_{ik} & \rho_{ii} \end{bmatrix}, \quad \mathcal{H}_0 = \begin{bmatrix} \hbar\omega_0 & 0 \\ 0 & 0 \end{bmatrix}, \quad \mathcal{V} = \begin{bmatrix} 0 & \hbar W e^{-i\omega t} \\ \hbar W e^{i\omega t} & 0 \end{bmatrix},$$

$$\Gamma\rho = \begin{bmatrix} -2\gamma\rho_{kk} & -\gamma\rho_{ki} \\ -\gamma\rho_{ik} & +2\gamma\rho_{kk} \end{bmatrix} \quad (16)$$

where

$$W = -\frac{\sqrt{I}}{\hbar\sqrt{2c\varepsilon_0}} \langle i|\hat{\mathbf{e}} \cdot \mathbf{P}|k\rangle \quad (17)$$

It is always possible to choose the relative phase between $|i\rangle$ and $|k\rangle$ so that W is real, and we shall assume this phase choice. The matrix elements of $\Gamma\rho$ have been evaluated by assuming that there is no collisional relaxation, and that the lifetime of the excited level $|k\rangle$, due to spontaneous decay, is τ. Under these conditions the natural linewidth in the absence of the light beam γ equals $\frac{1}{2}\tau$ owing to the quadratic dependence of the photon number on the electric field of the emitted wave. The time evolution of the matrix elements of ρ is thus

$$\frac{d}{dt}\rho_{kk} = -i(\rho_{ik}We^{-i\omega t} - \rho_{ki}We^{i\omega t}) - 2\gamma\rho_{kk}, \qquad \frac{d}{dt}\rho_{ii} = -\frac{d}{dt}\rho_{kk}$$

$$\frac{d}{dt}\rho_{ik} = i\omega_0\rho_{ik} - iWe^{i\omega t}(2\rho_{kk} - 1) - \gamma\rho_{ik}, \qquad \rho_{ki} = \rho_{ik}^* \qquad (18)$$

In our case, which is formally equivalent to the Bloch equations for a spin $\frac{1}{2}$ in the rotating wave approximation (optical Bloch equations), the steady-state solution for the density matrix has constant diagonal elements and off-diagonal elements oscillating at $\exp(\pm i\omega t)$:

$$\rho_{kk} = \sigma_{kk}, \qquad \rho_{ii} = \sigma_{ii} = 1 - \sigma_{kk}, \qquad \rho_{ik} = \sigma_{ik}e^{i\omega t}, \qquad \rho_{ki} = \rho_{ik}^* \qquad (19)$$

where σ is a constant Hermitian matrix.

If we substitute expressions (19) into equations (18), then we obtain

$$\sigma_{ik} = \frac{W(2\sigma_{kk} - 1)}{(\omega_0 - \omega + i\gamma)} \quad \text{and} \quad \rho_{kk} = \sigma_{kk} = \frac{W^2}{(\omega - \omega_0)^2 + \gamma^2 + 2W^2} \qquad (20)$$

At equilibrium, the rate at which photons are absorbed from the light beam, $-\alpha I/\hbar\omega$, must equal the photon emission rate due to spontaneous emission $2\gamma N_k$. If we denote by N the total atomic density, then $N_k = \rho_{kk}N$. Thus the gain coefficient α is derived in the form

$$\alpha = -\frac{2\gamma N\hbar\omega}{I} \frac{W^2}{(\omega - \omega_0)^2 + \gamma^2 + 2W^2}$$

Now we can replace W^2 by its expression in equation (17) to obtain

$$\alpha = -\gamma N\hbar\omega \frac{|\langle i|\hat{\mathbf{e}} \cdot \mathbf{P}|k\rangle|^2/(\hbar^2 c\varepsilon_0)}{(\omega - \omega_0)^2 + \gamma^2 + |\langle i|\hat{\mathbf{e}} \cdot \mathbf{P}|k\rangle|^2 I/(\hbar^2 c\varepsilon_0)} \qquad (21)$$

If we now define the *saturation intensity* I_s as

$$I_s = \frac{\hbar^2 c\varepsilon_0 \gamma^2}{|\langle i|\hat{\mathbf{e}} \cdot \mathbf{P}|k\rangle|^2} \qquad (22)$$

and the *saturated linewidth* γ_s as

$$\gamma_s = \gamma\sqrt{1 + I/I_s} \qquad (23)$$

then equation (15) finally becomes

$$\alpha = -\frac{|\langle i|\hat{\mathbf{e}} \cdot \mathbf{P}|k\rangle|^2 \pi\omega}{\varepsilon_0 c\hbar} \frac{N}{\sqrt{1 + I/I_s}} \frac{\gamma_s/\pi}{(\omega - \omega_0)^2 + \gamma_s^2} \qquad (24)$$

The term I/I_s, which in our case equals $2W^2/\gamma^2$, is called the *saturation parameter* and is usually denoted by S. With reference to equation (1), we see that in this case $\sigma = I/(\hbar^2 c\varepsilon_0 \gamma^2)$. The last fraction in equation (24) is a normalized Lorentzian curve of width γ_s. Equation (24) thus turns into equation (8), with $f(\omega)$ given by expression (9), at the limit $I \ll I_s$ (i.e., $S \ll 1$). The height of the Lorentzian curve (24) is seen to decrease with increasing intensities of the light beam, both because of the denominator of the second fraction and because of the normalization condition of the Lorentzian of the third fraction (since the area is normalized, the height must decrease as the width increases). The saturation intensity (22) is the intensity at which the height of the gain profile is reduced by a factor $\frac{1}{2}$. It is important to note that the saturation intensity is not constant for a given transition but, because of the matrix element in the denominator of equation (22), it depends also on the polarization of the exciting radiation. It is also important to note that, although we have again obtained a Lorentzian profile, the saturated linewidth γ_s, which is larger than the natural linewidth γ, is not simply obtained from the time–energy uncertainty relation.

Let us now consider the case of a Doppler-broadened gain profile. We shall assume a Boltzmann distribution of the velocity component of the atoms along the light beam direction, which we have denoted by z. The vapor layer between z and $z + dz$ can be represented by an ensemble averaged density matrix $\rho(v)\, dv$, which describes the atoms whose velocity has the z component in the $(v, v + dv)$ interval. By taking into account the Doppler shift, equation (24) becomes, in a nonrelativistic approximation,

$$\alpha(v) = -\frac{|\langle i|\hat{\mathbf{e}} \cdot \mathbf{P}|k\rangle|^2 \pi\omega}{\varepsilon_0 c\hbar} \frac{N(v)}{\sqrt{1 + I/I_s}} \frac{\gamma_s/\pi}{[\omega - \omega_0(1 + v/c)]^2 + \gamma_s^2} \qquad (25)$$

where $N(v) = (N/v_d\sqrt{\pi})\exp[-(v/v_D)^2]$, $v_D = \sqrt{2kT/m}$, m being the atomic mass. Equation (25) can also be obtained by writing the time evolution of the atoms with their z velocity component in the interval $(v, v + dv)$ as

$$\frac{\partial \rho(v)}{\partial t} + v\frac{\partial \rho(v)}{\partial z} = -\frac{i}{\hbar}[\mathcal{H}_0 + \mathcal{V}, \rho(v)] + \Gamma \rho$$

which has a steady-state solution of the kind

$$\rho_{kk} = \sigma_{kk}, \qquad \rho_{ii} = \sigma_{ii} = 1 - \sigma_{kk}, \qquad \rho_{ik} = \sigma_{ik} e^{i(kz - \omega t)}, \qquad \rho_{ki} = \rho_{ik}^*$$

where σ is a constant Hermitian matrix and $k = \omega/c$.

Integration of equation (25) over all velocities yields

$$\alpha = -\frac{|\langle i|\hat{\mathbf{e}} \cdot \mathbf{P}|k\rangle|^2 \omega}{\varepsilon_0 c \hbar v_D \sqrt{\pi}} \frac{N\gamma_s}{\sqrt{1 + I/I_s}} \int_{-\infty}^{+\infty} \frac{\exp[-(v/v_D)^2]}{[\omega - \omega_0(1 + v/c)]^2 + \gamma_s^2} dv \quad (26)$$

which, in the general case, is a Voigt-shaped curve. If γ_s is sufficiently small, equation (26) reduces to

$$\alpha = -\frac{|\langle i|\hat{\mathbf{e}} \cdot \mathbf{P}|k\rangle|^2 \omega}{\varepsilon_0 c \hbar \Delta\omega_D} \frac{N\sqrt{\pi \log 2}}{\sqrt{1 + I/I_s}} \exp\{-\log 2\,[(\omega - \omega_0)/\Delta\omega_D]^2\} \quad (27)$$

where $\Delta\omega_D = (\omega_0 v_D/c)\sqrt{\log 2}$.

Equation (27) is thus valid at the limit $\gamma_s \ll \Delta\omega_D$. Since under these conditions the width of the gain curve is practically determined by the Doppler width only, equation (27) can be rewritten as

$$\alpha(\omega, S) = \frac{\alpha_0(\omega)}{\sqrt{1 + S}} \quad (28)$$

where $\alpha_0(\omega)$ is the small signal gain coefficient obtained by inserting expression (10) into equation (8). Such a simple relation between the saturated gain and the small signal gain is not valid in the general case, where the width of the gain curve depends on the saturation parameter.

At the limit $I \ll I_s$ ($S \ll 1$) equation (28) reduces to equation (8) with $f(\omega)$ given by expression (10). It is important to note that, while any information on the natural linewidth γ is lost at the $I \ll I_s$ limit, equation (27) retains information on γ through the presence of I_s [see equation (22)]. Of course, it would be an extremely difficult experimental task to extract an accurate value of γ from measurements of the absorption coefficient of a two-level system at high intensities. The point is, however, that the saturated response is sensitive to γ even though the linear response is not.[4]

3. THE THREE-LEVEL CASE: HOMOGENEOUSLY BROADENED LINES

As noted earlier in Section 1, in order to observe the actual nonlinear Hanle effect we need the presence of coherence, and therefore an atomic or molecular transition involving at least three levels, two of which must be degenerate or near-degenerate. An example is given in Figure 6, which represents a case analogous to the $6\,^1S_0$–$6\,^3P_1$ transition of an even isotope of mercury (of course, it would not be easy to observe saturation effects on the 253.7 nm line!). The ground state is assumed to have $J = 0$ and the excited state $J = 1$. The quantization axis is chosen along the propagation direction of the beam, z. Since the NLHE is observed as a change in absorption, rather than in the spatial distribution of the fluorescence polarization, the experimental geometry required for monitoring the effect is simpler than the one of Figure 1. Therefore we shall assume a monochromatic,

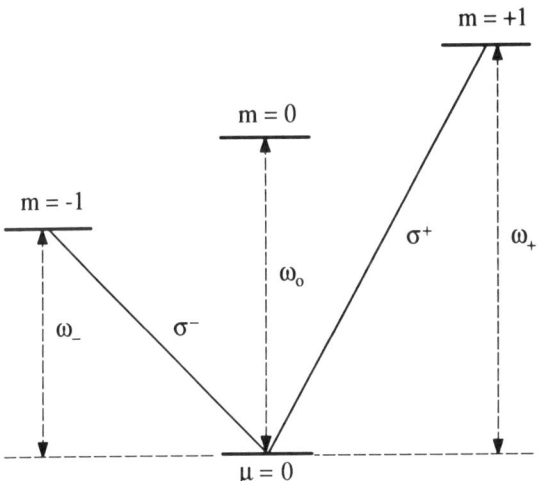

FIGURE 6. A transition from a $J = 0$ ground state to a $J = 1$ excited state. The energy of the excited state is assumed to be high enough so that the state is practically unpopulated at thermal equilibrium. A static uniform magnetic field $\mathbf{B}_0$ parallel to the propagation direction z of the exciting light beam, and a linear polarization of the light beam along the x axis are assumed. Under these conditions the radiation cannot excite the $|\mu = 0\rangle \to |m = 0\rangle$ transition, so that the system behaves as a three-level system.

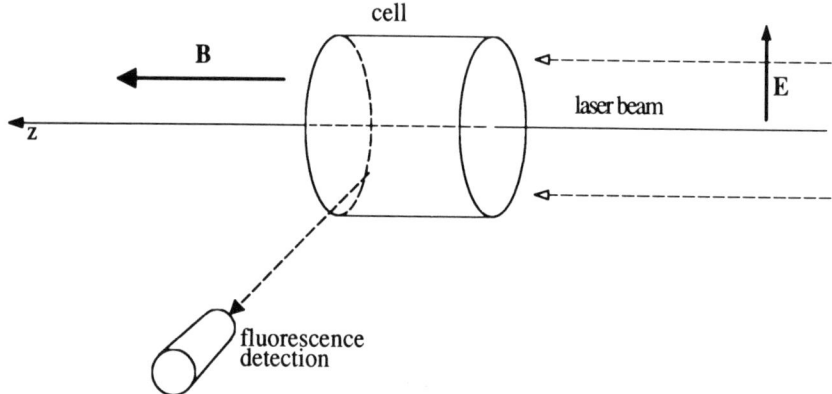

FIGURE 7. Experimental geometry for the fluorescence detection of the NLHE. A laser beam is propagating parallel to the direction of the static magnetic field, which defines the z axis. The beam is linearly polarized along the x direction, so that both σ transitions with $\Delta M = \pm 1$ can be induced simultaneously. The signal can be monitored, for instance, by measuring the intensity of the fluorescence light emitted in any direction.

linearly polarized light beam propagating along the z axis, with the electric vector parallel to the x axis, as shown in Figure 7. The effect can be monitored by measuring any quantity depending on the population distribution, such as the intensity of the fluorescence radiation emitted in any direction (of course, it is advisable to avoid interference with the strong excitation beam). Actually we have to deal with a four-level system, but if the light beam is linearly σ polarized, say in the $\mathbf{x}$ direction, the $|\mu = 0\rangle \to |m = 0\rangle$ transition cannot be induced, so that the level $|m = 0\rangle$ will not be populated (we are assuming that the energy difference between $|m = 0\rangle$ and $|\mu = 0\rangle$ is large compared to kT). Thus, practically, our vapor behaves as a V-type three-level system. If we apply an external field parallel to the beam direction, the sublevels of the excited state will be shifted with respect to each other. We shall denote by ω_+ the frequency of the $|\mu = 0\rangle \to |m = +1\rangle$ transition, and by ω_- the frequency of the $|\mu = 0\rangle \to |m = -1\rangle$ transition. Again, a density matrix treatment is mandatory for an exact, complete solution and, as in the two-level case, we shall exclude the presence of collisional relaxation and of a pumping mechanism. As stated above, we shall assume that $\hat{\mathbf{x}}$ is the polarization vector of the light beam. The time evolution of the density matrix is described by an equation formally identical to equation (11):

$$\frac{d\rho}{dt} = -\frac{i}{\hbar}[\mathcal{H}_0 + \mathcal{V}, \rho] + \Gamma\rho$$

The matrices are written explicitly in the form

$$\rho = \begin{bmatrix} \rho_{++} & \rho_{+0} & \rho_{+-} \\ \rho_{0+} & \rho_{00} & \rho_{0-} \\ \rho_{-+} & \rho_{-0} & \rho_{--} \end{bmatrix}, \quad \mathscr{H}_0 = \begin{bmatrix} \hbar\omega_+ & 0 & 0 \\ 0 & 0 & 0 \\ 0 & 0 & \hbar\omega_- \end{bmatrix},$$

$$\mathscr{V} = \begin{bmatrix} 0 & \hbar W e^{-i\omega t} & 0 \\ \hbar W e^{i\omega t} & 0 & \hbar W e^{i\omega t} \\ 0 & \hbar W e^{-i\omega t} & 0 \end{bmatrix},$$

$$\Gamma\rho = \begin{bmatrix} -2\gamma\rho_{++} & -\gamma\rho_{+0} & -2\gamma\rho_{+-} \\ -\gamma\rho_{0+} & 2\gamma(\rho_{++} + \rho_{--}) & -\gamma\rho_{0-} \\ -2\gamma\rho_{-+} & -\gamma\rho_{-0} & -2\gamma\rho_{--} \end{bmatrix} \quad (29)$$

The first three matrices of equations (29) are identical to those of equations (29) and (31) in Chapter 1. The difference between the last matrix of equations (29) here and of equation (33) in Chapter 1 is due to the fact that here we are assuming pure radiative decays. Under these conditions, the decay rates of the density matrix elements relative to populations (γ_p), Hertzian coherences (γ_c), and optical coherences (γ_o) appearing in equation (33) of Chapter 1 are all equal to one another, and are simple denoted by γ in the expression for $\Gamma\rho$. We have denoted by $|0\rangle$ the ground state, and by $|-\rangle$ and $|+\rangle$ the $|m = -1\rangle$ and $|m = +1\rangle$ sublevels of the excited state, respectively. The term W in the interaction Hamiltonian is formally identical to equation (17),

$$W = -\frac{\sqrt{I}}{\hbar\sqrt{2c\varepsilon_0}}\langle +|\hat{\mathbf{e}}\cdot\mathbf{P}|0\rangle = -\frac{\sqrt{I}}{\hbar\sqrt{2c\varepsilon_0}}\langle -|\hat{\mathbf{e}}\cdot\mathbf{P}|0\rangle \quad (30)$$

Again, it is possible to choose the relative phases of $|0\rangle$, $|+\rangle$, and $|-\rangle$ so that W is real. Its modulus squared is

$$|W|^2 = W^2 = \frac{I}{2\hbar^2 c\varepsilon_0}|\langle +|\hat{\mathbf{e}}\cdot\mathbf{P}|0\rangle|^2 = \frac{I}{2\hbar^2 c\varepsilon_0}|\langle -|\hat{\mathbf{e}}\cdot\mathbf{P}|0\rangle|^2 \quad (31)$$

Although this equation is formally identical to the modulus squared of equation (17), for which a σ^+ transition and a σ^+ light polarization had been assumed, equation (31) is actually smaller by a factor 2 if we assume equal reduced matrix elements for the two cases. In fact, now we have an $\hat{\mathbf{x}}$ polarization of the light beam, while an $\hat{\varepsilon}_{+1} = -(\hat{\mathbf{x}} + i\hat{\mathbf{y}})/\sqrt{2}$ polarization is required in order to induce the $|0\rangle \to |+\rangle$ transition, and an $\hat{\varepsilon}_{-1} = (\hat{\mathbf{x}} - i\hat{\mathbf{y}})/\sqrt{2}$ polarization is required for the $|0\rangle \to |-\rangle$ transition.

A general procedure for evaluating the steady-state solution of equation (11) at arbitrary light intensities is the development of the matrix elements of ρ into Fourier series in $\exp(\pm in\omega t)$. In fact, since both σ^+ and σ^- transitions can be induced by the light beam, multiple quantum transitions between $|0\rangle$ and $|+\rangle$ (as well as between $|0\rangle$ and $|-\rangle$), involving odd numbers of photons, are allowed by the angular momentum conservation. This is why the subdiagonal elements of ρ will oscillate at odd multiples of ω, while only even n appear in the diagonal elements, ρ_{+-} and ρ_{-+}. An exact solution is quite involved. However, since multiple quantum transitions induced by a light beam in the (optical) frequency region of the main resonance are negligible even at high intensities, it will be reasonable to assume that each steady-state matrix element oscillates at a single frequency. We shall thus look for approximate solutions to equation (11) of the form

$$\rho_{++} = \sigma_{++}, \qquad \rho_{--} = \sigma_{--}, \qquad \rho_{00} = 1 - \sigma_{++} - \sigma_{--},$$

$$\rho_{+0} = \sigma_{+0} e^{-i\omega t}, \qquad \rho_{0-} = \sigma_{0-} e^{+i\omega t}, \qquad \rho_{+-} = \sigma_{+-},$$

$$\rho_{0+} = \sigma_{+0}^*, \qquad \rho_{-0} = \sigma_{0-}^*, \qquad \rho_{-+} = \sigma_{+-}^* \tag{32}$$

where σ is, again, a constant Hermitian matrix. Inclusion of the components of the matrix elements oscillating at higher frequencies would lead to a Bloch-Siegert effect, which is negligible at optical frequencies. The steady-state gain coefficient is obtained, as for the two-level case, from the condition that the number of photons absorbed at equilibrium from the light beam equals the number of spontaneously emitted photons, i.e.,

$$-\frac{\alpha I}{\hbar \omega} = 2\gamma N(\rho_{++} + \rho_{--}) \tag{33}$$

where N is again the total number of atoms per unit volume. Substitution of expressions (32) into equation (33) yields the following set of simultaneous equations:

$$0 = -iW(\sigma_{+0}^* - \sigma_{+0}) - 2\gamma\sigma_{++} \tag{34}$$

$$0 = -iW(\sigma_{0-} - \sigma_{0-}^*) - 2\gamma\sigma_{--} \tag{35}$$

$$-i\omega\sigma_{+0} = -iW(1 - 2\sigma_{++} - \sigma_{--} - \sigma_{+-}) - i\omega_+\sigma_{+0} - \gamma\sigma_{+0} \tag{36}$$

$$i\omega\sigma_{0-} = iW(1 - \sigma_{++} - 2\sigma_{--} - \sigma_{+-}) + i\omega_-\sigma_{0-} - \gamma\sigma_{0-} \tag{37}$$

$$0 = -iW(\sigma_{0-} - \sigma_{+0}) - i(\omega_+ - \omega_-)\sigma_{+-} - 2\gamma\sigma_{+-} \tag{38}$$

Since equations (34)-(38) are linear, their solution is, in principle, straightforward, albeit rather involved. According to equation (33), we are interested in the linear combination $(\rho_{++} + \rho_{--}) = (\sigma_{++} + \sigma_{--})$. This has the form

$$\rho_{++} + \rho_{--} = 2W^2 \frac{A\Delta\omega^2 + B}{A\Delta\omega^4 + C\Delta\omega^2 + D} \tag{39}$$

where $\omega_0 = (\omega_+ + \omega_-)/2$, $\Delta\omega = \omega - \omega_0$ is the offset of the radiation frequency relative to the center of the doublet, $\delta\omega = \omega_+ - \omega_0 = \omega_0 - \omega_-$ is half of the splitting between the two upper levels and depends on the intensity of the applied static electric or magnetic field,

$$A = \delta\omega^2 + \gamma^2, \qquad B = (W^2 + \gamma^2 + \delta\omega^2)^2,$$

$$C = W^4 + 6W^2\delta\omega^2 + 6W^2\gamma^2 - 2\delta\omega^4 + 2\gamma^4$$

$$D = 4W^6 + 5W^4\delta\omega^2 + 9W^4\gamma^2 + 2W^2\delta\omega^4$$

$$+ 8W^2\gamma^2\delta\omega^2 + 6W^2\gamma^4 + (\delta\omega^2 + \gamma^2)^3$$

Equation (39) is the general steady-state solution of our three-level problem in the absence of Doppler broadening. According to equation (33), the gain coefficient is obtained by multiplying equation (39) by $-2\gamma N\hbar\omega/I$. The complexity of this expression does not allow a direct physical insight. Some physical insight, however, can be gained by considering the following three special cases:

(1) $\delta\omega = 0$, i.e., $\omega_+ = \omega_-$, the two upper levels are completely degenerate.

Since the frequency of the light beam is off-resonance by the same amount with respect to both the collapsed $|0\rangle \to |+\rangle$ and $|0\rangle \to |-\rangle$ transitions, we have $\rho_{++} = \rho_{--}$, and $|\rho_{+0}| = |\rho_{0-}|$, i.e., $|\sigma_{+0}| = |\sigma_{0-}|$. By examining equations (34)-(38) we see that these conditions imply $\sigma_{0-} = -\sigma_{+0}$. Equation (39) reduces to

$$\rho_{++} + \rho_{--} = \frac{2W^2}{\Delta\omega^2 + \gamma^2 + 4W^2}$$

which is identical to equation (20), with the substitution $W^2 \to 2W^2$. The gain curve is given again by equation (24), with γ_s given by expression (23), where

$$I_s = \frac{\hbar^2 c\varepsilon_0 \gamma^2}{2|\langle +|\hat{\mathbf{e}} \cdot \mathbf{P}|0\rangle|^2} \tag{40}$$

At this limit, in fact, the levels $|m = +1\rangle$ and $|m = -1\rangle$ are degenerate and, if we choose x as the quantization axis, the new $|m' = 0\rangle$ level is an eigenstate of the Hamiltonian, so that our system behaves as a two-level system. Since the light beam is polarized parallel to the new quantization axis, we have

$$|\langle m' = 0|\hat{\mathbf{e}} \cdot \mathbf{P}|\mu = 0\rangle|^2 = 2|\langle m = +1|\hat{\mathbf{e}} \cdot \mathbf{P}|\mu = 0\rangle|^2 = 2|\langle m = -1|\hat{\mathbf{e}} \cdot \mathbf{P}|\mu = 0\rangle|^2$$

so that equation (34) is actually identical to (16). This explains also the substitution of W^2 by $2W^2$ in the expression for the populations.

(2) $|\delta\omega|^2 \gg \gamma^2$, $|\delta\omega|^2 \gg W^2$.

Under these conditions there is no coherence between the two upper levels. We can disregard, both in the numerator and in the denominator, the terms whose order of magnitude is lower than $\delta\omega^4$. If we now confine ourselves to $|\Delta\omega| \approx |\delta\omega|$, equation (39) becomes

$$\rho_{++} + \rho_{--} \approx \frac{4W^2\delta\omega^2}{(\Delta\omega^2 - \delta\omega^2)^2 + 4\delta\omega^2(\gamma^2 + 2W^2)}$$

$$\approx \frac{W^2}{(\Delta\omega - \delta\omega)^2 + \gamma^2 + 2W^2} + \frac{W^2}{(\Delta\omega + \delta\omega)^2 + \gamma^2 + 2W^2} \quad (41)$$

The gain profile is thus approximated by the sum of two practically nonoverlapping Lorentzians, centered at ω_+ and ω_- respectively. The expression of each Lorentzian is again equation (24), with

$$I_s = \frac{\hbar^2 c\varepsilon_0 \gamma^2}{|\langle +|\hat{\mathbf{e}} \cdot \mathbf{P}|0\rangle|^2} \quad (42)$$

The factor 2 between the saturation intensities (42) and (40) is due to the fact that, for large $|\delta\omega|$, only one of the atomic $|0\rangle \to |+\rangle$ and $|0\rangle \to |-\rangle$ transitions can be excited at a time. Thus, only the accordingly polarized half of the beam intensity excites the transition, and twice the beam intensity as in case (1) is required for achieving the same saturation effects. Since equation (41) is practically zero when $|\Delta\omega|$ is not close to $|\delta\omega|$, we can consider this expression as valid for any $\Delta\omega$.

(3) $\Delta\omega = 0$, any value of $\delta\omega$.

Under these conditions ρ_{++} and ρ_{--} must be equal since, as in case (1), the light beam is off-resonance by the same amount with respect to both the $|0\rangle \to |+\rangle$ and the $|0\rangle \to |-\rangle$ transition. Again, we must have $\sigma_{0-} = -\sigma_{+0}$. Equation (39) reduces to

$$\rho_{++} + \rho_{--} = \frac{2W^2}{\delta\omega^2 + \gamma^2 + 4W^2(W^2 + \gamma^2)/(W^2 + \gamma^2 + \delta\omega^2)}$$

and the gain coefficient becomes

$$\alpha = -\frac{2\gamma N\omega_0 |\langle +|\hat{\mathbf{e}} \cdot \mathbf{P}|0\rangle|^2}{c\varepsilon_0 \hbar} \frac{1}{\delta\omega^2 + \gamma^2 + 4\gamma^2 S[(\gamma^2 S + \gamma^2)/(\gamma^2 S + \gamma^2 + \delta\omega^2)]} \quad (43)$$

where again S is the saturation parameter I/I_s, and

$$I_s = \frac{2\hbar^2 c\varepsilon_0 \gamma^2}{|\langle +|\hat{\mathbf{e}} \cdot \mathbf{P}|0\rangle|^2}$$

so that $S = W^2/\gamma^2$ in this case. The formal differences between this last expression for I_s, equation (40), and equation (42) are due to the fact that we have always used the matrix element $\langle +|\hat{\mathbf{e}} \cdot \mathbf{P}|0\rangle$ instead of the matrix element between the actual lower and upper state of each transition. Our choice has the advantage of allowing an easier comparison of the saturation intensities for the different cases.

Case (3) is the experimentally most interesting case. The computed profile of equation (43) as a function of $\delta\omega$ is shown in Figure 8 for different values of S. Here and in subsequent figures the sum of the populations of the sublevels of the upper state, rather than the gain coefficient, is shown as a function of the experimental parameters. Experimentally, $\delta\omega$ can be varied by sweeping the external magnetic or electric field around zero. It is seen from Figure 8 that the gain profile is a symmetrical bell-shaped curve at low field intensities. At higher intensities it remains symmetrical about $\delta\omega = 0$, but it displays a minimum at $\delta\omega = 0$ and two maxima at

$$\delta\omega_{\max}(S) = \pm\gamma[2\sqrt{S(1+S)} - (1+S)]^{1/2} \quad (44)$$

We see from the above expression that the two maxima are observed only if $S > \frac{1}{3}$. At the limit of high saturation parameters ($S \gg 1$) equation (44) becomes $\delta\omega_{\max}(S) \approx \pm\gamma\sqrt{S}$; this proportionality of $\delta\omega_{\max}$ to the square root of S at high S values is shown in Figure 9. Qualitatively, the behavior of Figures 8 and 9 can be interpreted as due to the competition of the following two phenomena: on the one hand, the increasing external field is causing a transition from a two-level to a three-level system which, alone, would lead to an increase in the gain coefficients. On the other hand, the external field is also shifting the two transitions out of resonance, and this latter phenomenon is dominant at $|\delta\omega| > |\delta\omega_{\max}(S)|$. An interesting saturation phenomenon can be observed in Figure 9. Here, for $S > 20$, the population of the upper state, $\rho_{++} + \rho_{--}$, has practically reached its theoretical maxima of 0.5 at $\delta\omega = 0$ and of $\frac{2}{3}$ at $\delta\omega = \pm\delta\omega_{\max}(S)$. The absorption coefficient

$\alpha(\delta\omega_{max})$ is thus larger by a factor $\frac{4}{3}$ than the absorption coefficient at zero field. For any fixed $\delta\omega$, with $0 < |\delta\omega| < |\delta\omega_{max}(20)|$, we see that the population of the upper state decreases with increasing radiation intensity. The explanation is that increasing radiation intensity broadens the levels; at fixed level spacing this causes a transition from a three-level system back toward a two-level system.

Since a transition from a two-level to a three-level system, or *vice versa*, is typical of the classical Hanle effect on the 253.7 nm Hg line, the saturation effect just described is known as the *nonlinear Hanle effect* (NLHE). Typically, the NLHE is observed in optically pumped molecular lasers emitting Stark-active lines, when an electric field is applied. An enhancement of the emitted laser radiation is observed at weak electric fields. The NLHE observed in laser emission is actually more complicated than the model outlined in this section by the following three factors: (1) there is a threshold for laser emission, (2) the lines are usually inhomogeneously broadened,

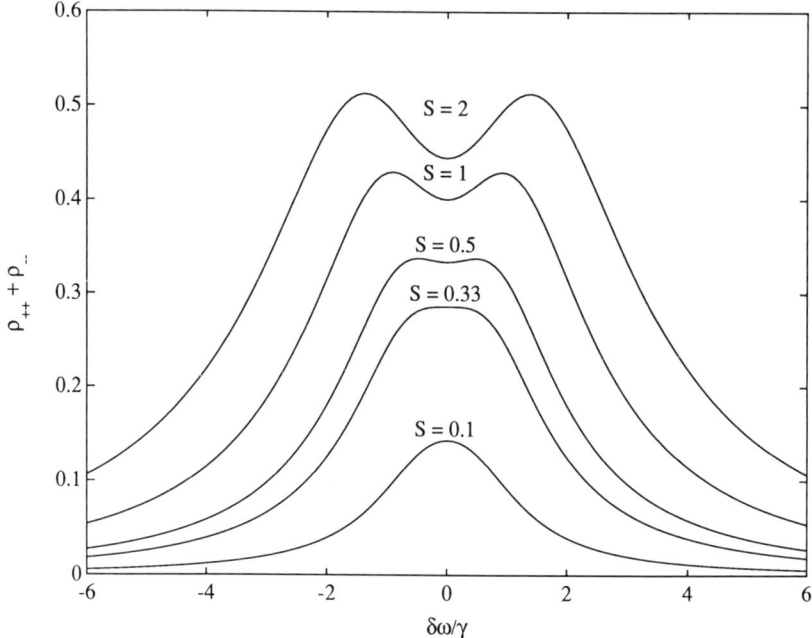

FIGURE 8. The total population of the excited state of the three-level system is reported as a function of $\delta\omega$, i.e., one half of the splitting between the two sublevels of the excited state, for various values of the saturation parameter $S = (W/\gamma)^2$. The linearly polarized exciting light beam is assumed to be tuned to the center of the doublet. A bell-shaped curve centered at $\delta\omega = 0$ is observed for $S < \frac{1}{3}$. At higher radiation intensities a relative minimum is observed at $\delta\omega = 0$, while two maxima are observed at $\delta\omega = \pm\gamma[2\sqrt{S(1+S)} - (1+S)]^{1/2}$.

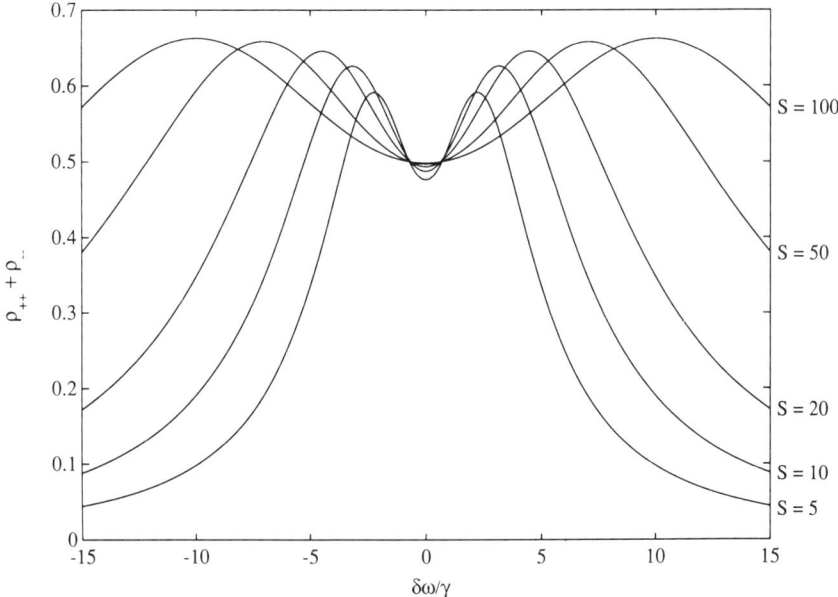

FIGURE 9. Population of the excited state of the three-level system at high saturation parameters. The theoretical maxima of 0.5 for a two-level system and of $\frac{2}{3}$ for a three-level system are reached practically at $\delta\omega = 0$ and $\delta\omega \approx \pm\gamma\sqrt{S}$, respectively. If $\delta\omega$ is fixed at an intermediate value, a population decrease in the excited level is observed as the intensity of the excited beam increases.

and (3) more than three levels are involved, since the J values of the upper and lower level may be higher than 1 and 0 and, in any case, the lower level of the laser transition is different from the ground state.

4. THE THREE-LEVEL CASE: DOPPLER-BROADENED LINES AND THE RATE EQUATIONS

As for the two-level system, in the case of Doppler-broadened absorption lines we represent the vapor layer between z and $z + dz$ by an ensemble-averaged density matrix $\rho(v)\,dv$ which describes the atoms whose velocity has the z component in the $(v, v + dv)$ interval. In a nonrelativistic approximation equation (39) becomes

$$\rho_{++}(v) + \rho_{--}(v)$$
$$= \frac{2W^2\{A[\omega - \omega_0(1 + v/c)]^2 + B\}}{A[\omega - \omega_0(1 + v/c)]^4 + C[\omega - \omega_0(1 + v/c)]^2 + D} \frac{e^{-(v/v_D)^2}}{v_D\sqrt{\pi}} \quad (45)$$

where, as in equation (25), $v_D = \sqrt{2kT/m}$, while A, B, C, D, ω, and ω_0 are given after equation (39). In order to evaluate $\rho_{++} + \rho_{--}$, we must integrate equation (45) over all velocities from $-\infty$ to $+\infty$. The exact integration of the general case has no analytical form and leads to an expression more complicated than a Voigt profile. Here, we shall confine ourselves to an experimentally interesting special case. As for equation (27), we shall assume a Doppler width much larger than the natural linewidth (and much larger than the splitting between the two excited levels), so that the exponential can be treated as constant in the region where the rest of the expression is sensibly different from zero. Moreover, we shall assume that $\omega = \omega_0$, i.e., that the frequency of the light beam coincides with the doublet center. Under these conditions we have, after the substitution $\xi = \omega_0 v/c$,

$$\rho_{++} + \rho_{--} = \frac{2W^2 c}{\omega_0 v_D \sqrt{\pi}} \int_{-\infty}^{+\infty} \frac{A\xi^2 + B}{A\xi^4 + C\xi^2 + D} d\xi$$

The integral here can easily be evaluated by means of the residue theorem:

$$\rho_{++} + \rho_{--} = 2W^2 \frac{\sqrt{\pi \log 2}}{\Delta\omega_D} \frac{A\sqrt{D} + B\sqrt{A}}{\sqrt{ACD + 2(AD)^{3/2}}} \qquad (46)$$

where

$$\Delta\omega_D = \frac{\omega_0 v_D}{c} \sqrt{\log 2}$$

Finally, insertion into equation (33) leads to

$$\alpha = -\frac{2\gamma N \omega_0 |\langle + |\hat{\mathbf{e}} \cdot \mathbf{P}|0\rangle|^2}{c\varepsilon_0 \hbar} \frac{\sqrt{\pi \log 2}}{\Delta\omega_D} \frac{A\sqrt{D} + B\sqrt{A}}{\sqrt{ACD + 2(AD)^{3/2}}} \qquad (47)$$

The computer-evaluated behavior of $\rho_{++} + \rho_{--}$, in the presence of a large Doppler broadening, as a function of $\delta\omega$, i.e., of the applied external field, is shown in Figure 10 for several values of the saturation parameter $S = W^2/\gamma^2$. The shapes of the Hanle signals of Figure 10 are shown on more convenient scales in Figure 11 for $S = 0.1$ and Figure 12 for $S = 10$. In both figures the vertical axis is in arbitrary units in order to normalize the height of the signal. The Lorentzian (dashed line) and Gaussian (dash-dotted line) curves of the same HWHM and same height as the signal are shown in both figures. It is interesting to note that the signal shape approaches a Gaussian curve as the saturation parameter is increased. The computed dependence of the signal width on $\sqrt{S}$ is shown in Figure 13; this dependence becomes linear at high saturation parameters, where $\Delta\omega \approx \gamma\sqrt{S} = W$. As

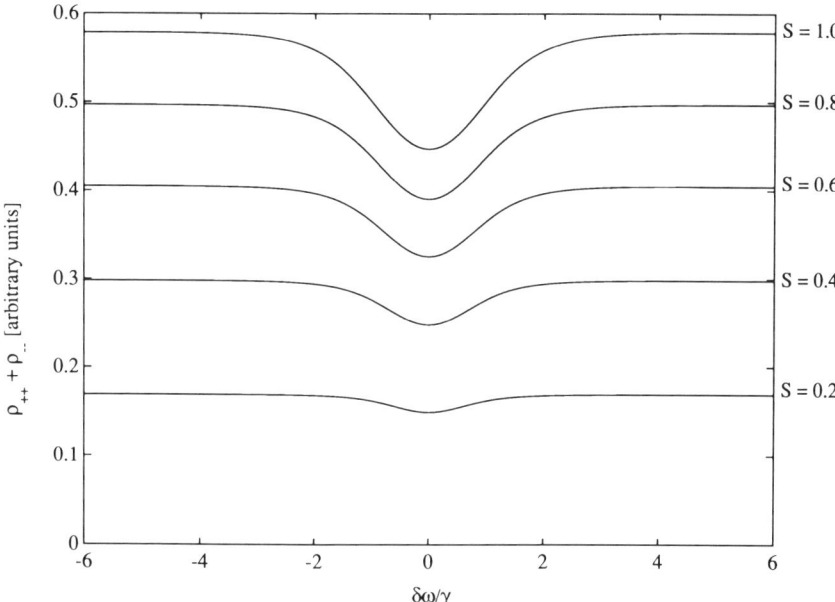

FIGURE 10. Population of the excited state of the three-level system at different values of the saturation parameter S, when the absorption line is Doppler broadened. A Doppler width ten times larger thant the natural width has been assumed.

for the classical Hanle effect, the width of the signal is determined practically by the homogeneous contribution to the linewidth only, and independent of the Doppler broadening.

Equation (47) itself is quite complex, but it has two simple, interesting limiting cases:

1. $\delta\omega = 0$, i.e., zero field,

$$\alpha = -\frac{2N|\langle +|\hat{\mathbf{e}} \cdot \mathbf{P}|0\rangle|^2 \omega}{\varepsilon_0 c\hbar\Delta\omega_D} \frac{\sqrt{\pi \log 2}}{\sqrt{1 + 2I/I_s}} \qquad (48)$$

2. $|\delta\omega|^2 \gg \gamma^2$, $|\delta\omega|^2 \gg W^2$,

$$\alpha = -\frac{2N|\langle +|\hat{\mathbf{e}} \cdot \mathbf{P}|0\rangle|^2 \omega}{\varepsilon_0 c\hbar\Delta\omega_D} \frac{\sqrt{\pi \log 2}}{\sqrt{1 + I/I_s}} \qquad (49)$$

which correspond to the minima and to the asymptotic parts of Figure 10. In both equations (48) and (49) I_s is given by expression (42). The simplicity of these expressions is due to the lack of coherence at both limits. Case (1) corresponds to a Doppler-broadened two-level system, as discussed in

Section 2, and case (2) is the Doppler-broadened equivalent of the homogeneously broadened nondegenerate three-level system, which was encountered in Section 3. In the case of molecules, where high J values are usually involved in both the upper and lower state of the transition, $\delta\omega$ can be large (and thus the coherence can be destroyed) also when no relevant line shift is observed. This is due to the simultaneous shift of the M sublevels in the upper and lower state.

At the limiting case of very high light intensities the ratio between α for the nondegenerate three-level system and α for the two-level system is $\sqrt{2}$, which is higher than the factor $\frac{4}{3}$ which we found for the homogeneously broadened case. The reason is that, while only one hole is burned into the velocity distribution at zero external field ($\delta\omega = 0$), two different holes are burned at high field. Thus a further important enhancing phenomenon is added to the population distribution between two or three levels. The same enhancing phenomenon is observed for a Doppler-broadened three-level system interacting with two electromagnetic waves of different frequencies.[2]

Equations (48) and (49) can be obtained directly by solving a system of rate equations, as we shall show in Section 6. Since only populations, and no coherences, appear in the rate equations, the number of unknowns

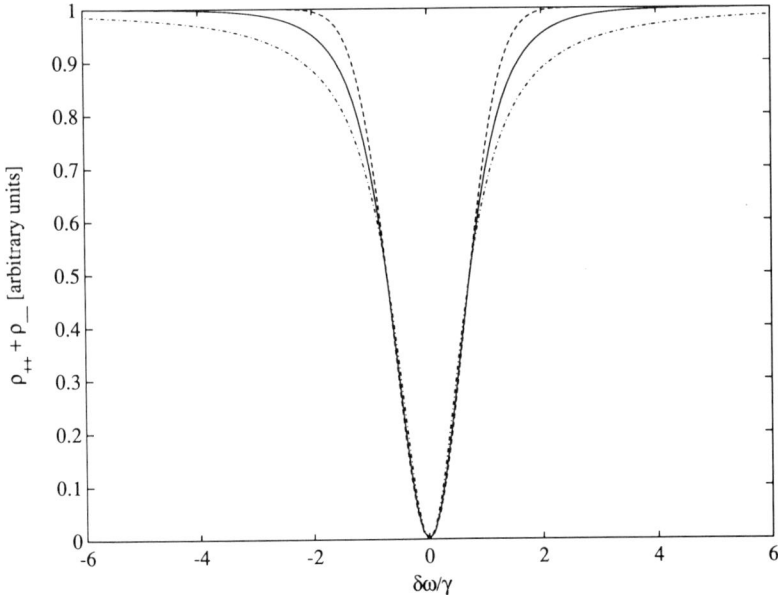

FIGURE 11. Calculated NLHE signal for a saturation parameter $S = 0.1$ (solid line). A Lorentzian (dash-dotted line) and a Gaussian (dashed line) curve having the same height and width of the signal are also reported.

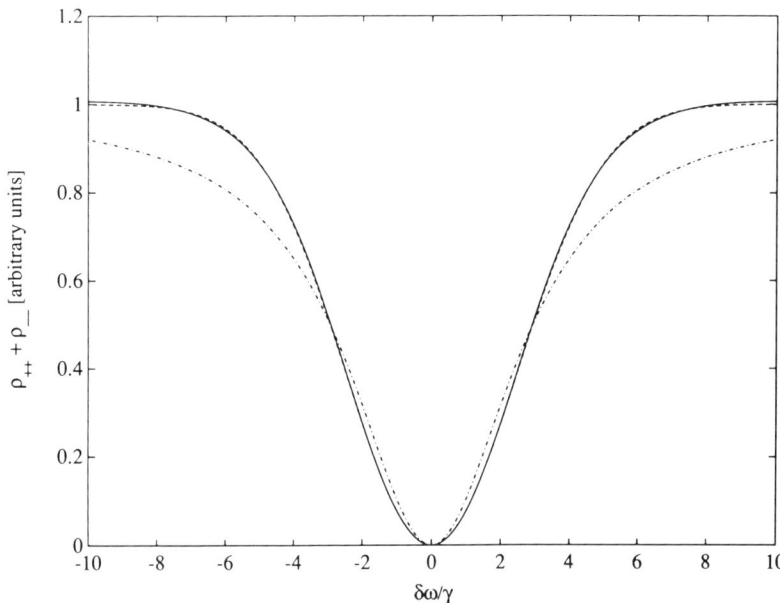

FIGURE 12. Calculated NLHE signal for a saturation parameter $S = 10$ (solid line). A Lorentzian (dash-dotted line) and a Gaussian (dashed line) curve having the same height and width of the signal are also reported. The signal is much closer to a Gaussian shape than the signal of Figure 11.

is correspondingly reduced and the solution is much simpler than for a density matrix treatment. Thus, in the case of Doppler-broadened lines with a Doppler width much larger than the natural linewidth, a more simple approach to the nonlinear Hanle effect is available. This approach, however, can provide only the height of the signal, which is determined by the difference between the behavior at zero field and the behavior at the high-field limit. The complete evaluation of the steady-state density matrix is still needed if the solution at intermediate fields is required. This is needed, for instance, for a precise evaluation of the signal width.

5. THE GENERAL CASE

Until now, the only cases that have been considered are the two-level system of Section 2 and the V-type three-level system of Sections 3 and 4. The extension of the method developed for homogeneously broadened lines to higher J values and to more complicated relaxation patterns is, in principle, straightforward. The method can also be easily modified in order

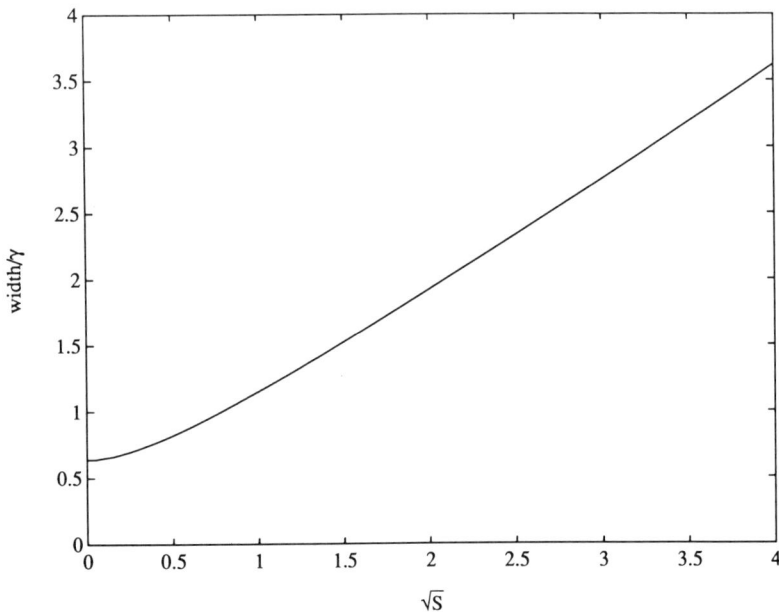

FIGURE 13. Width of the NLHE signal as a function of the square root of the saturation parameter, in the presence of a large Doppler broadening. At high light intensities the width of the signal is determined practically by the power broadening only.

to describe the NLHE on a laser transition. Here one must include terms describing the pumping mechanism into the matrix $\Gamma\rho$, which appears in equation (11). However, one should remember that the unknowns of the resulting system of linear equations are all the populations and half of the coherences of the density matrix, whose dimensions are $(2J' + 1 + 2J'' + 1) \times (2J' + 1 + 2J'' + 1)$. In the preceding expression, and in the rest of this chapter, we denote by J' the angular momentum of the upper level of the transition, and by J'' the angular momentum of the lower level, according to a convention which is of general use in molecular spectroscopy. It is thus apparent how fast the solution becomes more and more complicated with increasing J values.

It is also of interest that, while the $|J' = 1\rangle \leftarrow |J'' = 0\rangle$ transition can be treated as a V-type three-level system, the $|J' = 0\rangle \leftarrow |J'' = 1\rangle$ transition cannot be treated as a Λ-type three-level system. In fact, if we assume the experimental geometry of Figure 7 and absence of a redistribution mechanism within the ground state, all atoms are pumped into the $|\mu = 0\rangle$ sublevel, which cannot absorb σ radiation. As a result, no absorption is observed. Absorption can be restored by population redistribution within the ground state (which could be due, for instance, to collisional relaxation or to

RF-induced transitions). In any case, the presence of the $|\mu = 0\rangle$ sublevel of the ground state cannot be neglected, as we did for the $|m = 0\rangle$ state of the upper level in the V-type case. The number of simultaneous equations is thus increased by the presence of a further population and possible further coherences. Moreover, the necessary presence of a population redistribution mechanism in the ground state further complicates the solution. In conclusion, an analytical density matrix treatment of the NLHE is not practicable for a transition with both J' and J'' larger than 1, particularly at high saturation parameters, where a perturbative approach cannot be used.

However, in some experimental conditions, it is possible to introduce approximations which drastically simplify the evaluation of the magnitude of the NLHE signal.

In Section 3 we have seen that, for a homogeneously broadened line, saturation tends to equalize the populations of all the sublevels involved in the transition. Changing the external field from zero to high values changes the most convenient quantization axis and, in a sense, the effective numbers of sublevels involved in the transition. For instance, we have seen that the V-type system behaves as a two-level system at zero field, and as a three-level system at high fields; in any case, at complete saturation, all the sublevels tend to be equally populated. This remains true as long as no collisional relaxation and no important pumping effects (like those encountered for the $|J' = 0\rangle \leftarrow |J'' = 1\rangle$ case) are present. It is thus easy to evaluate the height of the signal, even if the evaluation of its width and shape may be very involved.

Doppler-broadened atomic or molecular lines occur much more often than homogeneously broadened lines. An important simplification is possible when the velocity dependence of the atom–laser interaction can be neglected, so that the evaluation of the convolution integral is not needed. Cohen-Tannoudji[54] pointed out that this is possible in two quite different experimental conditions.

The first case is the obvious one of a single-mode laser beam of frequency ω perpendicular to the investigated atomic beam. Under these conditions there is no first-order Doppler effect and no collisional relaxation. Since the correlation time of the interaction with the laser light is very long, the correlations between successive atom–light interactions must be taken into account. In general, if we monitor the fluorescence light, we expect a signal width of the order of the natural linewidth for the NLHE in the upper level of the transition, and of the order of the inverse of the transit time for the NLHE in the lower level (if $J'' > 0$). The NLHE in the upper state is reduced by the fact that the Zeeman (or Stark) effect, while removing the degeneracy, simultaneously shifts the absorption line out of resonance.

The second case is that of a Doppler-broadened transition, excited by a multimode laser beam oscillating on a large number N of modes with random phases and a spacing $\delta\omega$ between successive modes (Figure 14). If the two conditions

$$N\delta\omega \gg \Delta\omega_D, \gamma \quad \text{and} \quad \delta\omega \leq \gamma$$

apply, $\Delta\omega_D$ and γ being respectively the Doppler and the natural width of the transition, the *Bennett holes* burnt into the Doppler profile overlap. Thus the response of the atom does not depend on its velocity, and the elements of the density matrix are functions of the internal variables only. This remains valid also at nonzero magnetic fields, as long as the Larmor frequencies are much smaller than the laser spectral width: $\mu B \ll N\delta\omega$. This approximation is known as the *broad-line approximation* (BLA) and has been applied successfully to transitions involving low J values ($J \leq 2$). The treatment of the NLHE in the BLA approximation can be found elsewhere.[20,21,54]

In the rest of this chapter we shall be more interested in the intensity of the NLHE, rather than in its shape, because it is the size of the absorption (or gain) increase which determines the emitted power enhancement of a gas laser. General equations, which allow the evaluation of the size of the effect for any J' and J'' values, disregarding the shape of the signal, can be

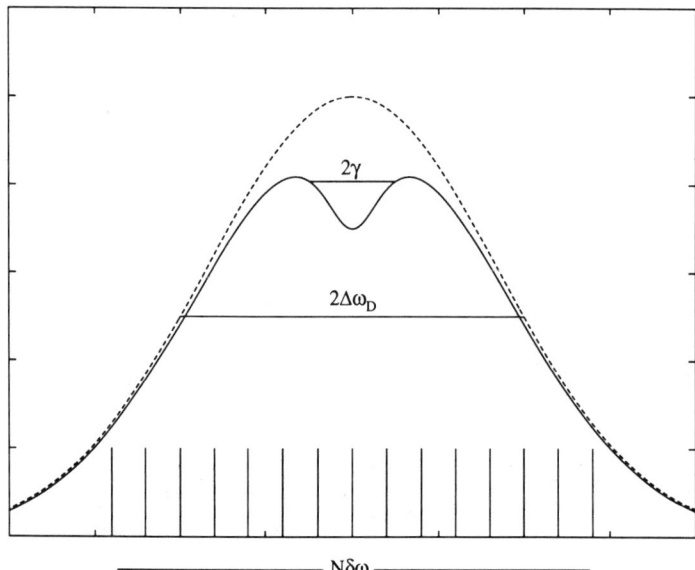

FIGURE 14. Broad-line approximation (BLA). When the frequency spacing of the laser modes $\delta\omega$ is smaller than the homogeneous linewidth γ ($\delta\omega < \gamma$) and the number of modes N is large enough to cover the Doppler linewidth ($N\delta\omega > \Delta\omega_D < \gamma$), the response of the single atom does not depend on its velocity.

obtained by a rate-equation approach. This treatment is the subject of the following two sections.

6. THE RATE-EQUATION APPROACH TO THE NONLINEAR HANLE EFFECT IN INHOMOGENEOUSLY BROADENED TRANSITIONS

The rate-equation approach is correct under many experimental conditions, particularly in the usual optoacoustic and optogalvanic experiments.[31,32,38]

Let us consider an inhomogeneously (Doppler) broadened transition from a lower level $|i, J''\rangle$ to an upper level $|k, J'\rangle$, where i and k stand for the remaining quantum numbers. It is assumed that our vapor is interacting with a linearly polarized laser beam propagating along the external static field, so that the $\Delta M = \pm 1$ transitions are excited simultaneously. We shall also assume a laser frequency close to the center of the absorption line ω_0. In this section the intensity of the light beam is taken constant over the whole interaction region (plane-wave approximation). The external field can be either a magnetic field $\mathbf{B}_0$ (Zeeman effect) or an electric field $\mathbf{E}_0$ (Stark effect, of particular interest in the case of polar molecules). The absorption enhancement R is defined as the ratio of the absorption coefficient at high field (i.e., high enough to remove degeneracy, but not enough to detune the Doppler-broadened absorption components from the laser frequency) to the absorption coefficient at zero external field (complete M_J degeneracy). We shall further assume that the relaxation between the M_J components of each level is fast enough to prevent important optical pumping effects.

If our vapor is interacting with unpolarized monochromatic radiation, equations (27) and (28) can be generalized as follows in order to take the M degeneracy into account:

$$\alpha(\omega, S) = \frac{\alpha(\omega, 0)}{\sqrt{1 + S}} \quad (50)$$

where

$$\alpha(\omega, 0) = \frac{\omega_0}{3\varepsilon_0 \hbar c} \left(\frac{N_k^0}{g_k} - \frac{N_i^0}{g_i} \right) \sum_{M,M'} |\langle kJ'M'|\mathbf{P}|iJ''m''\rangle|^2$$

$$\times \frac{\sqrt{\pi \log 2}}{\Delta\omega_D} \exp\left[-\log 2 \left(\frac{\omega - \omega_0}{\Delta\omega_D} \right)^2 \right] \quad (51)$$

where N_k^0 and N_i^0 are the unsaturated populations of the upper and lower levels, respectively, while g_k and g_i are the corresponding multiplicities. The factor $\frac{1}{3}$ is due to an average over the possible values of the scalar product $\hat{\mathbf{e}} \cdot \mathbf{P}$. The saturation parameter S is again defined as the ratio $S = I/I_s$ of the beam intensity to a saturation intensity yet to be defined. For an open system consisting of two degenerate levels, like that sketched in Figure 15, the saturation intensity can be written in the form[55]

$$I_s = \frac{A_{ki} 4\pi^2 \gamma}{B_{ki} \beta} \tag{52}$$

where γ is the homogeneous width (HWHM) and

$$\beta = \tau_k A_{ki} \left[1 + (1 - \tau_k A_{ki}) \frac{\tau_i g_k}{\tau_k g_i} \right] \approx \tau_k A_{ki} \left(1 + \frac{\tau_i g_k}{\tau_k g_i} \right) \tag{53}$$

In this expression τ_i and τ_k are the effective lifetimes (including all the decay processes, both spontaneous and collisional); A_{ki} is Einstein's spontaneous decay coefficient which, for an electric dipole transition, can be written as

$$A_{ki} = \frac{\omega_0^3}{3\pi\varepsilon_0 \hbar c^3} \frac{1}{g_k} \sum_{M'=-J'}^{J'} \sum_{M''=-J''}^{J''} |\langle kJ'M'|\mathbf{P}|iJ''M''\rangle|^2 \tag{54}$$

By applying the sum properties of the Clebsch-Gordan coefficients we have

$$\sum_{M''} |\langle kJ'M'|\mathbf{P}|iJ''M''\rangle|^2 = f(J' \to J'') |\langle kJ'\|P\|iJ''\rangle|^2$$

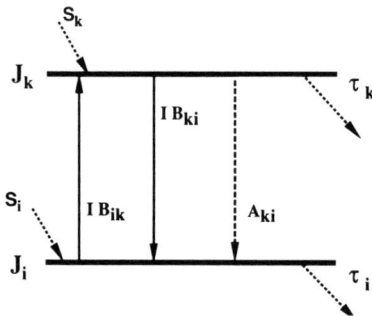

FIGURE 15. Open two-level system. The effective lifetimes τ_k and τ_i include all the incoherent decay processes, i.e., spontaneous emission and collisions. S_k and S_i are the filling rates of levels k and i, respectively.

with

$$f(J \to J) = J(J+1), \qquad f(J \to J+1) = (J+1)(2J+3),$$
$$f(J \to J-1) = J(2J-1) \qquad (55)$$

and

$$\sum_M |\langle kJ'M|\mathbf{P}|iJ''M+1\rangle|^2 = \sum |\langle kJ'M|\mathbf{P}|iJ''M\rangle|^2$$
$$= \sum |\langle kJ'M|\mathbf{P}|iJ''M-1\rangle|^2$$
$$= \tfrac{1}{3}(2J'+1)f(J',J'')|\langle kJ'\|P\|iJ''\rangle|^2$$

where $\langle kJ'\|P\|iJ''\rangle$ is the reduced matrix element of the transition, which is independent of M by definition.

The saturation intensity obtained by inserting this expression into equation (53), and then (53) into (52), is independent of the particular Zeeman (or Stark) component. This is valid for *unpolarized* light. If the light beam is *polarized*, the transition probabilities for the allowed $M'' \to M'$ components may be different from one another (while, in general, the lifetime of the single sublevel does not depend on M, because spontaneous emission is unpolarized). Thus, each line component can have a different saturation intensity. Under these conditions, we can no longer consider the vapor as a simple two-level system. Expression (51) must be rewritten deleting the factor $\tfrac{1}{3}$ and replacing each square matrix element $|\langle kJ'M'|\mathbf{P}|iJ''M''\rangle|^2$ by the corresponding $|\langle kJ'M'|\hat{\mathbf{e}} \cdot \mathbf{P}|iJ''M''\rangle|^2$. The sum is performed over all the transitions allowed by the selection rules. The saturated absorption will thus depend, in general, on the experimental geometry.

Let us first examine a linearly polarized light beam, interacting with an atomic (or molecular) vapor in the absence of external fields (completely degenerate Zeeman sublevels). The most convenient quantization axis is chosen along the electric field of the incoming radiation. In this reference frame only $\Delta M = 0$ transitions can be induced and, as discussed in Chapter 1, no coherences are created between the Zeeman sublevels. The total absorption is the sum of independent two-level processes, one for each M value, and the saturated absorption coefficient at zero field may be rewritten in the form

$$\alpha(\omega, S) = \frac{\omega_0}{3\varepsilon_0 \hbar c} \frac{\sqrt{\pi \log 2}}{\Delta \omega_D} \left(\frac{N_k^0}{g_k} - \frac{N_i^0}{g_i} \right) \exp\left[-\log 2 \frac{(\omega - \omega_0)^2}{\Delta \omega_D^2} \right]$$
$$\times \sum_{M=-J}^{+J} \frac{|\mu_M^0|^2}{\sqrt{1 + \sigma|\mu_M^0|^2}} \qquad (56)$$

where the saturation parameter S_M for each component has been written as a product $\sigma|\mu_M^0|^2$ of the square modulus of the dipole matrix element $|\mu_M^0|^2 = |\langle kJ'M|\hat{\mathbf{e}}\cdot\mathbf{P}|iJ''M\rangle|^2$ times the parameter σ, which depends on ω and M only through higher-order corrective terms. This is so because the coefficients β, defined in equation (53), contain terms proportional to $|\mu_M|^4$. In many experimental conditions, however, these terms are responsible for very small corrections, which decrease proportionally to the gas density (obviously we always have $A_{ik}\tau_k \leq 1$). By omitting these higher-order corrections, we can regard σ as independent of M.

A similar result is found for circularly polarized radiation; in this case $|\mu_M^0|^2$ in equation (56) is replaced by $|\mu_M^+|^2 = |\langle kJ'M|\mathbf{P}|iJ''M+1\rangle|^2$ or by $|\mu_M^-|^2 = |\langle kJ'M|\mathbf{P}|iJ''M-1\rangle|^2$.

Equation (56) and the $\Delta M = 0$ selection rule, obtained under the assumption of zero external field, remain valid for a nonzero external field parallel to the (linear) polarization of the light beam. In this case no coherence between the M sublevels can be induced, and no effect on absorption is observed as long as the Zeeman shift is smaller than $\Delta\omega_D$. A second case in which the degeneracy removal does not induce absorption changes is observed for a purely circularly polarized light beam and an external field parallel to the wave vector. In both cases, the system can be considered as the superposition of noninteracting two-level systems by choosing the appropriate quantization axis.

The situation is more involved for an arbitrary relative orientation of the polarization vector with respect to the external field. An interesting, yet still reasonably simple case, is found for a linearly polarized beam traveling along the external field, whose direction is chosen as quantization axis. Both $\Delta M = +1$ and $\Delta M = -1$ transitions can be excited simultaneously, so that coherences can be generated between states with their M values differing by $\pm 2, \pm 4, \pm 6, \ldots$. Two independent sets of mutually coherent sublevels are thus created (Figure 16a). An exact evaluation of the signal dependence on the external field requires the density matrix formalism. However, the nondiagonal elements of the density matrix coherences are rapidly destroyed if the external field is large enough to completely resolve the degeneracy. Thus, also the absorption coefficient at high fields can be written as a sum over independent transitions:

$$\alpha^F(\omega, S) = \frac{\omega_0}{3\varepsilon_0 \hbar c} \frac{\sqrt{\pi \log 2}}{\Delta\omega_D} \left(\frac{N_k^0}{g_k} - \frac{N_i^0}{g_i}\right) \exp\left[-\log 2 \frac{(\omega-\omega_0)^2}{\Delta\omega_D^2}\right]$$

$$\times \sum_{M=-J}^{+J} \left(\frac{|\mu_M^+|^2}{\sqrt{1+\sigma|\mu_M^+|^2}} + \frac{|\mu_M^-|^2}{\sqrt{1+\sigma|\mu_M^-|^2}}\right) \quad (57)$$

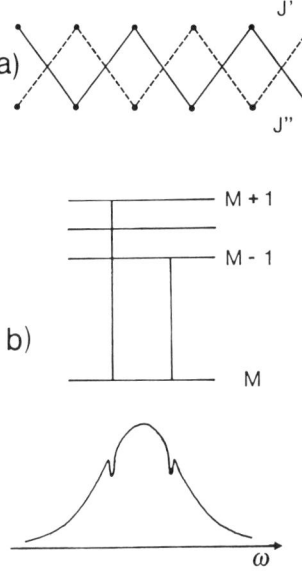

FIGURE 16. Heisenberg diagram for a $J' \leftarrow J''$ transition when the polarization vector of the light beam is perpendicular to the static field. Two separate holes are burned into the velocity distribution when the two absorption components are resolved with respect to their natural width.

This expression has been obtained under the same assumption of Section 4, i.e., that the Doppler width is *very large* compared to the natural linewidth of both levels, so that the Zeeman (or Stark) shifts of the single line components are negligible with respect to the Doppler width.

At this point, a general rule of the Zeeman and Stark effects is noteworthy. For each Zeeman (or Stark) component of a given line, the shift equals the difference between the shifts of the corresponding upper and lower sublevels. On the contrary, the level degeneracy can be completely removed at a field intensity of the order of $h\gamma/2\mu$, independently of the J value. Thus, if the line is inhomogeneously broadened and $\Delta\omega_D \gg \gamma$, all the $\Delta M = \pm 1$ transitions are excited simultaneously by a monochromatic laser beam, even if the line splitting induced by the external field is smaller than γ. In this case the $\Delta M = +1$ and $\Delta M = -1$ transitions starting from a given lower M sublevel involve atoms with different velocity components along the laser direction, i.e., different atoms (Figure 16b). This means that a given atom undergoes a pure $\Delta M = +1$ or $\Delta M = -1$ transition when the Zeeman effect is larger than the homogeneous linewidth.

The ratio

$$R(S, J', J'') = \frac{\alpha^F(\omega, S)}{\alpha(\omega, S)} \tag{58}$$

of expressions (57) to (56) states the enhancement of the saturated absorption for a linearly polarized light beam inducing $\Delta M = \pm 1$ transitions, due to an external field which completely resolves the transition components.

When the shift of the $\Delta M = \pm 1$ transitions induced by the external field is comparable to $\Delta\omega_D$, equation (57) should be completed by multiplying each term by a properly shifted exponential. This implies an obvious adjustment of equation (57), but we shall not consider this case here. Here we shall consider only the case $\Delta\omega_D \gg \delta\omega_M > \gamma$, where $R(S, J', J'')$ is independent of the frequency tuning of the laser frequency as long as it lies well within the Doppler profile.

An explicit expression for equation (58) can be obtained for any J' and J'' by using properly normalized values of $|\mu_M^0|^2$ and $|\mu_M^\pm|^2$.

If $J = J' = J''$ [such transitions are denoted by $Q(J)$ in the molecular notation, where it is customary to denote the lines by the letter P, Q, or R when $J' = J'' - 1$, J'', or $J'' + 1$, respectively, followed by the value J'' of the angular momentum in the lower level] the square moduli of the dipole matrix elements are given by

$$|\mu_M^0|^2 = |\langle kJM|\hat{\mathbf{e}} \cdot \mathbf{P}|iJM\rangle|^2 = |\mu|^2 M^2/J(J+1)$$

$$|\mu_M^\pm|^2 = |\langle kJM|\hat{\mathbf{e}} \cdot \mathbf{P}|iJM \pm 1\rangle|^2$$

$$= |\mu|^2[J(J+1) - M(M \pm 1)]/2J(J+1) \tag{59}$$

where $|\mu|^2$ is the average dipole moment, defined as

$$|\mu|^2 = |\langle kJ\|P\|iJ\rangle|^2 \frac{f(J \to J)}{2J+1} \tag{60}$$

so that $\sum (|\mu_M^0|^2 + |\mu_M^+|^2 + |\mu_M^-|^2)$ is normalized to $g_k|\mu|^2$.

For an arbitrary experimental geometry, equations (58) and (59) depend on the orientation of the light polarization vector with respect to the quantization axis z. There are two cases of particular interest for a linearly polarized light beam: (1) the light beam is propagating along the static field (quantization axis), and (2) the static field is orthogonal both to the beam wave vector and to the beam polarization vector. In both cases the saturation of the $\Delta M = \pm 1$ transitions is the same.

For the sake of simplicity, we shall introduce an *average* saturation parameter S, defined as the saturation parameter of the transition at zero external field (complete M degeneracy). By taking the normalization of equation (60) into account and substituting the dipole moments of equation (59) into the ratio (58), we obtain the relative absorption increase R in the form

$$R(S, J \to J) = \frac{\sum_{M=-J}^{+J} \dfrac{J(J+1) - M(M+1)}{\left[1 + 3S \dfrac{J(J+1) - M(M+1)}{4J(J+1)}\right]^{1/2}}}{\sum_{M=-J}^{+J} \dfrac{2M^2}{\left[1 + 3S \dfrac{M^2}{J(J+1)}\right]^{1/2}}} \qquad (61)$$

where the symmetry between $|\mu_M^+|^2$ and $|\mu_M^-|^2$ has been taken into account.

For a transition from an initial level with angular momentum J to a final level with angular momentum $J - 1$, which, in the molecular spectroscopy notation, could be either a $P(J)$ line in absorption ($J' = J - 1, J'' = J$) or an $R(J - 1)$ line in emission ($J' = J, J'' = J - 1$), we have

$$|\mu_M^0|^2 = |\langle kJM|\mathbf{P}|i, J-1, M\rangle|^2 = |\mu|^2 (J^2 - M^2)/J(2J - 1)$$

$$|\mu_M^\pm|^2 = |\langle kJM|\mathbf{P}|i, J-1, M \pm 1\rangle|^2$$
$$= |\mu|^2[(J \mp M)^2 - (J \mp M)]/4J(2J - 1) \qquad (62)$$

and

$$R(S, J \to J - 1) = \frac{\sum_{M=-J}^{+J} \dfrac{(J+M)^2 - (J+M)}{\left[1 + 3S \dfrac{(J+M)^2 - (J+M)}{4J(2J-1)}\right]^{1/2}}}{\sum_{M=-J}^{+J} \dfrac{2(J^2 - M^2)}{\left[1 + 3S \dfrac{J^2 - M^2}{J(2J-1)}\right]^{1/2}}} \qquad (63)$$

Finally, for a transition from an initial level with angular momentum J to a final level with angular momentum $J + 1$, which, in the molecular spectroscopy notation, could be either an $R(J)$ line in absorption or a $P(J + 1)$ line in emission, we have

$$|\mu_M^0|^2 = |\langle kJM|\mathbf{P}|i, J-1, M\rangle|^2 = |\mu|^2 \frac{(J+1)^2 - M^2}{(J+1)(2J+3)}$$

$$|\mu_M^\pm|^2 = |\langle kJM|\mathbf{P}|i, J-1, M \pm 1\rangle|^2 = |\mu|^2 \frac{(J \mp M)^2 + 3(J \mp M) + 2}{4(J+1)(2J+3)} \qquad (64)$$

and

$$R(S, J \to J+1) = \frac{\sum_{M=-J}^{+J} \dfrac{(J+M)^2 + 3(J+M) + 2}{\left[1 + 3S\dfrac{(J+M) + 3(J+M) + 2}{4(J+1)(2J+3)}\right]^{1/2}}}{\sum_{M=-J}^{+J} \dfrac{2[(J+1)^2 - M^2]}{\left[1 + 3S\dfrac{(J+1)^2 - M^2}{(J+1)(2J+3)}\right]^{1/2}}} \quad (65)$$

Obviously $R(S, J+1 \to J)$ and $R(S, J \to J+1)$ refer to the same transition with the initial and final level interchanged and

$$g_k|\mu_{ki}|^2 = g_i|\mu_{ik}|^2 \quad (66)$$

Since the saturation intensity is different for the different M components, also the power broadening, at fixed light intensity, will affect differently the various line components.

In Figures 17 and 18 the value of R is plotted vs. S for different $J \to J$ and $J \to J+1$ transitions. The enhancement factor R for a $\Delta J = 0$ transition

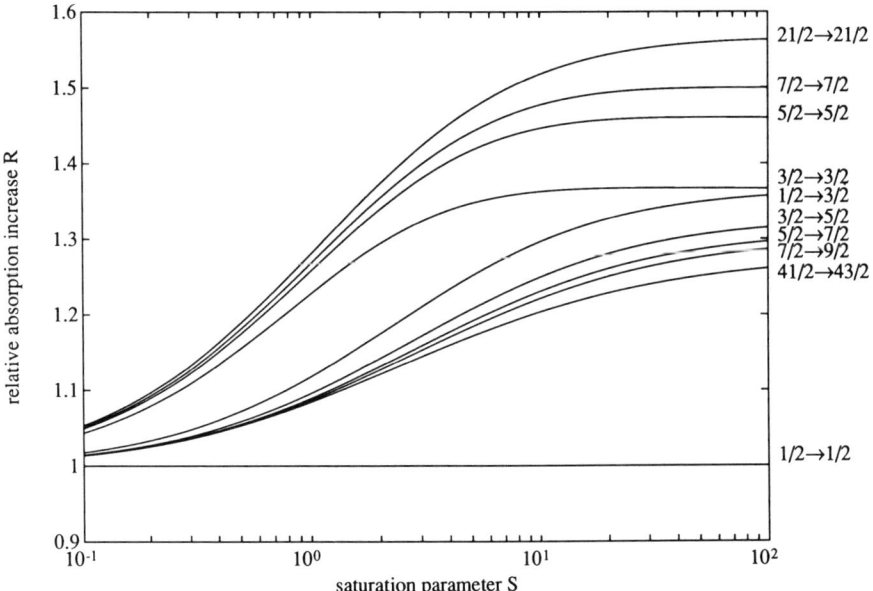

FIGURE 17. Enhancement factor $R(S)$ vs. the saturation parameter S. The case of a plane wave and half-integer values of J', J'' is shown. Note that when the wave polarization is orthogonal to the static field, no enhancement is observed for the $\frac{1}{2} \to \frac{1}{2}$ transition.

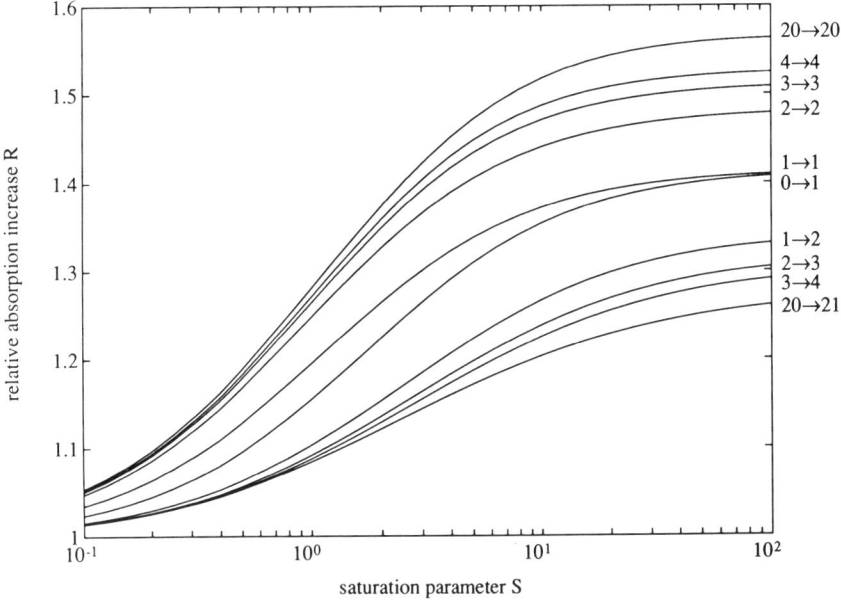

FIGURE 18. As Figure 17, but for integer values of J', J''. At large J values the $\Delta J = 0$ and $\Delta J = \pm 1$ transitions converge toward two limit curves with a maximum asymptotic value of $R(\infty) = 1.57$ and 1.27, respectively.

shows rapid convergence, for increasing J values, to a curve giving a maximum asymptotic enhancement value of 1.57. On the contrary, the enhancement for $\Delta J = \pm 1$ transitions decreases to a limiting curve giving an asymptotic value of about 1.27.

For transitions involving low J values, it can be seen that $R(S, 1, 0)$, $R(S, 0, 1)$, and $R(S, 1, 1)$ have the same maximum absorption increase of $\sqrt{2}$ for large values of S. Of course, the simplest case, $J' = J'' = \frac{1}{2}$, produces no enhancement for a light beam polarized orthogonal to the static field. It is worth noting that the rate-equation approach can be applied, by using equations (5) and (6), also to the evaluation of the intensity of saturated level-crossing experiments (Hanle effect at field different from zero).

7. THE NONLINEAR HANLE EFFECT WITH A GAUSSIAN LASER BEAM

The results of the preceding section have been obtained under the assumption of a light intensity constant over the whole interaction region. This assumption is no longer valid as soon as the transverse linear dimensions of the interaction region are of the order of, or greater than, the diameter of the laser beam. The intensity of the laser beam decreases with increasing distance from the beam axis, and usually it is reasonable to assume a Gaussian profile (TEM_{00} mode)

$$I(r) = I_0 \exp(-2r^2/w^2) \tag{67}$$

where w is the *waist* of the laser beam and r is the distance from the beam axis. Equation (67) is valid for an optically thin medium, otherwise the beam intensity depends also on the axial coordinate z. If the condition $r \ll w$ is not verified, the saturation parameter S is not constant over the interaction volume. Under these conditions, it is convenient to define an effective absorption enhancement (or effective emission enhancement for an active medium) as the ratio of the total power absorbed (or emitted) by the sample at high field to the total power absorbed (or emitted) at zero field. The numerator is thus obtained by multiplying the high-field absorption (gain) coefficient $\alpha^F(\omega, S(r))$, where the dependence of expression (57) on r is taken into account, by the beam intensity $I(r)$ and integrating their product over the interaction volume. The denominator is obtained by the same procedure, replacing $\alpha^F(\omega, S(r))$ by the position-dependent zero-field absorption coefficient $\alpha(\omega, S(r))$, as defined by equation (56).

We shall now develop an expression for the absorption (emission) enhancement valid for an optically thin homogeneous sample crossed by a cylindrical laser beam with a Gaussian profile (cylindrical symmetry about the axis of the laser beam is assumed). The position-dependent saturation parameter can be written as

$$S(r) = I(r)/I_s = S_0 \exp(-2r^2/w^2) \tag{68}$$

where $S_0 = I_0/I_s$ is the saturation parameter on the beam axis. The enhancement factor R is thus

$$R = \frac{P^F}{P} = \frac{\int_{z_1}^{z_2} \int_0^\rho \int_0^{2\pi} \alpha^F[\omega, S(r)] I(r) r \, dr \, d\theta \, dz}{\int_{z_1}^{z_2} \int_0^\rho \int_0^{2\pi} \alpha[\omega, S(r)] I(r) r \, dr \, d\theta \, dz} \tag{69}$$

where z is oriented along the laser beam axis ($L = z_2 - z_1$ is the sample length) and ρ is the radius of the interaction region, assumed cylindrical. Assuming $\rho \gg w$, and with our hypothesis of an optically thin sample, we obtain

$$P^F = 2\pi L \int_0^\infty \alpha^F[\omega, S(r)] I_0 \exp(-2r^2/w^2) r \, dr$$

$$P = 2\pi L \int_0^\infty \alpha[\omega, S(r)] I_0 \exp(-2r^2/w^2) r \, dr \tag{70}$$

These integrals can be resolved analytically. The solution has the form

$$\frac{w^2 I_0}{2S_0} \sum_M \frac{A}{B} (\sqrt{1 + S_0 B} - 1) \qquad (71)$$

where A and B are constants depending on the quantum numbers and selection rules involved in the particular transition.

In the case of a Gaussian beam, the final expressions for R are thus, for the $J \to J$, $J \to J - 1$, and $J \to J + 1$ lines, respectively,

$$R(S_0, J \to J) = \frac{2 \sum_{M=-J}^{+J} \left[1 + 3S_0 \frac{J(J+1) - M(M+1)}{4J(J+1)}\right]^{1/2} - 2(2J+1)}{\sum_{M=-J}^{+J} \left[1 + 3S_0 \frac{M^2}{J(J+1)}\right]^{1/2} - (2J+1)} \qquad (72)$$

$$R(S_0, J \to J - 1) = \frac{2 \sum_{M=-J}^{+J} \left[1 + 3S_0 \frac{(J+M)^2 - (J+M)}{4J(2J-1)}\right]^{1/2} - 2(2J+1)}{\sum_{M=-J}^{+J} \left[1 + 3S_0 \frac{J^2 - M^2}{J(2J-1)}\right]^{1/2} - (2J+1)} \qquad (73)$$

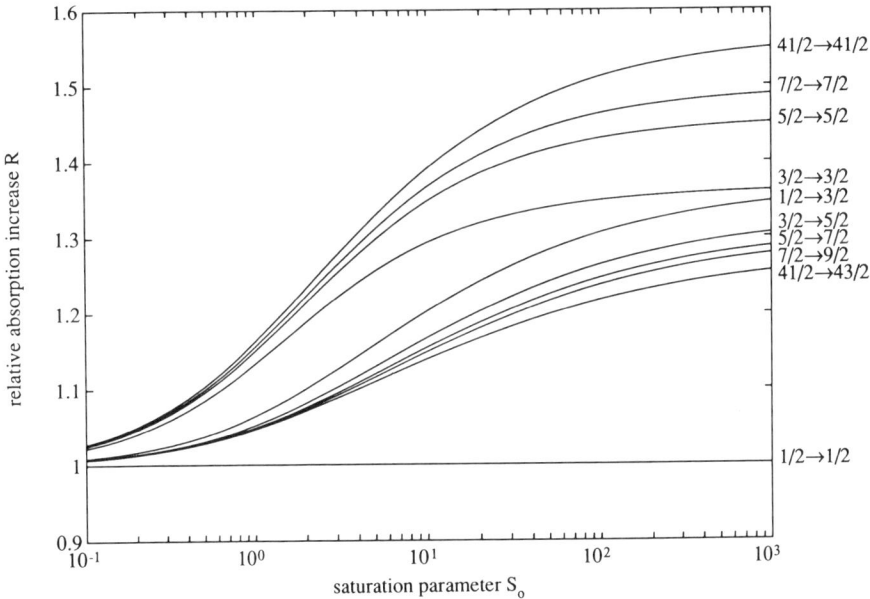

FIGURE 19. As in Figure 17, but for a Gaussian light beam. Here S_0 is the saturation on the beam axis.

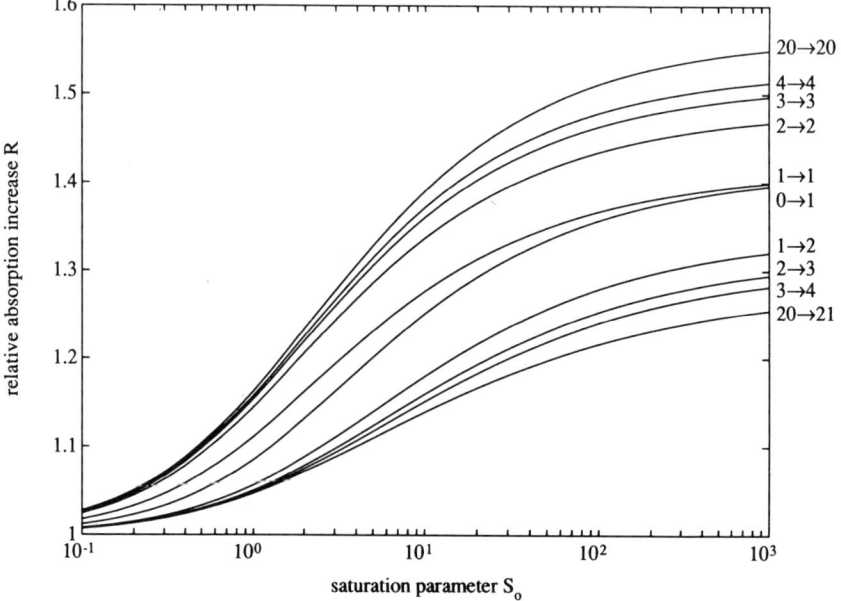

FIGURE 20. As in Figure 19, but for integer values of J' and J''.

$$R(S_0, J \to J+1) = \frac{2 \sum_{M=-J}^{+J} \left[1 + 3S_0 \frac{(J+M)^2 + 3(J+M) + 2}{4J(J+1)(2J+3)}\right]^{1/2} - 2(2J+1)}{\sum_{M=-J}^{+J} \left[1 + 3S_0 \frac{(J+1)^2 - M^2}{(J+1)(2J+3)}\right]^{1/2} - (2J+1)}$$

(74)

These expressions are plotted in Figures 19 and 20; they have the same asymptotic values as equations (61), (63), and (65).

8. THE NONLINEAR HANLE EFFECT IN ABSORPTION

The NLHE, like the classical Hanle effect, is a powerful tool to achieve sub-Doppler (and in some case also subnatural) resolution in atomic and molecular spectroscopy.

The first experiments were performed on the laser-active Ne and Xe atoms[1-11] whose Landé factors and hyperfine structure splittings were measured by sweeping an external field applied to the laser cavity, and on molecular absorbers, like CH_4 and CH_3F, contained in intracavity or extracavity cells.[12-14]

The shape and height of the NLHE signal in the He-Ne laser was investigated extensively by Decomps, Dumont, and Ducloy.[20] Javan[1] pointed out that the NLHE on the He-Ne laser can be easily detected by monitoring the fluorescence radiation emitted from the upper or lower level of the laser transition toward lower-lying levels, not directly involved in the laser cycle. This was confirmed experimentally.[1,5] When the fluorescence and the laser frequency are sufficiently apart, the fluorescence signal can be easily isolated and detected with a good signal/noise ratio. Since the signal depends on the J value of the lower level of the fluorescence transition, the theoretical treatment is somewhat involved. Several experimental arrangements are possible (Figure 21). The ratio

$$P = \frac{F_{\sigma y} - F_{\sigma x}}{F_{\sigma y} + F_{\sigma x}}$$

($F_{\sigma x}$ and $F_{\sigma y}$ being the fluorescence intensities observed orthogonally to the z direction and polarized along the y and x directions, respectively) is proportional to the precession of the induced dipole moment (Hanle effect) and is independent of the total fluorescence intensity. Therefore, P is also independent of the sample absorption, while the single polarized fluorescence components $F_{\sigma x}$ and $F_{\sigma y}$, as well as the fluorescence component polarized along z, F_π, are proportional to the NLHE according to the description of Sections 1-7. In this sense the behavior of P corresponds to

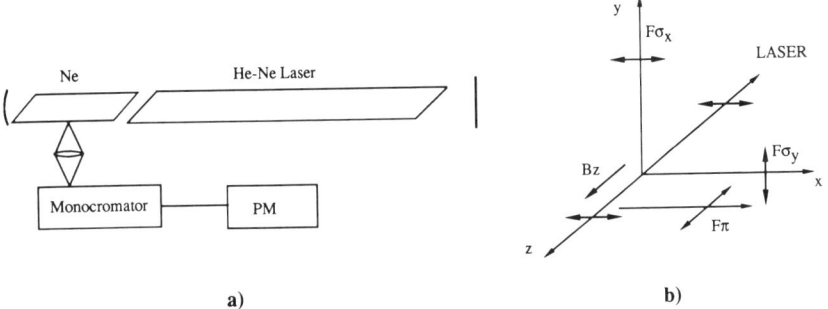

FIGURE 21. Typical experimental apparatus for the observation of the NLHE on the fluorescence inside a laser cavity. (a) Detection scheme. (b) The difference $F_{\sigma y} - F_{\sigma x}$ gives a signal directly associated to the induced dipole motion as in the linear Hanle effect. Quantity F_π is associated to the pure saturation effect, proportional to the change in the absorption of the sample.

a usual Hanle effect in the presence of saturation, while the dependence of F_π on the applied external field corresponds to a new effect, which has no analogy in the linear regime. Decomps, Dumont, and Ducloy[20] named this effect *zero-field saturation resonance*. However, we feel that NLHE is a more appropriate terminology, because of the role of degeneracy removal and of the presence of other important differences from the usual saturation effects. Actually, a change in the total absorption (and therefore in the total fluorescence intensity) of the sample is observed in any optical pumping experiment, as soon as the intensity of the pumping radiation becomes sufficiently high. However, the saturation of an optical pumping cycle always induces an absorption decrease. Moreover, optical pumping by linearly polarized light can only induce an alignment of the sample, which corresponds to a small second-order effect, negligible when high J values are involved. Alignment can be large only in the case of a $J = 0 \leftarrow 1$ transition, $J = 1$ being the lower state. Also in this case, however, if $\Delta\omega_D \gg \gamma$ the absorption is unaffected by the presence of an external field ($\Delta\omega_D \gg \mu B/h \gg \gamma$), independently of the laser power. On the contrary, the fluorescence signal F_π can be sensitive to alignment depending also on the J value of the third level.

A detailed description of the experimental results obtained with the experimental scheme of Figure 21 and their theoretical interpretation are given by Decomps, Dumont, and Ducloy[20] and references cited therein. Here it is worth noting that in these experiments coherence effects as high as the hexadecapole moment ($\Delta M = 4$ coherence in a $J = 2 \leftarrow 1$ transition) have been observed.[56] Figure 22 shows the experimental results and the corresponding theoretical prediction. The agreement is very good and proves the validity of the interpretation.

Decomps, Dumont, and Ducloy[20] describe several experiments on the He–Ne laser operated in a multimode regime, and develop a theoretical interpretation of these experiments in the frame of the broad-line approximation, as explained in Section 5.

The BLA model has been extended by Ducloy[56,57] to transitions involving large J values, like the usual molecular transitions. According to Ducloy's model, the hexadecapole moment in the fluorescence signal should be observed also in this case. However, a very small total fluorescence (F_π) signal is predicted, tending to disappear at high saturation parameters. These predictions are in disagreement with the experimental results obtained for both fluorescence[60] and absorption.[32-34] This disagreement is due to the fact that the BLA model is not applicable to single-mode lasers, which were used elsewhere[32-34] nor when the frequency spacing between the laser modes is much larger than γ, which is the case of Ref. 60.

More recently the NLHE in the visible lines of Ne has been investigated by means of a single-frequency laser. McLean, Ballagh, and Warrington

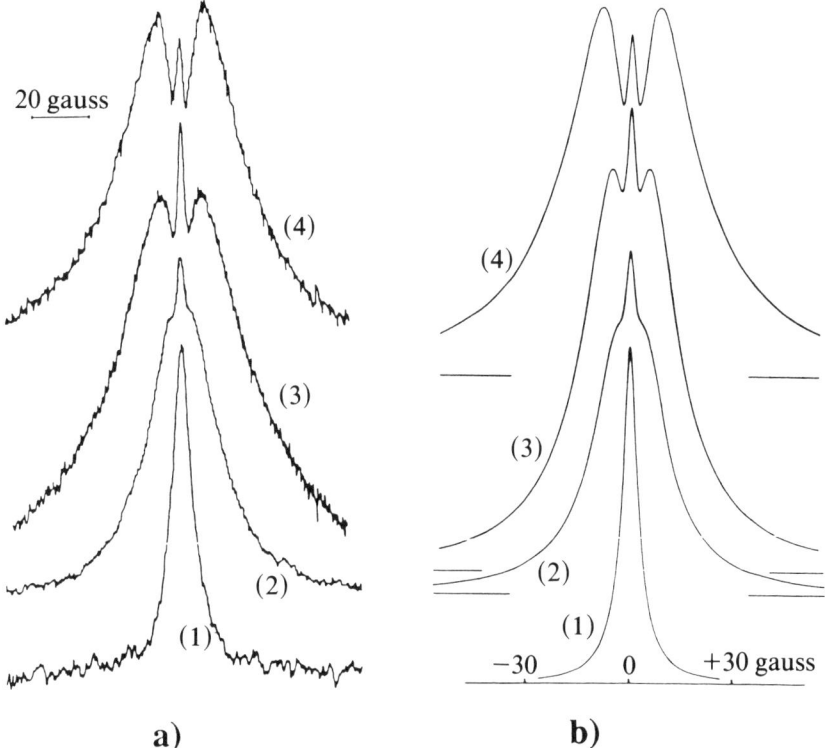

FIGURE 22. NLHE observed on the fluorescence of the 730.5 nm Ne line for laser irradiation at 632.8 nm. (a) Curves 1, 2, 3, and 4 correspond to laser intensities in the ratios 0.2, 1, 2, and 3.75, respectively. (b) Corresponding calculated shape of the NLHE in the broad-line approximation (from Ref. 56).

used a single-mode dye laser to excite the 607.4 nm $2p_3 \leftarrow 1s_4$ and the 585.2 nm $2p_1 \leftarrow 1s_2$ transitions (both $J' = 1 \rightarrow J'' = 0$)[85] and the 597.5 nm transition ($2p_5 \leftarrow 1s_5$; $J' = 1 \leftarrow J'' = 2$).[86] In both experiments the fluorescence emitted orthogonally to the magnetic field (F_σ or F_π signals) was monitored. The NLHE signal of the first experiment had an approximately Lorentzian shape. It is noteworthy that in these experiments the Gaussian profile of the laser beam can significantly deform the shape of the signal, especially at high saturation parameters.[87] Also in the second experiment the signal corresponding to π-polarized fluorescence was approximately Lorentzian, the ratio between the high-field and zero-field intensities being of the order of $R \approx 1.2 \rightarrow 1.3$. This value is in good agreement with the equations of Sections 6 and 7, but larger than what was expected by the BLA model. The σ-polarized fluorescence signal gave clear evidence of the hexadecapole resonance. A density matrix treatment was also presented which, however, could be carried out only numerically and under drastic approximations. One of the approximations, for instance, consisted in neglecting the velocity changing collisions, which are known to play an important role, especially when the lower state is metastable, as is the case for $1s_2$ and $1s_5$.

The NLHE on the 632.8 nm laser transition ($J' = 2 \to J'' = 1$) was investigated in a Ne discharge by monitoring the absorption signal.[88,89] The signal was very close to a single Lorentzian, since the homogeneous widths of the upper and lower level of the transition are close to each other. Other Ne lines have been investigated in hollow cathode lamps.[38,40] These are the 607.4 nm ($J' = 0 \to J'' = 1$), 594.5 nm ($J' = 2 \to J'' = 2$), and 603 nm ($J' = 1 \to J'' = 1$) lines. The corresponding NLHEs were measured by monitoring the optogalvanic signal of the discharge (the discharge impedance is affected by the absorption increase of the sample). Figure 23 shows the NLHE observed at fixed values of the magnetic field, while the Doppler profile is scanned by sweeping the frequency of the dye laser. Figure 23(Ia)

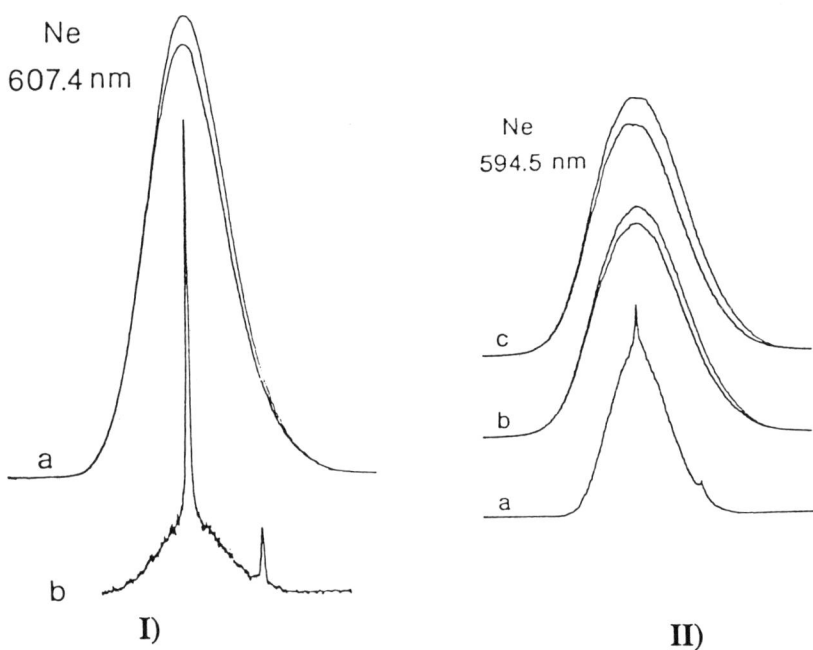

FIGURE 23. NLHE absorption increase in the optogalvanic signal for the 607.4 nm and 594.5 nm Ne lines vs. the frequency of a single-mode dye laser. The Ne is excited in a hollow cathode lamp (redrawn from Ref. 38).

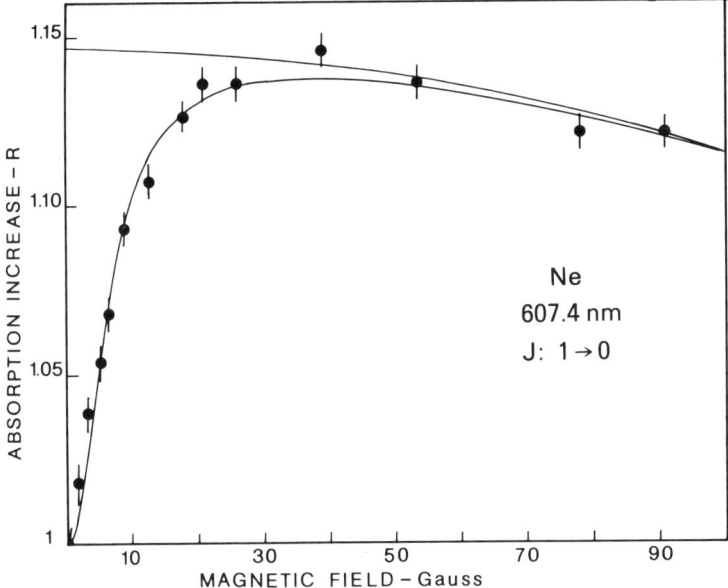

FIGURE 24. Relative enhancement of the optogalvanic signal for the 607.4 nm Ne transition as a function of the magnetic field (from Ref. 40).

shows the Doppler curve at zero magnetic field and in the presence of a magnetic field (2.6 mT) oriented along the laser beam ($\sigma^+ + \sigma^-$ exitation). The absorption enhancement is detected for a wide range of laser tunings, as predicted. The asymmetry of the signal is due to the presence of two distinct isotopes in natural Ne (20 and 22, respectively), which give origin to two overlapping Doppler curves. The two line centers are displayed in Figure 23(Ib) using Doppler-free intermodulated optogalvanic spectroscopy. Analogous results are shown in Figure 23(II) for the 594.5 nm line. In this case the magnetic field is again constant (11.7 mT), but the laser intensity in curve c is three times larger than in curve b. The corresponding relative power enhancements are 13% and 8%, respectively. Curve a is again a Doppler-free intermodulated optogalvanic signal. The background due to the elastic velocity-changing collisions is larger for this line because its lower state is metastable. This result stresses the role played by the elastic collisions when the NLHE signal has to be convoluted with the velocity distribution.

An example of NLHE at fixed laser frequency as a function of the magnetic field intensity is shown in Figure 24. The experimental points are fitted by a Lorentzian minus a Gaussian curve, the latter corresponding to the Zeeman scanning of the Doppler profile. This Gaussian, as obtained from the data of the fit, is also shown in Figure 24. Its width of 2160 ± 210 MHz is in good agreement with the value of 2190 ± 40 MHz obtained by recording the Doppler-broadened absorption optogalvanic curve. The zero-field height of the Gaussian gives the value R of the absorption increase for a given laser intensity and, in turn, the value of the saturation parameter S by using the equations of Sections 6 and 7.

The NLHE absorption increase for transitions involving large J values can be observed in molecular vapors. A first experiment was performed on I_2 molecules in the visible region.[60] Unfortunately, the interpretation of the effect for this molecule is not straightforward because of the presence of several hyperfine structure components, whose spacing is smaller than the Doppler width. Nevertheless, a fluorescence increase as large as 30% has been observed.

The experiments on the vibrational infrared transitions of polar molecules like CH_3F or CH_3OH are more easily interpretable. These molecules do not display structures in their Doppler profiles, and small electric fields are sufficient to originate large linear Stark shifts. The accidental coincidence of several vibrational transitions of these molecules with the rich spectrum of the CO_2 laser makes powerful laser lines around 10 μm available for investigation. Contrary to the atomic transitions, these molecular transitions are unaffected by magnetic fields. Moreover, as expected by the parity rule and demonstrated experimentally,[61,62,83] the collisional mixing of the M sublevels is negligible. The effect of collisions results in a fast rotational relaxation within each vibrational manifold leading to a linear pressure broadening, of the same order of magnitude in the lower and upper states, and a quadratic dependence of the saturation intensity on the gas density, since $\Gamma_{coll} \gg \gamma$.

The optoacoustic technique provides a sensitive, and the only possible, linear detection of the NLHE for these molecules. In fact, the optogalvanic technique cannot be applied because it would require an electric discharge, there is no fluorescence radiation, and absorption is very small. Linearly polarized radiation from a stable CO_2 laser is fed into a cell containing the investigated molecular vapor, two parallel conducting plates which create the static electric field, and a microphone used as a detector. The polarization of the laser radiation is oriented either parallel or orthogonal to the static field, in order to induce $\Delta M = 0$ or $\Delta M = \pm 1$ transitions. The sensitivity of this technique allows operation at low pressure (a few Pa) with a sample length of a few cm.[32-34] The absorption signal displayed no field dependence when the $\Delta M = 0$ transitions were excited.[32] The dependence of absorption on the electric field in the case of the $\Delta M = \pm 1$ transitions is shown in Figure 25 for two different CH_3OH lines. The upper part of the figure refers to irradiation with the $9P(36)$ CO_2 laser line, which induces a transition belonging to the CO-stretch Q-branch ($J = 16$, $K = 8$, $\Delta J = 0$).

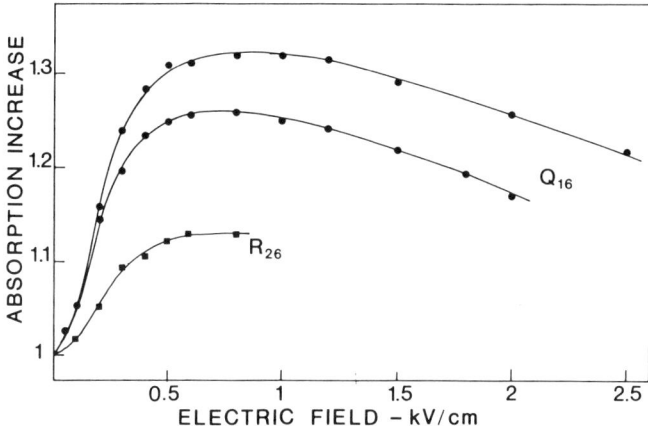

FIGURE 25. NLHE observed as an increase in the acoustooptic absorption signal on the rovibrational Q ($K = 8$, $J = 16$) transition of CH_3OH excited by the $9P(36)$ line of a CO_2 laser, and on the R ($K = 10$, $J = 27$) transition excited by the $R(10)CO_2$ line (from Ref. 33).

The upper and lower curves were measured at 6.25 and 12 Pa, respectively. At low electric fields (below some 600 V/cm) a large absorption increase, due to the NLHE, is observed. The relative absorption increase is lower at the higher pressure, as expected by theory since $S \propto 1/p^2$. The signal decrease at higher fields is due to the Stark effect on the absorption transitions, which is shifted out of resonance. The lower part of Figure 25 shows a CH_3OH transition belonging to the R branch ($J' = 27$, $J'' = 26$, $K = 10$) excited by the $9R(10)$ CO_2 laser line. The gas pressure was 9.3 Pa; although the laser intensity was higher by a factor 1.6 than in the upper curves, the NLHE absorption increase is lower than that of the $Q(16)$ line, as expected by the rate-equation theory of Section 6.

The S value can be varied by changing the laser power density and the gas pressure, and a maximum value $R = 1.48$ was observed for the $Q(16)$ transition. On the contrary, the maximum observed value for the $R(26)$ transition was only $R = 1.15$.

The measurement of the relative absorption increase R in a NLHE experiment is a simple and convenient method for a direct determination of the saturation parameter S. Figure 26 reports the experimental values of S vs. p^{-2}, as obtained from the measurement of R and by using the equations of Section 7; the behavior is linear, as expected.

The experimental dependence of R on S over a wide range is shown in Figure 27. The experimental points refer to the $Q(16)$ transition and were measured at 12.2 Pa (dots) and 17.7 Pa (squares), respectively. The solid line has been evaluated from the formulas of Section 7; its good agreement with the experimental points proves the reliability of the rate-equation

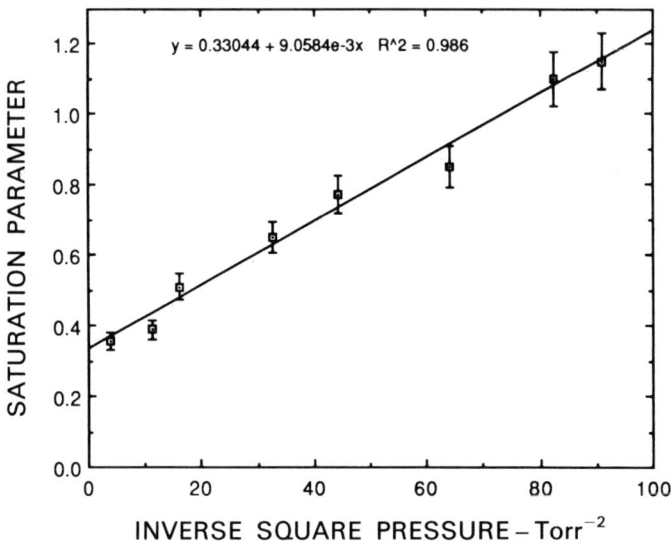

FIGURE 26. Saturation parameter S vs. the inverse square pressure (p^{-2}), as obtained from the acoustooptic absorption in a cell filled with pure CH_3OH excited by the $9P(36)$ CO_2 laser line and by using equation (61). The linear dependence on p^{-2}, as expected from the definition of I_s, confirms the validity of the rate-equation approach (redrawn from Ref. 34).

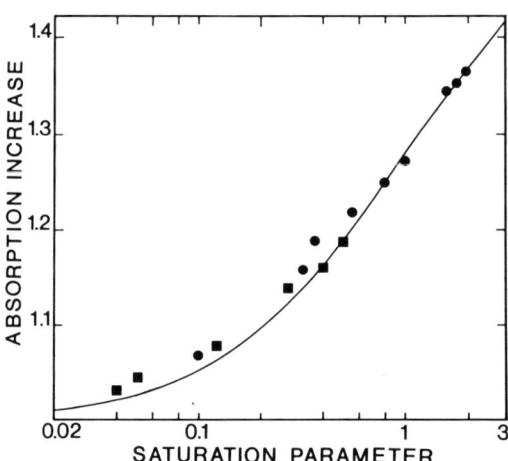

FIGURE 27. Experimental dependence of the enhancement factor R on the saturation parameter S for the $Q(8, 16)$ line of CH_3OH. S is assumed proportional to the laser beam intensity. The continuous line corresponds to the behavior expected from equation (61) (from Ref. 33).

model in predicting the size of the NLHE absorption increase at high J values.

A density matrix treatment for infrared molecular transitions, based on a perturbative development up to the third order, has been presented by Sokabe et al.[41] The validity of this treatment is limited to small saturation parameters ($S < 1$). Other experimental results[32-34,41] are in agreement with this perturbative approach applied to the $Q(16)$, 8 CH_3OH line.

The high sensitivity of the optogalvanic technique applied to hollow cathode experiments opened the possibility to investigate also refractory atoms, which are easily produced inside a hollow cathode discharge at room temperature. Several transitions of the Zr, Yb, and Y atoms were investigated in a series of papers.[37,39,42,63,64,84] Both the NLHE at zero magnetic field and high-field level-crossing resonances were observed (Figure 28), and the Landé factors and hyperfine interaction constants could be determined from these experiments. Narrow NLHE signals were obtained by using also a pulsed broad-band dye laser instead of a CW single-mode laser as in the previous experiments. The experiment, carried out in a hollow cathode Ne discharge,[65] proves the insensitivity of the NLHE to the frequency profile of the laser source.

We have seen that the collisional broadening is of the same order of magnitude for all levels of polar molecules, and that for these molecules radiative broadening is negligible. On the contrary, the lower level of an allowed atomic transition can be narrower than the upper level by orders

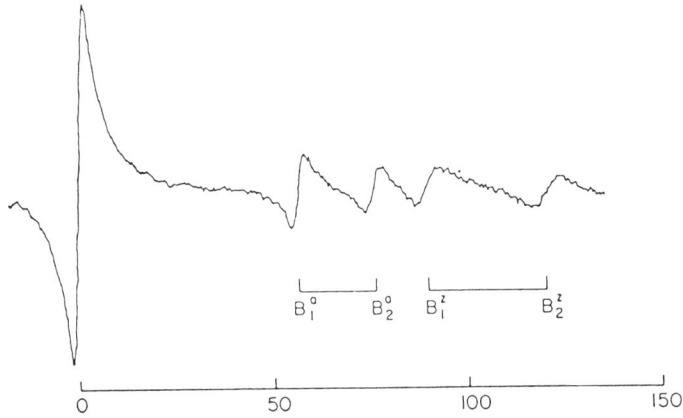

FIGURE 28. Example of zero-field and level-crossing NLHE in the optogalvanic signal for the 619.2 nm ($a\,^2D_{3/2} \leftarrow z\,^2D_{3/2}$) transition in yttrium I. Argon discharge, 66 Pa, single-mode laser operation. Level crossings in both the ground states (B_1^a and B_2^a) and upper states (B_1^z and B_2^z) can be observed. In all crossings, as in the zero-field signal, the degeneracy restoration causes an absorption decrease (from Ref. 63).

of magnitude. This happens, for instance, when the lower level of the transition is the atomic ground state and the collisional broadening of the line is smaller than the natural width of the upper state. In this case, if the angular momentum of the lower state is $J > 0$, a very small magnetic field is enough to remove degeneracy, and the NLHE line shape is the sum of two Lorentzians with very different linewidths. The narrower Lorentzian corresponds to the ground state, or to a metastable state. It is thus possible to measure the widths of the two levels involved in the optical transition directly and independently, and the resolution for the narrower line is higher than the natural linewidth of the transition, which equals the sum of the widths of the two levels: $\gamma_N = \gamma' + \gamma''$; $\gamma'' \ll \gamma'$. This remarkable resolution beyond the natural linewidth of the transition is also independent of the stability of the laser frequency, and the measurement requires a single laser beam crossing the sample. The resolution is thus largely independent of the actual distribution of the wave front of the laser beam. In these conditions the real limit to the spectroscopic resolution is set only by the interaction time between atoms and radiation.

Of course, the same resolution can also be achieved in an RF-visible double-resonance experiment. However, a NLHE experiment can be performed even at very small external fields, for which double-resonance experiments become unfeasible in the absence of fine or hyperfine structures.

As mentioned in Section 5, these narrow resonances had been predicted by Cohen-Tannoudji[54] for an atomic beam with homogeneous linewidth and in the absence of collisional relaxation. The effect was observed in a Na atomic beam crossed by a laser beam, the interaction region being of the order of 2 mm.[66] The same effect has also been detected in the optogalvanic signal of a hollow cathode lamp experiment, i.e., in a Doppler-broadened transition. A first experiment[82] was carried out on the metastable triplet levels of Ca (Figure 29). In this case the upper state is the $^3S_1$ state while the lower state can be one of the $^3P_{0,1,2}$ metastable triple states (whose natural lifetime is much longer than the interaction time). The corresponding transitions are at 610.2 nm ($J' = 1 \to J'' = 0$), 612.2 nm ($J' = 1 \to J'' = 1$), and 616.2 nm ($J' = 1 \to J'' = 2$), respectively, and, in the absence of collisional broadening, $\gamma_N = \gamma' = 10.4$ MHz. Figure 30 shows the corresponding NLHE signals observed in a hollow cathode discharge in the presence of 150 Pa of Ar as buffer gas. In the case of the $^3S_1 \to {}^3P_0$ transition, only the upper level is degenerate and the experimental results can be fitted by the difference between a Lorentzian a (Zeeman effect in the upper level) and a Gaussian b (Zeeman tuning of the Doppler-broadened absorption line). The height of the Lorentzian corresponds to an enhancement of $R = 1.15$, and to a saturation parameter of $S = 1.26$ for the Gaussian laser beam. The corresponding pressure-broadening coefficient is $\Gamma''(\text{Ar}) = 518$ kHz/Pa, in good agreement with the already published value of

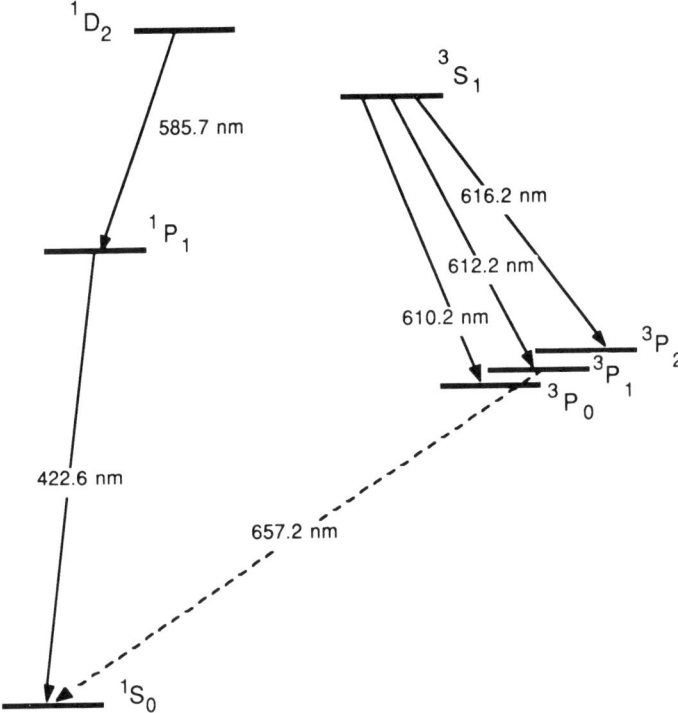

FIGURE 29. Simplified level scheme of the Ca I atom. The lifetime of the $^3P_1$ state is 0.5 ms, the lifetimes of the $^3P_0$ and $^3P_2$ states are much longer. The lifetimes of the $^1P_1$, $^3S_1$, and $^1D_2$ states are a few nanoseconds.

562 kHz/Pa.[67] For the $^3S_1 \to {}^3P_1$ transition both the upper and lower level split simultaneously, and the NLHE line shape is the sum of two Lorentzians of the same sign and height, but of different widths, corresponding to the different g-factors and broadenings of the upper and lower levels. The experimental signal has been fitted by fixing the parameters of the larger Lorentzian to the values obtained for the 610 nm line, and a FWHM of 8 MHz has been obtained for the narrower Lorentzian c. This corresponds to a pressure-broadening coefficeint of 45 kHz/Pa for the $^3P_1$ level, which is smaller by one order of magnitude than the coefficient for $^3S_1$. It is interesting to note that the width of the narrower Lorentzian is smaller than the natural width of the transition. For comparison, a homogeneous linewidth of 130 MHz was measured by intermodulated saturated spectroscopy under the same experimental conditions. This latter value, which is due to saturation and pressure broadening and to the natural linewidth, is in agreement with the Lorentzian width of about 100 MHz of the 610 nm line.

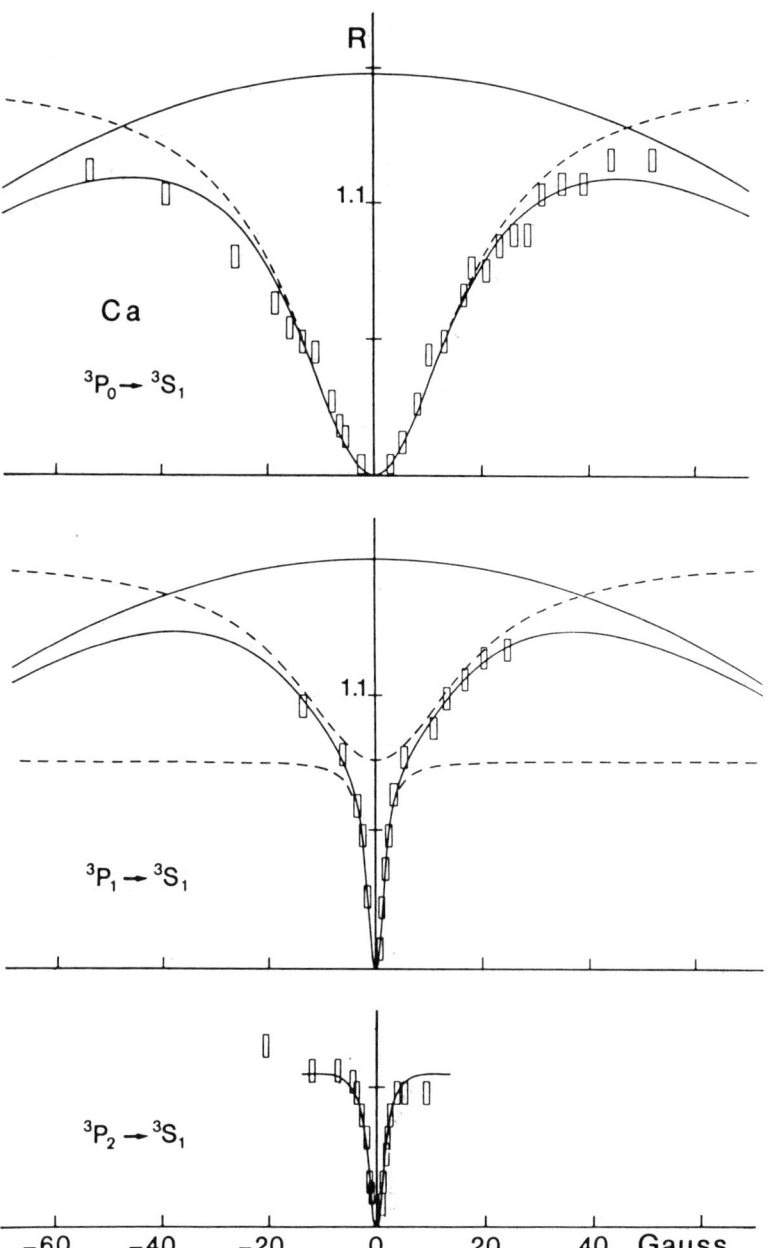

FIGURE 30. NLHE observed as a function of the magnetic field for the 610, 612, and 616 nm Ca lines, having the metastable ^3P states as lower levels. The width of the NLHE in the lower state is much smaller than the homogeneous linewidth of the transition (from Ref. 82).

For the 616 nm line only one narrow Lorentzian with a smaller height was observed. The width of 7 MHz is comparable to the width observed for $^3P_1$, while the absence of the larger Lorentzian can be explained by the rate-equation model. In fact, the contributions to the absorption increase due to degeneracy removal in the upper and lower level can be calculated separately, and it is found that about 90% of the NLHE signal is due to the degeneracy removal in the lower level only. Therefore, the larger Lorentzian signal is smaller than the signal/noise ratio of Figure 30. From the FWHM value of 7 MHz, again smaller than the natural linewidth, a pressure-broadening coefficient of 58 kHz/Pa can be estimated for the $^3P_2$ state.

Figure 31 shows the NLHE observed for the 585.7 nm line of Ca ($^1D_2 \rightarrow {}^1P_1$; $J' = 2$, $J'' = 1$). Here the natural lifetime and the g_J-factors are approximately equal for both levels and, as expected, the NLHE is very well fitted by a single Lorentzian curve.

The 570.7 nm line of samarium, which corresponds to the $^7F_1 \rightarrow {}^7F_0$ transition within the ground state, has been investigated in a hollow cathode discharge by similar experiments. Even in the presence of 93 Pa of Ne in the discharge, the observed resonance was very narrow, its width corresponding to a rate of collisional misalignment in the ground state as low as 50 kHz.[68]

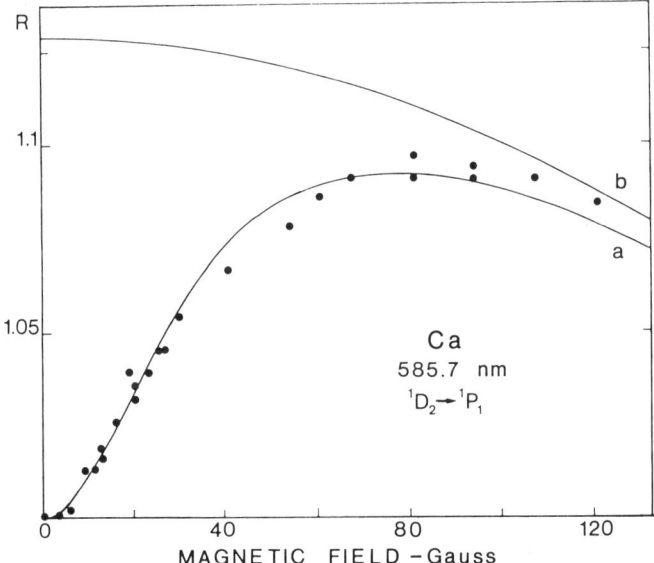

FIGURE 31. Enhancement of the optogalvanic signal for the Ca transition $^1D_2 \leftarrow {}^1P_1$ at 585.7 nm vs. the applied magnetic field. Hollow cathode lamp with Ne buffer gas at a pressure of 186 Pa (from Ref. 40).

The NLHE on a metastable Ca atomic beam has been investigated by monitoring the fluorescence radiation, as was done elsewhere for Na.[66] In a first experiment[69] a single laser beam excited $\sigma^+ + \sigma^-$ transitions in the 612 nm and 616 nm lines. Narrow resonances, corresponding to degeneracy removal in the ground state, were observed for both lines. The dependence of the signal width on the intensity of the exciting laser radiation is in agreement with the relation $\gamma = \gamma_0\sqrt{(1 + S)}$, the width extrapolated to zero laser intensity corresponding to the inverse transit time of the atoms through the laser–atomic beam interaction region. The width of the NLHE signal was as small as 1.4 μT (corresponding to 31 kHz). A subsequent experiment was performed introducing two separate interaction regions, in a Ramsey-like configuration.[70,71] For a separation of 3.9 cm between the two interaction regions, a width of only 0.29 μT was observed, the lower limit for the width being set only by the unavoidable magnetic field inhomogeneities (Figure 32). This is equivalent to a resolution of the order of 6 kHz, more than three orders of magnitude narrower than the natural radiative transition width, and two orders of magnitude narrower than the laser frequency jitter (about 1 MHz).

Hemmer and co-workers obtained the resonant Raman excitation of an atomic sodium beam by using a resonant laser beam, optically modulated at the frequency of the Na ground-state hyperfine splitting, and could thus observe the Ramsey fringes.[29,30,72] This experiment is a version of the OODR scheme of Figure 1d. A similar experiment was performed also on a samarium atomic beam[73]; in this case the modulation frequency of the laser side band was brought into resonance with the Zeeman splitting in the ground state.

More recently, an experiment was undertaken in order to investigate how the laser frequency noise spectrum affects the NLHE in the case of the very narrow linewidths typical of the atomic beams. The experiment is in progress on an ytterbium beam.[74]

The NLHE can also be detected in the forward scattered light. The experimental apparatus is similar to that described in Figure 2a, but a linear polarizer, whose axis is perpendicular to the polarization of the laser beam, is inserted between the sample and the detector. Thus, the direct laser light is suppressed and the detector receives only the spontaneously emitted light from the laser excited sample. A first experiment was performed on a Doppler-broadened Na vapor excited by a pulsed dye laser, in order to have an excitation linewidth much broader than the Doppler width (BLA approximation).[75] The behavior of the absorption signal was almost identical to the signal arising from the creation of the hexadecapole moment observed in neon atoms.[56] However, a later theoretical analysis[76,77] established that the observed line shapes are not necessarily due to hexadecapole couplings ($\Delta M = 4$ coherence). A further experiment was performed on

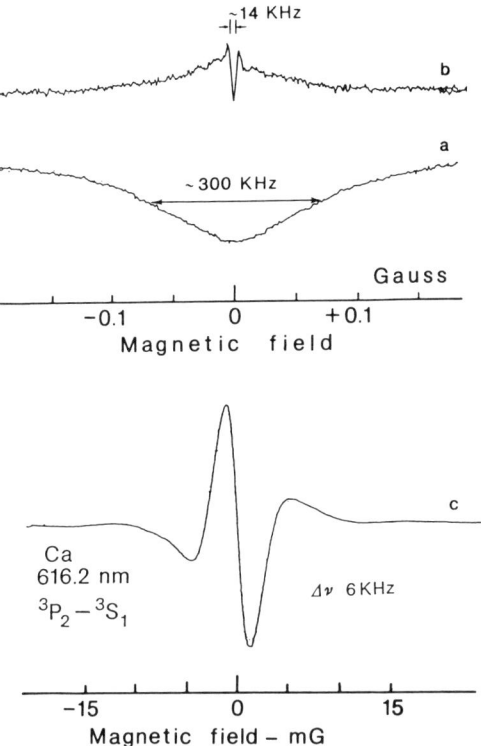

FIGURE 32. Typical recordings of NLHE as observed on an atomic beam of metastable Ca. Curves (a) and (b) correspond to the effect on the $^3P_1 \leftarrow {}^3S_1$ transition: (a) single interaction region ($d = 6$ mm), (b) two separate interaction regions ($L = 32$ mm, $d = 6$ mm). (c) Derivative signal observed on the $^3P_2-{}^3S_1$ transition ($L = 39$ mm). The natural linewidth is about 10 MHz (from Ref. 70).

the 570.7 nm line of Sm ($J' = 0 \to J'' = 1$) and the results were interpreted by using the perturbative treatment.[78]

More recently the forward scattering NLHE was investigated in the excited state of the 607.4 nm ($J' = 0 \leftarrow J'' = 1$) and 594.5 nm ($J' = 1 \leftarrow J'' = 2$) lines of Ne.[79,80] Again, the experimental results were interpreted by a semiclassical perturbative development up to second order in the laser intensity. Although this model does not allow the treatment of higher-order coherences, the agreement between experiment and theory is quite good also for the $J = 2 \leftarrow 1$ transition, as shown in Figure 33. It is thus apparent that the higher-order coherences do not contribute significantly to the signals observed in forward scattering experiments. In a further experiment on the forward scattering signal for the 633 nm line of Ne, the polarizer axis was not perpendicular to the laser beam polarization. Also in this experiment

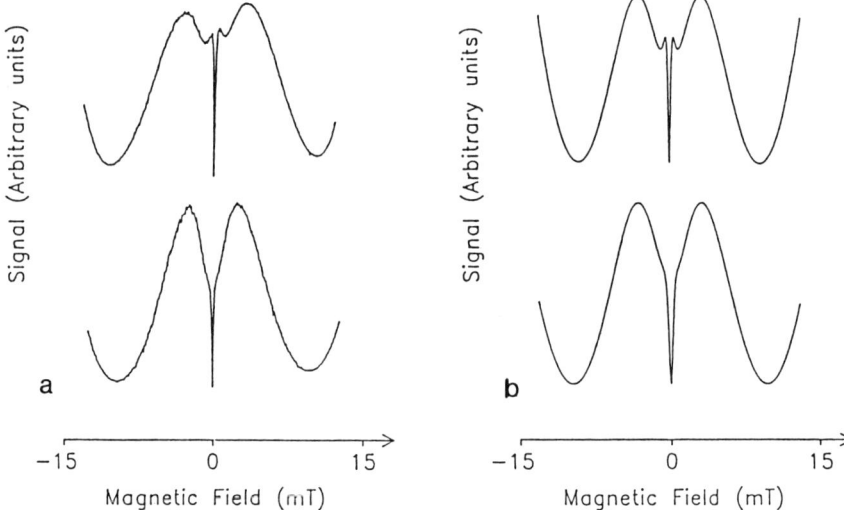

FIGURE 33. NLHE observed in forward scattering in a Ne discharge excited by a single-mode dye laser. (a) Experimental line shape obtained for the $1s_5$ ($J = 2$) ← $2p_4$ ($J = 2$) transition. Lower curve: $P_{\text{laser}} = 20$ mW, $p(\text{Ne}) = 40$ Pa. Upper curve: $P_{\text{laser}} = 50$ mW, $p(\text{Ne}) = 15$ Pa. (b) Calculated line shape for a $J = 1$ ← $J = 1$ transition with a perturbative treatment up to the second order (from Ref. 79).

the line shapes were found in good agreement with the perturbative approach.[81]

In a series of elegant papers by the Giessen group (Scharmann, Abt, Graubner, Hermann, Lasnitschka, and Schleicher) the zero-field crossing, the level crossings, and the mode crossings in the He-Ne and He-Cd lasers have been directly monitored as changes in the laser output power.[11,90-100,190] In particular, high-order mode crossings were observed when the applied external magnetic field was large enough to split the laser emission into two groups of longitudinal modes corresponding to the $\Delta M = \pm 1$ lines (Figure 34), allowing accurate measurements of the Landé factors and of the collisional broadening. In the case of the 442 nm CdII laser line also the level crossings in the hyperfine structure of the odd isotopes were detected.[98-100] In a similar experiment[191] the NLHE was measured for the He-Ne and He-SeII laser, yielding again precise measurements of the collisional broadening cross section of the natural linewidth. It is noteworthy that in all cases (zero field, level crossing, and mode crossing) a power decrease in the laser output was observed whenever the degeneracy of a level was restored. This effect on the laser power will be considered in the last section of this chapter.

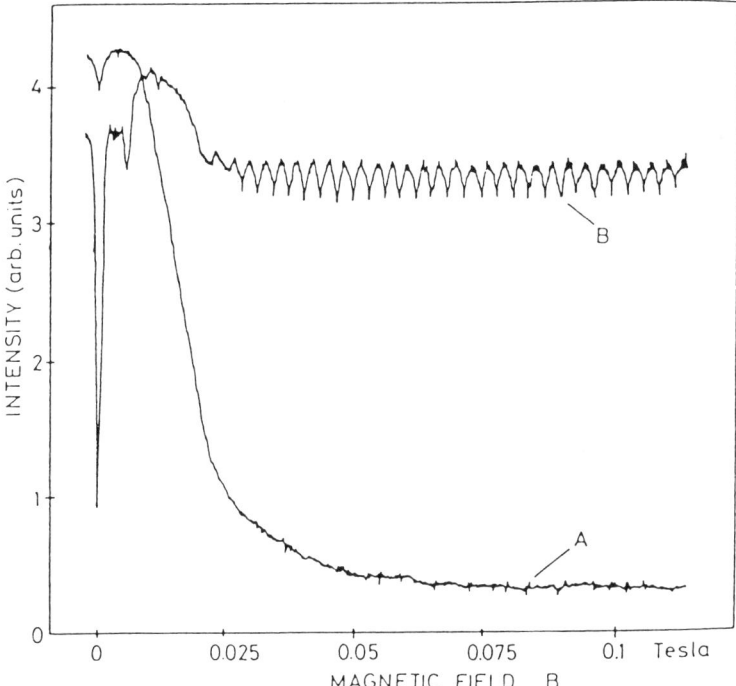

FIGURE 34. Laser intensity of the Ne line at 1.523 μm vs. the axial magnetic field intensity. Curve A shows the directly measured power, curve B the phase-sensitive detection. The zero-field NLHE and a number of high-order mode crossings are observed (from Ref. 95).

9. THE NONLINEAR HANLE EFFECT IN LASER-ACTIVE MEDIA

In Sections 6 and 7 we have obtained expressions for the ratio R of the high-field absorption (or gain) coefficient to the zero-field absorption (or gain) coefficient. While the NLHE induces an absorption increase at thermal equilibrium, it induces a gain increase in the presence of population inversion. This effect can sensibly affect the output power of gas lasers, since their homogeneous linewidth is usually smaller than the Doppler width, and since the degeneracy removal for these lasers can be achieved already at a small or intermediate external field.

The actual magnitude of the output power increase due to the NLHE depends on several factors. Some of them hinder the effect. For instance, some important laser-active molecules, like CO_2 or N_2O, display negligible Zeeman and Stark effects. In other cases the best laser operation is achieved at densities large enough to reduce the importance of the Doppler broadening. It should also be noted that the Stark effect, which is very efficient in removing the level degeneracy in polar molecules, cannot be used whenever the population inversion is achieved by an electrical discharge. On the other hand, the NLHE can increase both the gain of the lasing transition and the efficiency of the pump absorption in optically pumped lasers.

The gain increase for an inhomogeneously broadened transition is due to the increment in the velocity profile interval which can interact with the radiation field. This effect is maximum for a single frequency laser with $\Delta\omega_D \gg \gamma$. In multimode conditions, the size of the NLHE is reduced as the mode frequency spacing approaches the homogeneous linewidth γ (broad-line approximation).

NLHE power enhancement has been observed in quite different CW lasers such as: (1) the He-Ne laser, both on visible and near-infrared transitions; (2) the Xe discharge laser in the IR; (3) the He-Cd and He-Zn lasers; (4) the noble-gas ion lasers; (5) the optically pumped far-infrared molecular lasers. The effect is of practical relevance in the last two cases. In particular, all high-power commercial Ar^+ and Kr^+ lasers are equipped with a solenoid providing an axial magnetic field of appropriate intensity. Actually, also the magnetic field confinement of the plasma contributes to the power enhancement for these lasers, as was soon realized. However, the NLHE still plays an important role.[50]

An external magnetic, or electric, field can affect not only the output power of a laser, but also its output frequency. This effect was carefully investigated, both experimentally and theoretically, in the early works on the He-Ne laser, with a view to exploiting it in order to tune and stabilize the laser frequency. The theoretical results are described in Refs. 46 and 47, and references cited therein. In some cases, as for the infrared Xe laser[101] and the visible Ar^+ laser,[102] it has been possible to use the Zeeman effect for tuning the laser frequency beyond the Doppler limit. The progress in laser physics and in quantum electronics has superseded the Zeeman lasers, especially as tunable sources. On the other hand, the Stark effect remains a practical tool for tuning and modulating the frequency of the FIR optically pumped lasers (see Ref. 61 and references cited therein).

The theory presented in Refs. 46 and 47 is based on a perturbative density matrix approach, valid only if $I \ll I_s$. This theory is of particular interest for describing the polarization and frequency of the laser beam when *flat* resonators are used, i.e., when there is no constraint to the polarization of the electromagnetic field inside the laser cavity. As expected, the theory becomes more and more complex with increasing J values. Here we shall give a simplified description of the power enhancement (PE) effect by means of the rate-equation approximation, assuming single-mode laser operation and large Doppler broadening. We shall use only the basic parameters of the laser, i.e., the small signal gain α_0 and the saturation intensity I_s. In the following analysis the Lamb dip has been neglected, in the sense that the power enhancement is calculated as the ratio between the best output powers at high and at zero field. The best output power at zero field is obtained by detuning the laser frequency from the central frequency of the transition, which is affected by the Lamb-dip power decrease. At high field the laser frequency is assumed not to be tuned in resonance with any level crossing, mode crossing, or saturation dip. A possible frequency is now in the vicinity of the central frequency of the unperturbed (zero external field) line. The PE thus defined has a direct practical significance: it is the real power increase which we expect from the addition of an external field. It is, of course, smaller than that reported in Ref. 46, where both the Lamb-dip removal by the Zeeman shift and the NLHE contributed to the gain increase. This model can give an easy estimate of the size of the power enhancement as a function of J' and J''.

For an inhomogeneously broadened transition, the steady-state lasing condition is

$$\frac{2\alpha_0 l}{\sqrt{1+S}} = a + T \tag{75}$$

where a represents the roundtrip intracavity losses, T is the transmission of the output mirror, and l is the length of the active medium ($l \leq L$, where L is the cavity length). Equation (75) yields the saturation parameter and output intensity in the forms

$$S = G^2 - 1 \tag{76}$$

and

$$I_{out} = IT = TI_s(G^2 - 1) \tag{77}$$

where G is the excitation parameter (i.e., gain above threshold) defined by

$$G = 2\alpha_0 l/(a + T) \tag{78}$$

If an external field is added, the NLHE increases the gain by a factor $R(S)$, thus increasing the saturation parameter from S to S' and, according to Section 7, we expect a further relative gain increase $R(S')$. The NLHE originates a positive feedback: the saturation increases the gain and the gain increase increases the saturation. A new steady state is obtained,

$$GR(S') = \sqrt{1+S'} \tag{79}$$

assuming that the laser cavity is terminated by the same output mirror as before. Of course, the transmission coefficient T can be optimized only for one specific condition.

From equation (78) we obtain

$$S' = R^2(S')G^2 - 1 \tag{80}$$

and

$$I'_{out} = TI_s[R^2(S')G^2 - 1] \tag{81}$$

where $R(S')$ is given by equations (61), (63), and (65) for a plane wave, and by equations (72)-(74) for a Gaussian laser beam. We note that the approximation of an optically thin sample is here alleviated, because the laser optical cavity imposes the mode of the field.

The new steady-state saturation S' can be calculated by solving equation (80). For a $J = 0 \to 1$ transition in the plane-wave approximation we have the simple result

$$S' + 1 = G^2(1 + S')/(1 + S'/2) \tag{82}$$

and the power is expected to be doubled by the NLHE ($I'_{out} = 2I_{out}$) independently of the actual value of the excitation parameter G.

Unfortunately, for J', J'' values different from 1 and 0, $R(S)$ has a complex expression. Thus, for high J values, it is more convenient to solve equation (80) numerically or graphically by rewriting it in the form

$$(1/G^2)(S' + 1) = R^2(S') \tag{83}$$

The laser power enhancement PE is defined as the ratio of the saturation intensity in the presence of the NLHE (i.e., at high field) to the saturation intensity at zero field,

$$\text{PE} = \frac{S'}{S} = R^2(S') + \frac{R^2(S') - 1}{S} = \frac{R^2(S')G^2 - 1}{G^2 - 1} = \frac{S'}{G^2 - 1} \tag{84}$$

and is always larger than $R^2(S')$.

Equation (83) can be solved graphically as shown in Figure 35, where the value of S corresponds to the abscissa of the crossing point of the straight line, representing the left-hand side of equation (83), with the horizontal line $R = 1$, and the value of S' corresponds to the abscissa of the crossing point with the functions $R^2(S)_{J'J''}$. This gives a visual perception of the PE effect. Of course, equation (83) can also be solved numerically,

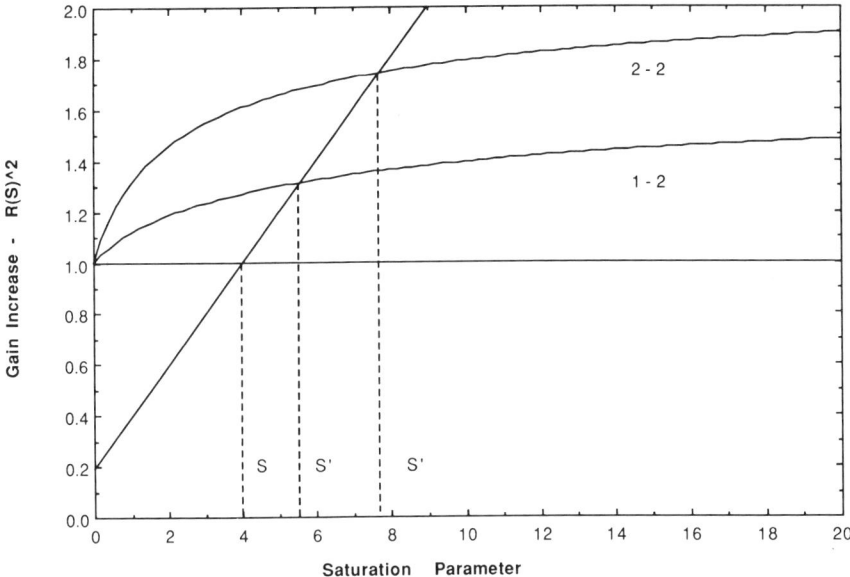

FIGURE 35. Graphical representation of the NLHE saturation increase in a gas laser. The ratio S'/S gives the power enhancement factor.

and the corresponding PE as a function of G is shown in Figures 36 and 37 for several values of J' and J'', and for both a plane wave and a Gaussian beam, respectively.

Because of the positive feedback the PE is larger than the R^2 factor also at small values of the excitation parameter G, depending strongly on the slope of the $R^2(S)$ function. Roughly speaking, we can conclude that, when $\gamma \ll \Delta\omega_D$, the NLHE will produce a power enhancement of the order of 1.6–2.2 for a gas laser operating in a single-mode regime (independently of the actual small signal gain).

Equations (83) and (84) must be modified, both in the presence of additional effects which modify α_0, and when the approximations under which the equations have been obtained are not satisfied.

An additional increase in α_0 can be observed in two important cases:

1. If the laser is excited by an electrical discharge, then the effect of the external magnetic field on the plasma must also be taken into account. In favorable geometrical and electrical arrangements the excitation efficiency, and consequently α_0, are increased. This is, in particular, the case of the noble-gas ion lasers which operate at high current densities in the presence of an axial magnetic field.

2. If the laser is optically pumped, the NLHE can affect also the pump transition. This effect is of particular relevance for the optically pumped far-infrared molecular lasers, since polar molecules are used as active laser media. Many of these molecules exhibit a linear Stark effect.

However, the PE is reduced in the following two instances:

1. The approximation

$$\gamma_s = \gamma\sqrt{1 + S} \ll \Delta\omega_D \tag{85}$$

is no longer valid (γ or S too large). Under these conditions the convolution between the velocity distribution and the Lorentzian of half-width γ_s must be taken into account, and this leads to a gain decrease.
2. The laser is operating multimode and there is substantial overlapping between the different saturation dips in the velocity distribution ($\delta\omega \approx \gamma_s$). In this case, a large fraction of the available population inversion is used and the addition of an external field cannot significantly increase the gain. This is the case of the BLA approximation discussed in Section 5.

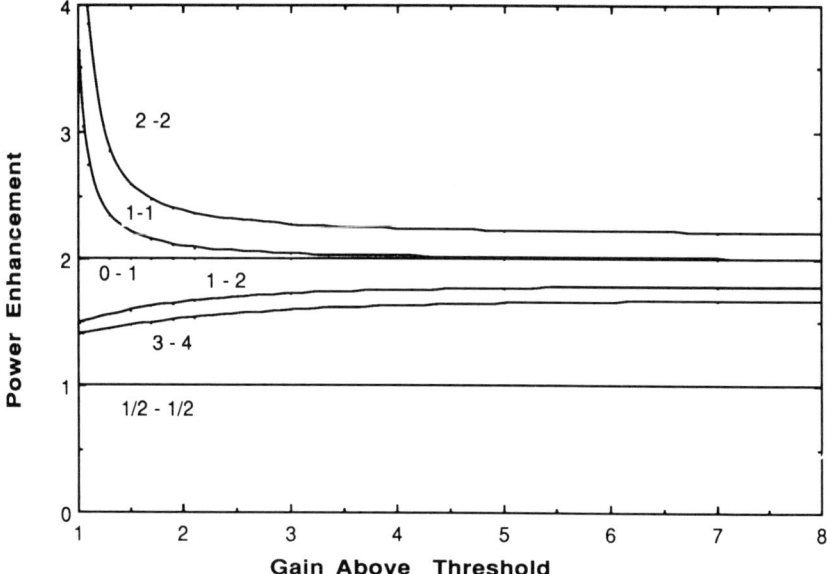

FIGURE 36. NLHE power enhancement in a gas laser vs. the excitation parameter G. Plane-wave approximation.

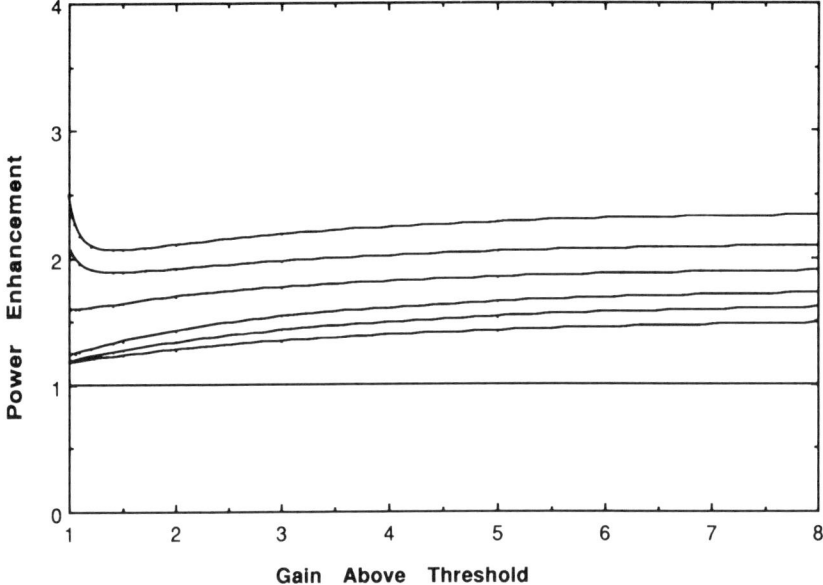

FIGURE 37. Power enhancement as in Figure 36, but for a Gaussian laser beam. From the top the curves correspond to the following transitions: $J' = 20 \to J'' = 20$, $J' = 2 \to J'' = 2$, $J' = 1 \to J'' = 1$, $J' = \frac{1}{2} \to J'' = \frac{3}{2}$, $J' = \frac{3}{2} \to J'' = \frac{5}{2}$, $J' = 20 \to J'' = 21$, $J' = \frac{1}{2} \to J'' = \frac{1}{2}$.

Let us now consider the gain reduction of the latter instance (1). If equation (85) is not satisfied, then the factor $R(S)$ is reduced because the saturated gain decreases as γ_s increases, and also because the frequency shift necessary for the removal of the degeneracy is a substantial fraction of $\Delta\omega_D$ and the exponentials of equation (2) cannot be approximated by unity.

When equation (85) is not satisfied, the saturated gain becomes

$$\alpha_s = \frac{\alpha_0 \gamma}{\pi} \int_0^\infty \frac{\exp[-(\nu_c - \nu_0)^2/\Delta\nu_D^2]}{(\nu - \nu_c)^2 + \gamma_s^2} d\nu_c$$

$$= \frac{\alpha_0}{\sqrt{1+S}} \mathcal{R} W\left[(\nu - \nu_0)/\Delta\nu_D + i\left(\frac{\gamma\sqrt{1+S}}{\Delta\nu_D}\right)\right] \quad (86)$$

where $\mathcal{R}W(z)$ is the real part of the error function for complex arguments, which is defined by

$$W(z) = [1 + \text{erf}(iz)] \exp(-z^2) \quad (87)$$

as defined and tabulated, for instance, in Ref. 103.

Since at the limits $S \to 0$ and $\nu \to \nu_0$ we have

$$\alpha(S = 0, \nu = \nu_0) = \alpha_0 \mathscr{R} W(0 + i\gamma/\Delta\nu_\mathrm{D})$$

the saturated gain can be conveniently normalized to

$$\alpha_\mathrm{s}(\nu) = \frac{\alpha_0}{\sqrt{1+S}} \frac{\mathscr{R}W[(\nu - \nu_0)/\Delta\nu_\mathrm{D} + i\gamma\sqrt{1+S}/\Delta\nu_\mathrm{D}]}{\mathscr{R}W(0 + i\gamma/\Delta\nu_\mathrm{D})} \qquad (88)$$

where $\alpha_0/\sqrt{1+S}$ is the saturated gain at the center of the line at the limit $\gamma \ll \Delta\nu_\mathrm{D}$. A physical insight into the PE effect, when the correction of equation (88) is taken into account, can be obtained by introducing some approximations. At zero external field the laser power is affected by the saturation dip at the line center ν_0. This saturation dip can be neglected in the evaluation of the PE, if the laser power is calculated at a frequency $\nu_0 + \Delta\nu$ corresponding to the output power maximum. If $\gamma \ll \Delta\nu_\mathrm{D}$, no offset is necessary, provided that the saturation dip is neglected. Otherwise $\Delta\nu$ depends on the $\gamma/\Delta\nu_\mathrm{D}$ ratio. For the Ar^+ and He-Ne lasers ($\gamma/\Delta\nu_\mathrm{D} \approx$ 0.05–0.15) a good approximation can be[104-106]

$$\Delta\nu = \gamma\sqrt{1+S}/\Delta\nu_\mathrm{D} \qquad (89)$$

At zero external field, the saturation S of the laser is thus given by

$$S + 1 = G^2 \left[\frac{\mathscr{R}W(\Delta\nu/\Delta\nu_\mathrm{D} + i\gamma\sqrt{1+S}/\Delta\nu_\mathrm{D})}{\mathscr{R}W(0 + i\gamma/\Delta\nu_\mathrm{D})} \right]^2 \qquad (90)$$

If the NLHE for a $J = 0 \to 1$ transition is now considered, then equations (1) and (2) must be modified as follows:

$$\alpha_\mathrm{s}(\Delta M = 0) = \frac{\alpha_0}{\sqrt{1+S}} \left[\frac{\mathscr{R}W(\Delta\nu/\Delta\nu_\mathrm{D} + i\gamma\sqrt{1+S}/\Delta\nu_\mathrm{D})}{\mathscr{R}W(0 + i\gamma/\Delta\nu_\mathrm{D})} \right] \qquad (91)$$

and

$$\alpha_\mathrm{s}(\Delta M = \pm 1) = \frac{\alpha_0/2}{\sqrt{1+S/2}} \left\{ \frac{\mathscr{R}W[(\Delta\nu + \Delta\nu_\mathrm{F})/\Delta\nu_\mathrm{D} + i\gamma\sqrt{1+S/2}/\Delta\nu_\mathrm{D}]}{\mathscr{R}W(0 + i\gamma/\Delta\nu_\mathrm{D})} \right.$$

$$\left. + \frac{\mathscr{R}W[(\Delta\nu - \Delta\nu_\mathrm{F})/\Delta\nu_\mathrm{D} + i\gamma\sqrt{1+S/2}/\Delta\nu_\mathrm{D}]}{\mathscr{R}W(0 + i\gamma/\Delta\nu_\mathrm{D})} \right\} \qquad (92)$$

where $\pm\Delta\nu_\mathrm{F}$ is the frequency shift of the $\Delta M = \pm 1$ transition due to the Zeeman or Stark effect. The external field is assumed to have an intensity such that $\gamma < \Delta\nu_F < \Delta\nu_D$. If $\Delta\nu_F > \Delta\nu_D$, each $\Delta M = \pm 1$ transition is lasing independently when in resonance with a cavity mode. If $\Delta\nu_F < \Delta\nu_D$, we can assume $\Delta\nu = 0$ in the presence of an external field and the expression for the gain increase reduces to

$$\frac{\alpha^F(\Delta M = \pm 1)}{\alpha(\Delta M = 0)} = \frac{\sqrt{1+S}}{\sqrt{1+S/2}} \frac{\mathcal{R}W(\Delta\nu_F/\Delta\nu_D + i\gamma\sqrt{1+S/2}/\Delta\nu_D)}{\mathcal{R}W(\Delta\nu/\Delta\nu_D + i\gamma\sqrt{1+S}/\Delta\nu_D)} \quad (93)$$

where the offsets $\Delta\nu_F$ and $\Delta\nu$ are to be determined according to the above criteria and the actual experimental apparatus. The above correction to the gain enhancement is small for $\gamma/\Delta\nu_D \leq 0.1$, as is the case for the noble-gas ion lasers, but it is significant for the FIR lasers, where the ratio $\gamma/\Delta\nu_D$ can be as large as 0.5. The corrections become more complex for transitions involving higher J values, because of the presence of exponential terms involving the different detunings of each component of the Zeeman or Stark multiplet. The detunings depend on the actual dipole moment and on the J value (and on the quantum number K for the symmetric-top molecules) of the upper and lower level of the transition. A favorable situation, approaching the classical Zeeman effect, is observed when the dipole moments of the upper and lower state are equal or nearly equal. In this case all the $\Delta M = \pm 1$ transitions are grouped together with a frequency shift of almost $\Delta\nu_F = \pm\mu B$, respectively. Equation (93) can still represent a satisfactory approximation to the gain enhancement if $S/2$ is considered as an appropriate average saturation parameter. Of course, equation (92) can also be written for any J' and J'' values by adding the correction for the convolution to equations (61), (63), and (65) for the plane-wave approximation. Equation (92) becomes more mathematically involved in the case of a Gaussian beam.

Fortunately, the approximation $\mu' \approx \mu''$ is verified for many inportant laser lines, such as the 633 nm line of Ne, the 488 and 514 nm lines of the Ar$^+$ ion, and several strong far-infrared laser lines.

In conclusion, equation (83) can be generalized in the following way:

$\Delta M = 0$ transitions

$$S + 1 = G^2 F \left[\frac{\mathcal{R}W(\Delta\nu/\Delta\nu_D + i\gamma\sqrt{1+S}/\Delta\nu_D)}{\mathcal{R}W(0 + i\gamma/\Delta\nu_D)} \right]^2$$

$\Delta M = \pm 1$ transitions

$$S + 1 = G^2 FR^2(S) \left[\frac{\mathcal{R}W(\Delta\nu_F/\Delta\nu_D + i\gamma\sqrt{1+S/2}/\Delta\nu_D)}{\mathcal{R}W(0 + i\gamma/\Delta\nu_D)} \right]^2 \quad (94)$$

where F is a correction, independent of S', and takes into account additional phenomena affecting the gain, like the magnetoplasma effect or the NLHE in the pump transition; the last term is the correction when the ratio $\gamma/\Delta\nu_D$ is not negligible, and $\Delta\nu_F$ is to be taken of the order of μB.

Equation (94) is a useful guideline for evaluating the relevance of the PE for a single-mode laser; it can be applied also to multimode laser operation provided that the interference between the modes is negligible. On the other hand, as soon as the interference increases, the system evolves toward the BLA and the power enhancement effect is expected to be strongly reduced.

The results obtained for different lasers will be examined below.

9.1. The He-Ne Laser

The He-Ne gas discharge was the first medium to produce CW laser emission and the first to be investigated in the presence of an external magnetic field. A magnetic field can affect the operation of a gas laser in three ways. First, the Zeeman effect influences the gains and the frequency of the components of the laser transitions. Second, a sufficiently high magnetic field can affect the gas-discharge plasma characteristics (i.e., the electron density and the electron temperature), thus influencing the excitation rate and the gain of the lasing transitions. The effects becomes noticeable at field intensities of the order of a few tens mT, the minimum field intensity increasing with the gas density. Third, the magnetic field can change the dispersion properties of the laser medium around the laser transition. The separation of the $\Delta M = +1$ and $\Delta M = -1$ transitions by an axial field induces a rotatory power of the medium, so that the polarization axis of a linearly polarized light beam crossing the medium along the field direction experiences a rotation (Macaluso-Corbino effect). This effect can introduce additional cavity losses if the laser resonator contains Brewster windows. However, it was soon realized that the rotation of the polarization plane is negligible when the gain of the laser medium is reduced by saturation under the steady-state oscillation conditions.[128]

The neutral Ne atoms are selectively excited into the $3s_2$ and $2s_2$ levels by collisions with the metastable He atoms. Several laser transitions are possible from the s_2 levels toward the lower-lying $3p$ and $2p$ levels. The most intense laser lines are observed at 633, 1152, and 3390 nm, corresponding to the $3s_2 \to 2p_4$, $2s_2 \to 2p_4$, and $3s_2 \to 3p_4$ transitions, respectively (Figure 38). All these transitions occur from a $J = 1$ upper level to a $J = 2$ lower level. Since all the involved Landé factors are approximately equal, the observed Zeeman effect is very close to the classical case, with the shifts for all the $\sigma^\pm$ transitions being practically coincident with $\pm g_J \mu_B B$. It is also worth noting that the high-gain 3.39 μm laser line is in strong competition with the red laser line because of the common upper level. Thus, if the oscillation at 3.39 μm is not suppressed, this competition can affect

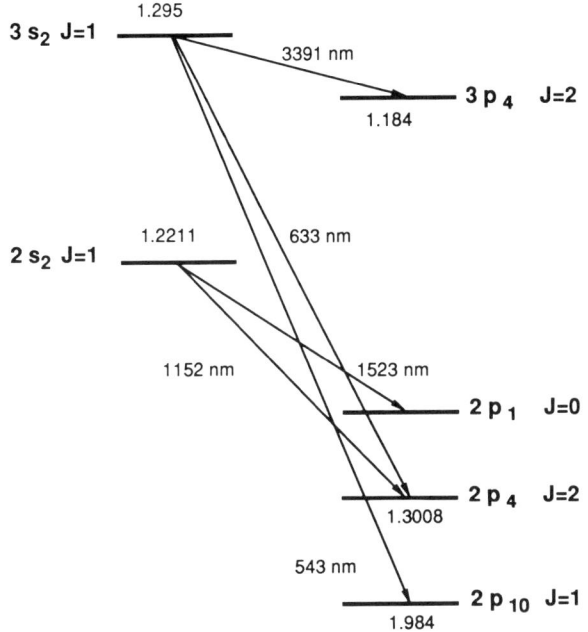

FIGURE 38. Scheme of the Ne levels and of the Landé factors relevant to laser emission.

many experimental results, including the NLHE. The natural linewidth of the Ne laser lines is of a few MHz only. Thus, the ratio $\gamma/\Delta\nu_D$ is small ($\ll 1$) and the laser is expected to operate independently on all the longitudinal TEM$_{00}$ cavity modes that are above threshold. This actually happens in lasers with a large bore tube operating at low pressure. The population inversion is exploited more efficiently in lasers operating at a relatively higher pressure in tubes whose internal diameter has been reduced to the minimum allowed by the diffraction losses. Under these conditions, at a pressure of a few hundred Pa, the collisions increase the homogeneous linewidth ($\gamma/p \approx 0.5$ MHz/Pa) up to a substantial fraction of the frequency spacing of the longitudinal modes. The saturation holes in the velocity distribution overlap and the population inversion are used more efficiently.[105-108] Thus, the commercial He-Ne lasers are designed to operate on all the longitudinal modes above threshold, with the saturation holes largely overlapping in a regime of broad-line approximation (BLA), by using an appropriate combination of pressure broadening, discharge current intensity, and bore diameter.[107]

If the pressure is further increased above the optimum value, the homogeneous linewidth becomes larger than the mode spacing and the efficiency decreases. The competition between the modes becomes so large that some of them are suppressed. By further increasing the pressure, all the modes but the central one are eventually suppressed, and single-frequency operation can be obtained.[109]

Two other He–Ne laser lines, at 1523 and 543 nm, respectively, are of interest because they involve J values more favorable to the NLHE ($J' = 1 \rightarrow J'' = 0$ and $J' = 1 \rightarrow J'' = 1$, respectively). Lasers operating on these lines are also commercially available.

The operation of gas lasers can be strongly affected by the presence of Brewster windows, or other polarization-dependent optical components, inside the laser cavity. In the absence of Brewster windows ("*flat* lasers") the polarization of the laser field is free to accommodate to the modes of largest gain. Such modes are circularly polarized in the presence of a longitudinal magnetic field. On the other hand, the laser field is forced to be linearly polarized in the presence of Brewster windows, so that the electromagnetic field interacts simultaneously with both $\Delta M = \pm 1$ transitions. Thus, the presence of the Brewster windows can strongly affect the operation of a laser in a magnetic field. If the Brewster windows are present, the magnetic field effects can be observed only in the emitted power (zero field, level crossing, and mode crossings) when the Zeeman splitting is smaller than $\Delta \nu_D$. However, a rich variety of additional new effects can be observed in flat lasers, whose polarization is free. Among these effects are frequency splitting and frequency modulation, rotation of the polarization plane, etc.

These effects are essentially due to the anisotropy induced in the medium by the field, and to the positive feedback effect of the laser amplification, and are not directly related to the NLHE. They are therefore beyond the scope of the present review and will not be discussed here. Excellent reviews are available,[44,46,47] providing also an almost complete bibliography.

The first power enhancement of a gas laser in a magnetic field was observed on the 1.153 μm line of the He–Ne laser by Buser, Kainz, and Sullivan[43] and by Culshaw and Kannelaud,[110,111] In both experimets the He–Ne laser was operating multimode with a longitudinal mode spacing of about 120 MHz. The power enhancement observed in Ref. 43 was of the order of 1.11. In Refs. 110 and 111 the laser had Brewster windows and the maximum observed PE was between 1.2 and 1.3. More extensive measurements on the 1.15 μm laser line were reported by Fodiati and Fridrikhov.[112] The laser tube, 8 mm internal diameter and terminated by Brewster windows, was placed inside a 1 m long semiconfocal resonator. At a pressure of 130 Pa, and in multimode operation, a power enhancement between 8% and 15% was observed by increasing the magnetic field from 0 to 1.5 mT (Figure 39). The PE was followed by a flat plateau up to 11 mT, and then by a Gaussian power decrease for $B > 11$ mT. The PE is smaller than expected for single-frequency laser operation. This can be explained

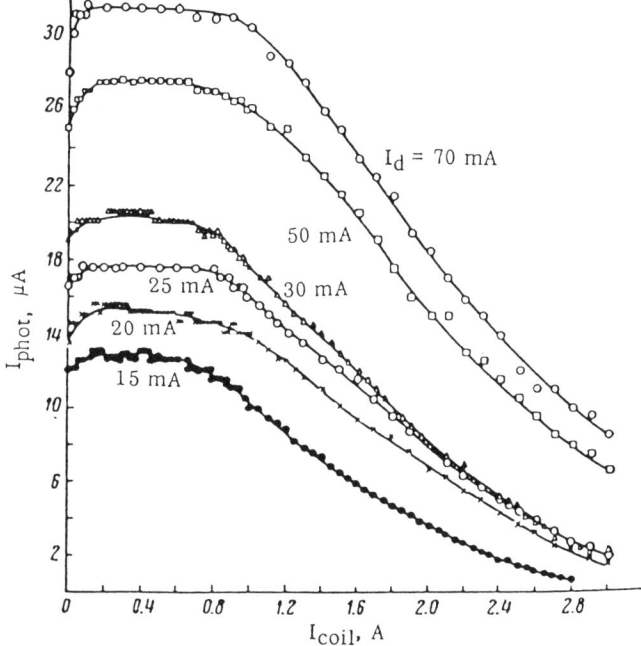

FIGURE 39. NLHE power enhancement for the 1.15 μm laser line of Ne, as a function of the magnetic field (1 A = 0.0145 T) at different discharge currents (from Ref. 112).

by the interference between the saturation dips inside the velocity distribution. In fact, the Zeeman shift between the σ^+ and σ^- components is about 36 MHz/mT which, at 1.5 mT, corresponds to a splitting smaller than the homogeneous linewidth γ. Under these conditions, the BLA is valid, so that the laser power remains constant, as expected, until the Zeeman splitting becomes of the same order as the Doppler width.

The power enhancement for the 0.6328 μm laser line was first investigated by Terekhin and Fridrikhov.[113] Their experimental apparatus was essentially the same as that of Fodiati and Fridrikhov,[112] but the gas pressure was somewhat lower (between 80 and 105 Pa). A PE of about 1.5-1.6 was observed when the field was raised from 0 to 1-1.5 mT; this PE was followed by a plateau, by a further power increase between 5 and 15 mT, and by a final power decrease. The first PE can be interpreted as almost entirely due to a NLHE, while the second one was correctly attributed to the suppression of the simultaneous and competitive laser line at 3.30 μm (see Figure 38). This interpretation was confirmed by a successive experiment by Alekseeva and Gordeev,[114] who observed the lines at 0.632 and 3.39 μm, both in simultaneous and in separated laser emission as functions of the magnetic field. These authors used a shorter laser cavity (70 cm) and investigated the laser operation with and without Brewster windows, in multimode oscillation, in longitudinal multimode only (TEM$_{00}$), and in single-frequency oscillation. The PE observed in single-line operation at 0.6328 μm was found to increase by reducing the number of laser modes.

On the other hand, when the laser was oscillating only at 3.39 μm, the emitted power showed no PE. In the absence of Brewster windows the power was almost independent of the magnetic field, while a strong power decrease with increasing magnetic-field intensity was observed in the presence of Brewster windows. Similar results are also reported by Ladygin and Tsarkov.[186] These results can be explained by taking into account the different values of the $\gamma/\Delta\omega_D$ ratio at different wavelengths ($\gamma/\Delta\omega_D \approx 0.07$, 0.12, 0.16, and 0.36 at 0.6328, 1.15, 1.52, and 3.39 μm, respectively). At 3.39 μm the large $\gamma/\Delta\omega_D$ ratio completely counterbalances the gain increase of the NLHE predicted by equation (94) (see Figure 40). The Zeeman degeneracy is removed only when the Zeeman splitting is of the order of the Doppler width. Under these conditions the σ^+ and σ^- lines oscillate independently and the gain is maximum for a circularly polarized wave. In absence of Brewster windows the gain and saturation intensity are the same as for a linearly polarized wave at zero field. Therefore no significant change in the laser power is expected, provided that the laser cavity is frequency-

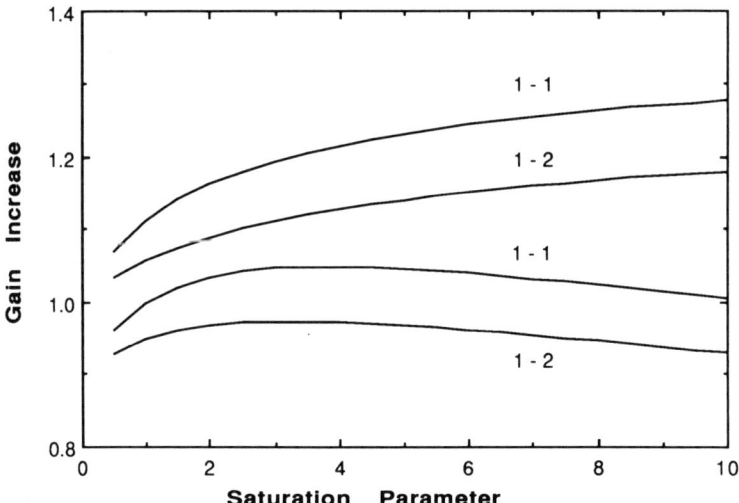

FIGURE 40. Example of the reduction in the NLHE gain increase with a finite $\gamma/\Delta\nu_D$ ratio. The vertical scale corresponds to the quantity $R(S)\mathcal{R}W(\delta\nu/\Delta\nu_D + i\gamma\sqrt{1+S/2}/\Delta\nu_D)/\mathcal{R}W(0 + i\gamma/\Delta\nu_D)$ for a Gaussian beam. Two transitions are considered: $J' = 1 \to J'' = 1$ and $J' = 1 \to J'' = 2$. The two upper curves correspond to $\gamma = \delta\nu = 0$, the two lower curves to $\gamma/\Delta\nu_D = 0.3$ and $\delta\nu/\Delta\nu_D = 0.4$.

tuned in resonance with the Zeeman shift. However, the Brewster windows introduce an additional large cavity loss for circularly polarized modes, and a power decrease with increasing magnetic field is expected.

In a different arrangement the laser was set to oscillate simultaneously on both the 0.633 and 3.39 μm lines. In this case the PE observed for the 0.633 μm line was large due to the suppression of the 3.39 μm line. A PE of factor 8 was observed at a field of about 10 mT. This result is in agreement with equation (94), if an additional gain increase is introduced in order to take into account the suppression of the 3.39 μm line. The absence of a NLHE power enhancement in the case of the 3.39 μm line had been previously reported by Bell and Bloom,[115] who observed a steady decrease in the output power with increasing intensity of the axial magnetic field. On the contrary, a power increase was reported for the visible He–Ne laser lines at 0.6328 and 0.6118 μm ($J' = 2 \rightarrow J'' = 2$). Other observations of power enhancement in flat cavities are also reported by Terekhin, Fridrikhov, and Antonov for the 0.632 μm line,[187] and by Gudelev and Yasinskii[188] and Lenstra[189] for the 1.15 μm line.

Menzies, Dienes, and George[129] investigated the NLHE gain increase in a He–Ne laser amplifier both for the $3s_2 \rightarrow 3p_4$ ($J' = 1 \rightarrow J'' = 2$) at 3.3923 μm neon transition discussed above, and for the $3s_2 \rightarrow 3p_2$ ($J' = 1 \rightarrow J'' = 1$) neon transition at 3.3912 μm, which displays a much larger NLHE. The amplifier tube was filled at 130 Pa with a 9:1 mixture of ^{3}He and ^{20}Ne ($\gamma/\Delta\nu_D = 0.3$), and the amplification for single-frequency radiation tuned to the line center was measured. A small gain increase ($R \approx 1.08$) was still observed for the $J' = 1 \rightarrow J'' = 1$ transition, even smaller than the value $R \approx 1.2$ expected by the rate equations for $\gamma/\Delta\nu_D = 0$, in the saturation range $1.2 < S < 2.8$ (see Figure 40).

At zero external field the output power of a gas laser displays a saturation dip (Lamb dip) when the laser frequency is exactly tuned to the center of the Doppler profile ν_0. We have seen that, if an external field is applied, we can observe a zero-field dip (NLHE), and level-crossing and mode-crossing dips. Level-crossing dips can be observed in a single-mode laser, while mode crossings require multimode laser operation. All these dips appear at fixed field values and are independent of the exact tuning of the laser. Additional saturation dips were predicted by Dyakonov and Perel[197] when the detuning $\delta\nu = \nu_L - \nu_0$ of the laser frequency from the center of the Doppler profile is such that the saturations of the σ^+ and σ^- are in coincidence. Such a *magnetic* Lamb dip appears, for instance, when $\delta\nu = \mu B$, but also under many other conditions when the laser is running multimode. All the above kinds of dips have been observed in the emission of the 633 nm He–Ne laser line by Brun, Le Floch and Tache.[192–194] In Ref. 192 the laser was linearly polarized due to the presence of the Brewster windows, but the 3.39 μm line was not suppressed. In fact, a large PE (between 50% and 100%) was observed for magnetic fields between 1 and 3 mT, followed by a number of inverted dips which were interpreted as Zeeman dips and mode crossings occurring in the 3.39 μm line, thus enhancing the 633 nm line.[193] In one of the experiments[193] the authors were able to suppress the 3.39 μm line. Under these conditions only regular dips (i.e., decreases in the output power) were observed. In a further experiment[194] they modified the laser in order to obtain either single-frequency or multifrequency operation. In single-frequency operation they could observe Zeeman dips only, while both Zeeman dips and mode crossings were observed in multifrequency operation. Of course, no level crossing is possible in the Ne levels. Figure 41 shows the experimental results. Curves 1, 2, and 3 of Figure 41a correspond to single-frequency operation with $\delta\nu = 0$, $\delta\nu$ small, and $\delta\nu$ large, respectively. Thus, the zero-field dip of curve 1 is a superposition of the Lamb dip and of the NLHE dip. The resolved NLHE dip and Zeeman dip are present in curve 2, while the Zeeman dip is shifted far away in curve 3. The results for multifrequency operation are shown in Figure 41b, where the two curves correspond to different $\delta\nu$ values. The NLHE and two mode-crossing dips are observed at fixed frequencies, while the weaker Zeeman dips occur at different field values. It was demonstrated that Zeeman saturation dips can be used for the frequency stabilization of the laser as the Lamb dips, with the additional possibility to tune the frequency as a function of the field.[167,195,196]

Durand[116] investigated the effect of an external magnetic field perpendicular to the laser axis on the $2s_2 \rightarrow 2p_1$, 1.52 μm line of ^{20}Ne ($J' = 1 \rightarrow J'' = 0$). The experiment was performed in a cavity without Brewster windows. By proper adjustment of the cavity it was possible to obtain single-frequency operation and a linear polarization either parallel (π) or orthogonal (σ) to

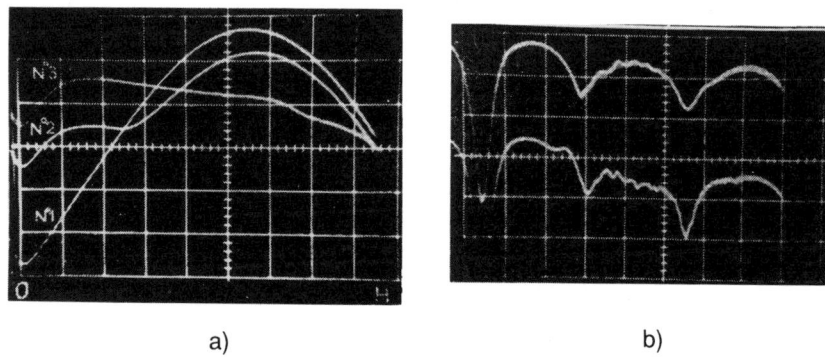

a)   b)

FIGURE 41. Output power of the 633 nm He-Ne laser line vs. the magnetic field: Lamb dip, NLHE dips, Zeeman dips, and mode crossings are present (from Ref. 194).

the magnetic field. Power increase was observed in the latter case, reaching its maximum at a field of 1.5 mT. The value PE ≈ 2 is in good agreement with the prediction of the rate equations in the plane-wave approximation (see Figure 36). Hermann and Scharmann measured a NLHE power increase both for the 0.6327 μm line[90,91] and, in collaboration with Lasnitschka, for the 1.523 μm line.[95] The reported PE are small (1.07 for the 0.6328 μm line) because the laser was operating under BLA conditions, due to the length of the resonator (1.3 to 1.9 m).

Ahmed, Kocher, and Gerritsen investigated the influence of an external magnetic field on the gas-discharge excitation efficiency (*magnetoplasma effect*) for the 3.39 μm line in resonators both with and without Brewster windows.[117] An increase in the electron density of the He-Ne plasma is expected at fields higher than 10-20 mT, and the results of these authors are in agreement with this prediction. However, an effective power increase was observed only when the discharge current was below the optimum value. In fact, the dependence of the output power of the He-Ne laser on the excitation current displays a maximum. If the current is further increased, the output power decreases because of excess ionization in the plasma. An additional magnetic field can only change the value for the optimum output, but cannot improve the overall efficiency at the best excitation rate. The situation is different for the noble-gas ion lasers, where the saturation of the output power as a function of the excitation current can be reached.

Recently Cartaleva, Gateva, and Kolarov reported an interesting application of the NLHE on the 0.6328 μm line.[118] Saprikin, Yudin, and Atutov[109] had shown that single-frequency operation can be obtained if the homogeneous linewidth is broadened by increasing the gas pressure above a critical value. Cartaleva and co-workers demonstrated that in a laser cavity with Brewster windows the critical pressure is substantially decreased, thus increasing the output power, by the presence of an axial magnetic field of about 3-7 mT. The output power in single-mode operation with the magnetic field was about the same as that obtained in a multimode operation regime at zero magnetic field.

As noted before, a large NLHE is expected for the He-Ne laser line at 543 nm because it corresponds to a $J' = 1 \rightarrow J'' = 1$ transition (see Figure 36). This line was obtained in CW operation by Perry[202] and now commercial lasers emitting this line are available. We have investigated the effect of an axial magnetic field on a Melles-Griot model 05-LGR-171. Unfortunately, this is a flat laser running on two TEM$_{00}$ modes separated by 380 MHz and with a rather low output power. Under these conditions only a very small PE is expected, as already observed in similar He-Ne lasers for the 633 nm line, and in He-CdII lasers for the 441 nm line (see later in this chapter). Nevertheless, we could obtain the interesting signals shown in Figure 42. A NLHE PE (of the order of 5%) was observed at low magnetic

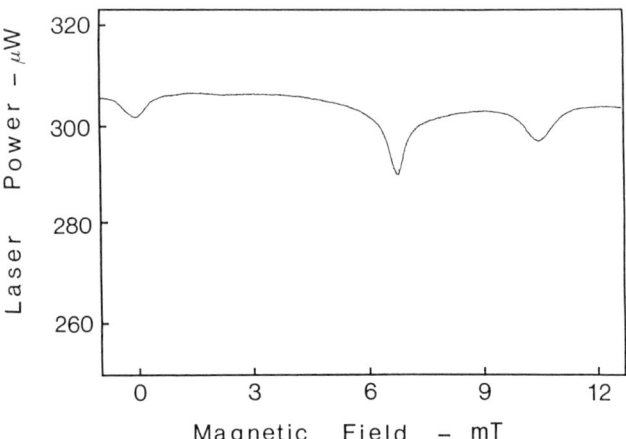

FIGURE 42. NLHE and mode crossings observed on the green 543 nm line of a commercial He-Ne laser.

field. Two mode crossings could also be observed, connected to the splitting of the upper ($g = 1.300$) and lower ($g = 1.984$) laser level. It is interesting to note that the width of the mode crossing in the lower state is 1.75 times larger than the width for the upper state, as expected for a CW operating laser.

In conclusion the NLHE, the mode crossings, and the Zeeman dips can be easily observed in a He-Ne laser. However, the NLHE is not of practical interest for increasing the output power in a well-designed laser, where the saturation of the excitation, the pressure broadening, and the multimode operation almost completely utilize the population inversion. In addition, there is experimental evidence that the NLHE is strongly reduced when the ratio $\gamma/\Delta\nu_D$ is larger than 0.3-0.4. On the other hand, the magneto-plasma effect is almost absent and also of no practical interest.

9.2. The Xe Laser

The He-Xe gas discharge is known to lase on several infrared lines with large unsaturated gain. The output power of these lines is small because of their low saturation intensities.[119] Fork and Patel[120] measured the magnetic-field dependence of the gain of the 2.026 μm line ($J = 1 \rightarrow J = 1$). The gain for linearly polarized radiation was found to decrease slightly with the field between 0 and 30 mT. Kannelaud and Culshaw[121] investigated the 2.65 μm line ($J = 1 \rightarrow J = 0$). They could easily obtain single-frequency operation and a tunability within the Doppler width at the partial pressures of 370 Pa He and 13 Pa Xe in a resonator without Brewster windows. In the presence of a weak magnetic field (<0.1 mT) the laser beam was circularly polarized, and the polarization could be switched from σ^+ to σ^- by tuning the laser frequency across the line center. No power enchancement was observed. The absence of PE even for the favorable $J = 1 \to J = 1$ and $J = 1 \to J = 0$ transitions is in agreement with the absence of PE for the 3.39 μm laser line of Ne, an can be explained by the fairly large ratio between the homogeneous and Doppler linewidths ($\gamma/\Delta\nu_D \approx$ 0.35–0.6).

9.3. The He-CdII and He-ZnII Lasers

The Penning effect is responsible for the excitation of several metal ion laser lines in the visible and UV regions.[119] The He-CdII laser is of particular interest because it provides two CW laser lines at 441.6 nm and 325.0 nm with output powers of several mW. Other laser lines in the green and red regions can also oscillate in the CW regime. Therefore the He-CdII lasers have become a convenient, commercially available source of CW blue and UV coherent radiation. The operation of some of the laser lines in the presence of an external magnetic field has been investigated. Contrary to the He-Ne laser, the power of the He-CdII laser does not reach saturation when the discharge current is increased. Thus, in this case, the unsaturated gain is expected to increase because of the plasma confinement at sufficiently large magnetic fields ($B > 20$ mT or more, depending on the He pressure).

A Grotrian scheme of the CdII laser transitions is shown in Figure 43. This scheme is limited to the laser lines for which a NLHE investigation is available. The most investigated line is the 441.6 nm, which is the strongest emitted line and corresponds to the $5s^2\,^2D_{5/2} \to 5p^2\,^2P_{3/2}$ transition. The upper level has a long lifetime and a correspondingly narrow width ($\gamma \approx$ 1–4 MHz), while the lifetime of the lower $^2P_{3/2}$ level is much shorter (about 3 ns). Thus the NLHE in the upper state is easily observable, especially at low He pressures (below 500 Pa), since the resonance has a width of a few gauss. On the other hand, the NLHE in the lower state gives origin to a much broader signal which can be distorted by Zeeman and mode crossings, by the magnetoplasma effect, and by the convolution with the Doppler profile. In addition, the natural Cd metal consists of several even and two odd isotopes, the odd isotopes displaying a hyperfine structure. Both the isotopic shifts and the hfs splittings are of the order of the Doppler linewidth.[98] The largest gain is obtained in correspondence of the even isotopes 112 and 114.

The first investigation of magnetic effects on the He-CdII laser output power was performed by Hernquist on the 441 nm line.[122] He used a short laser tube (42 cm) with Brewster windows, filled with He at 900 Pa. The

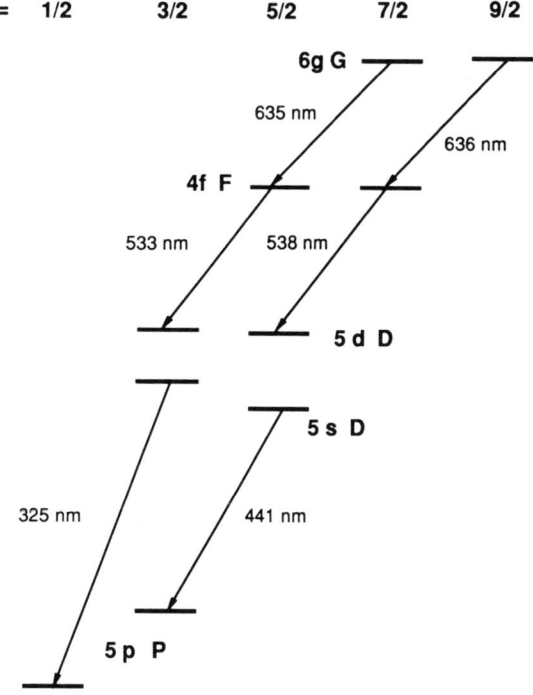

FIGURE 43. Scheme of the CdII levels involved in the He-CdII laser lines.

magnetic field could be oriented either parallel or orthogonal to the polarization imposed by the Brewster windows. For parallel orientation ($\Delta M = 0$ transitions) a power increase was observed for $B \geq 20$ mT; this effect is a consequence of the improved excitation efficiency due to the plasma confinement. For orthogonal orientation ($\Delta M = \pm 1$ transitions), a small PE was observed for magnetic fields up to 20 mT, followed by a power decrease for $B > 40$ mT. This power decrease occurs when the separation between the $\Delta M = +1$ and $\Delta M = -1$ transitions becomes larger than the Doppler linewidth. In fact, at higher fields (of the order of 0.08 T) the laser output is split into two components, separated by about 34 MHz/mT.[198]

The enhancement of the 441 nm laser line in an axial magnetic field was investigated by Brown and Swift.[199] The laser had Brewster windows, a He pressure of 900 Pa, an active region of 50 cm, and a cavity length of 88 cm. The operation was always on TEM_{00} modes only, with output powers between 10 and 20 mW. The maximum field strength investigated was 12.2 mT. A sharp power increase (~20%) was observed in the first mT, followed by some oscillations with a maximum PE of 37%. Owing to the natural isotopic mixture and multimode operation the oscillation of the power as a function of the field in the region between 1 and 12.2 mT is probably due to several level crossings, mode crossings, and Zeeman dips. The spontaneous emission intensity of the CdII 441.6 nm and of other lines of neutral Cd and He was also investigated as a function of the field intensity, but no change was observed. This result is important because it disproves the possibility that the PE might be related to an increase in the excitation rate, thus strongly supporting the interpretation as a pure NLHE.

The effect of an orthogonal magnetic field on the He–CdII laser lines at 533.7 and 537.8 nm was investigated by Cristescu, Lupescu, Popescu, and Preda.[126] Their results were similar to those obtained by Hernquist.[122] For magnetic fields up to 10 mT a PE of the order of 1.2 was observed for $\Delta M = \pm 1$ transitions. Above 20 mT the magnetoplasma effect was also observed. Again, the laser oscillation occurred simultaneously on several longitudinal modes. The Giessen group investigated the mode crossing for the even Cd isotopes and the level crossings for the odd isotopes in the $^2D_{5/2}$ level of the 441 nm line, providing accurate values for the Landé factors and for the hyperfine-splitting factors.[97-99] The pressure broadening and the spontaneous decay rate of the upper level of the CdII 441.6 and 537.8 nm lines, and of the ZnII 492.4 nm line, were measured from the width of the NLHE by Deki, Takenaka, Matsura, and Ohta.[200] Similar measurements were also done on several blue-green lines of the He–Se laser.[191]

Sasaki, Ueda, and Ohta investigated the NLHE on the 537.8 and 636.0 nm He–CdII laser lines and on the 492.4 nm He–ZnII laser line in a hollow cathode laser.[127] The corresponding transitions are $J = \frac{7}{2} \to J = \frac{5}{2}$ and $J = \frac{9}{2} \to J = \frac{7}{2}$ for CdII and $^2F_{7/2} \to {}^2D_{5/2}$ for ZnII. The laser tubes had rotatable Brewster windows in order to separate the $\Delta M = \pm 1$ and $\Delta M = 0$ components of the oscillation. The laser oscillates on several longitudinal modes with a spacing of 208 MHz, so that mode interaction was only partial. The 537 and 636 nm lines oscillated simultaneously, but they were separated by a monochromator and their respective powers were measured independently. Since the two lines are in cascade, some interaction in the NLHE is expected. In addition, the NLHE power enhancement is expected to be small because of the large J value and of the $\Delta J = -1$ transition. The magnetic field could be oriented either parallel or orthogonal to the laser tube. The results on the power output are in agreement in the predictions for all lines. Figure 44 shows the results for the 537.8 nm line. A power increase is observed for both axial and orthogonal magnetic fields and for the $\Delta M = \pm 1$ selection rule. A power decrease is observed for the $\Delta M = 0$ selection rule when the field is orthogonal. The maximum PE, of the order of 1.2, is observed between 6 and 8 mT. The effect of an axial magnetic field on the 441 nm laser line was investigated also in a flat cavity by Qiu and Zhou.[201] In this case no PE was observed. Instead, the polarization

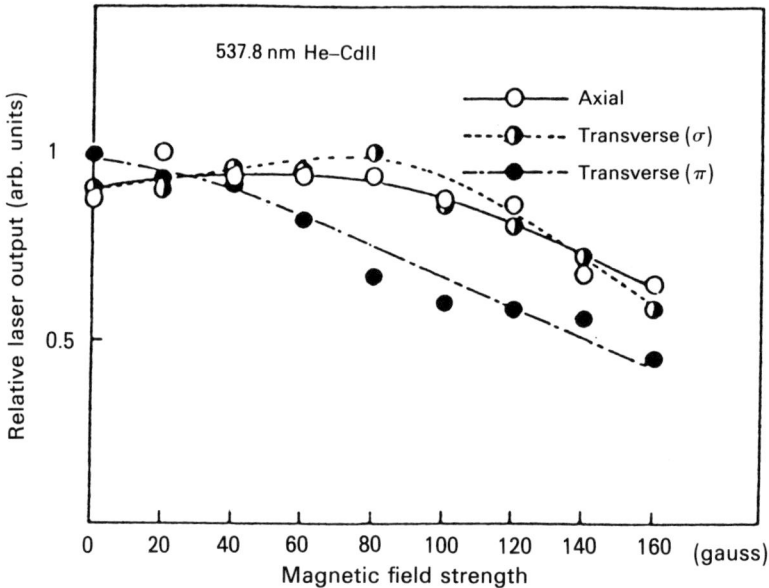

FIGURE 44. Output power of the He-CdII 537.8 nm laser line vs. the intensity of a transverse or axial magnetic field (from Ref. 127).

of the laser beam was affected by the magnetic field, in agreement with the third-order theory of the Zeeman laser.[47]

In order to check the efficiency of the NLHE for enhancing the power output of a He-Cd laser, we have investigated the PE on a commercial laser, designed for achieving maximum efficiency. The laser, a Liconix model 4207NB, has Brewster windows and interchangeable mirrors for operation at either 325 nm (nominal power 1 mW) or 441 nm (nominal power 7 mW). An axial magnetic field up to about 13 mT was generated by a pair of large-diameter (80 cm) Helmholtz coils. Typical results are shown in Figures 45 and 46. The power increased following a Lorentzian curve for fields up to 8-10 mT for both lines. The relative increase was larger for the 325 nm line (PE = 1.15) than for the 441 nm line (PE = 1.05); this result is in agreement with what is expected for the J values involved in the two transitions. In both cases the PE is smaller than predicted for single-frequency operation. In fact, the laser was simultaneously oscillating on several longitudinal modes. The width of the NLHE is in agreement with the lifetime of the lower levels of the laser transitions (unfortunately, the He pressure was not known) and the expected collisional broadening, thus indicating that the observed NLHE power enhancement was present also

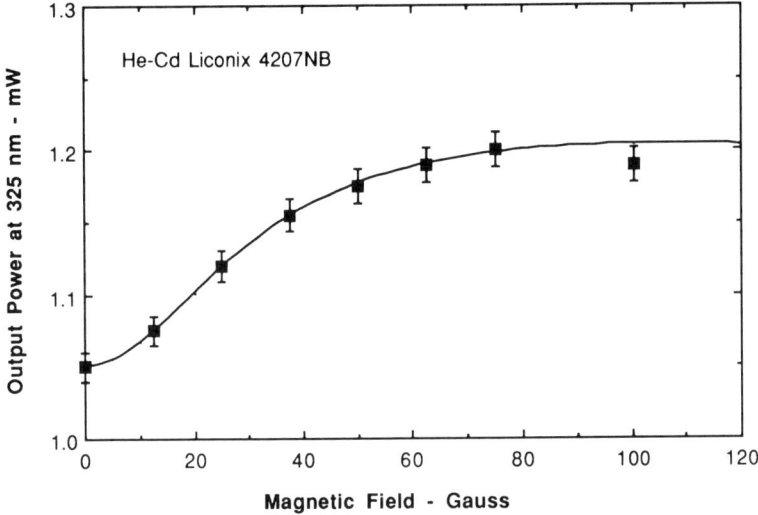

FIGURE 45. NLHE power increase in the 325 nm laser line observed on a commercial He-CdII laser.

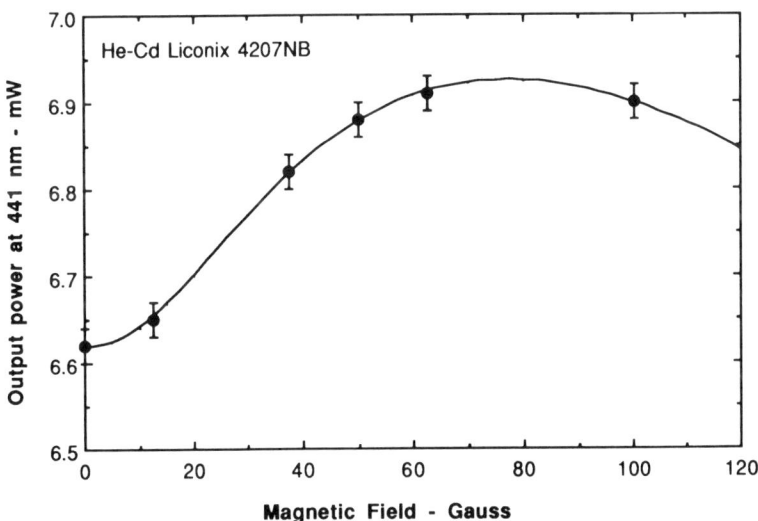

FIGURE 46. NLHE Power increase in the 441 nm laser line observed on a commercial He-CdII laser.

in the lower levels. The PE observed for the 325 nm line is sufficiently large to be of practical interest.

9.4. The Noble-Gas Ion Lasers

The noble-gas ion lasers (expecially the Ar^+ ion laser) are the most powerful and popular sources of CW laser radiation in the visible and UV spectral regions. Systems with output power between 2 and 25 W are commercially available, and laboratory lasers can reach powers of several tens of watts. Since their discovery, systematic efforts have been dedicated to the investigation of their lasing mechanism and to the improvement of their performance. The effect of an external magnetic field has also been thoroughly investigated, and now the discharge tubes of all the large power commercial systems ($P > 2$ W) are located inside solenoids providing axial magnetic fields of the order of 0.08–0.12 T. Typical power enhancements of the order of four are thus obtained for most of the laser lines, in particular for the strongest ones, like the 488.0 and 514.5 nm of Ar^+ and the 647.1 nm of Kr^+.

Not surprisingly, a large number of papers has been dedicated to the noble-gas ion lasers, and extensive review papers are available.[48,49,119,130,132,151,232] Nevertheless, at first the contribution of the NLHE to the large power enhancement observed in a magnetic field was not properly understood. This enhancement was attributed almost exclusively to the magnetoplasma effect, even if this explanation was not entirely satisfactory, as pointed out elsewhere.[55] Only in Ref. 50 was it demonstrated that the NLHE must be taken into account for explaining the different enhancements observed for laser transitions involving different J' and J'' quantum numbers.

The noble-gas ion laser is a gas laser, like the He–Ne or He–Cd laser. However, the excitation mechanism and the operation are quite different, and deserve some comment in order to understand why the NLHE can strongly affect the output power also in cases where the effect is small for other gas lasers.

First of all, in a noble-gas ion laser the upper level of a laser transition is populated by two or more successive inelastic collisions with the electrons of the discharge. Thus, the laser gain depends strongly on the discharge current density J (it is proportional to J^2 or to higher powers of J). Second, the population inversion is not obtained via a selective excitation transfer, but only because the lifetimes of the lower laser states are much shorter ($<10^{-9}$ s) than the lifetimes of the upper levels (about 10^{-8} s). This is why noble-gas ion lasers require much higher current densities than other gas lasers. At a typical current density of some hundreds A/cm^2 and a pressure of some tens Pa the ion temperature is quite high ($T > 3000$ K), the Doppler linewidth is correspondingly large ($\Delta \nu_D \geq 3.5$ GHz), and serious technological problems are encountered in the cooling and protection of the tube. A further difference from the He–Ne laser is that the output power of the CW noble-gas ion lasers continues to increase with increasing excitation power. This is because the saturation of the pump inversion would become significant only at current densities so high that they cannot be reached in practice.

The addition of a static magnetic field parallel to the laser tube axis has several beneficial effects on the laser operation. The Lorentz force reduces the diffusion rates of the electrons toward the walls and confines the discharge in the central region of the tube. Wall damages are thus attenuated, the voltage drop across the tube is reduced, and the electron density is increased, thus increasing the excitation rate and the laser gain. In addition the NLHE induces a further and independent gain increase. A partial diagram of the energy levels of Ar$^+$, including the CW laser transition, is shown in Figure 47. The angular momentum of singly ionized Ar is half-integer, and there are two laser lines (457.9 and 488.9 nm) which, corresponding to $J' = \frac{1}{2} \to J'' = \frac{1}{2}$ transitions, are not affected by the NLHE. The two most powerful laser lines (488.0 and 514.5 nm) correspond to $J' = \frac{5}{2} \to J'' = \frac{3}{2}$ transitions and display a moderate NLHE gain increase. Other laser lines (454.5 and 472.7 nm) are $J' = \frac{3}{2} \to J'' = \frac{3}{2}$ transitions with a larger NLHE gain increase. This different behavior of lines involving different angular momenta is typical of the NLHE. On the other hand, the

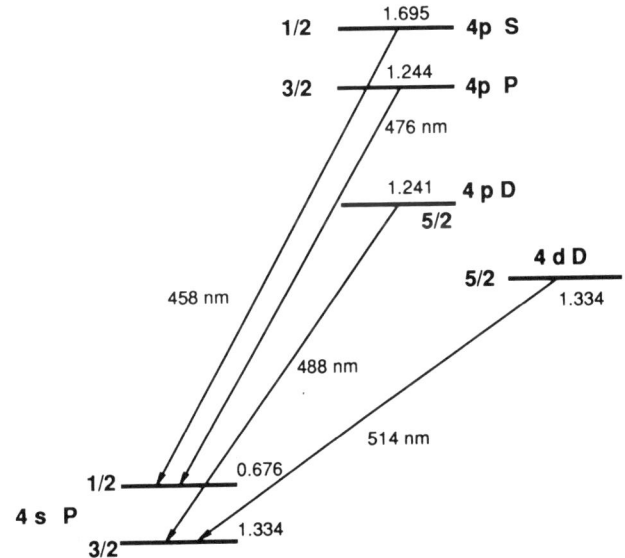

FIGURE 47. Scheme of the ArII levels involved in the CW laser emission.

gain increase due to the magnetoplasma effect is expected to be practically the same for all the laser lines. This different response to the magnetic field provides the possibility to experimentally discriminate between the two effects.

The Kr^+ ion has analogous levels. Again, we have two laser lines (468.0 and 676.4 nm) corresponding to $J' = \frac{1}{2} \to J'' = \frac{1}{2}$ transitions, while the strongest line (647.1 nm) corresponds to a $J' = \frac{5}{2} \to J'' = \frac{3}{2}$ transition.

Another important feature of the Ar^+ laser is its very large homogeneous linewidth, due to the very short lifetime of the lower laser levels. The natural linewidth is about 460 MHz, and the homogeneous linewidth at the laser operating conditions is of the order of 600-700 MHz.[134,135] Thus the ratio to the Doppler linewidth is of the order of $\gamma/\Delta\nu_D \approx 0.1\text{-}0.15$. In addition, this large homogeneous linewidth induces a large competition between the axial modes of the laser cavity, whose typical spacing is between 70 and 200 MHz. It is thus to be expected that a number of modes will be completely suppressed and that the Ar^+ laser will operate on only a fraction of the cavity modes above threshold. On the contrary, He-Ne and the other neutral gas lasers satisfy the condition $\gamma < c/L$ in most experimental conditions, so that the laser is operating on all the cavity modes and the BLA is more or less valid.

At first sight, one would expect a mode interaction also for the Ar^+ laser, which would substantially reduce the NLHE gain increase. On the contrary, a different experimental behavior is observed because of the large homogeneous linewidth. Already in 1965 Bridges and Rigrod[136] observed that the Ar^+ laser line at 488 nm has quite different behavior from the lines emitted by neutral gas lasers. When the excitation is moderately above threshold one observes either stable single-frequency operation near the center of the gain profile, or two-frequency operation with the two frequencies located symmetrically with respect to the center of the gain profile, and separated by several cavity mode spacings. Above a critical discharge current the stable single- or two-frequency operation is suddenly replaced by an unstable multimode pattern of oscillations appearing in random sequence at all the longitudinal cavity mode frequencies, $c/2L$ apart. The transition to fluctuating frequency emission is not accompanied by an abrupt change in the output power, thus suggesting that only a few modes are oscillating simultaneously. This interpretation is also supported by the observation that the power emitted on a given mode fluctuates violently, at times even reaching zero. The fluctuation time is of the order of several μs.

The results of Bridges and Rigrod[136] were confirmed in other experiments. Forsyth[137,138] and Grimblatov, Ostapchenko, and Teselkin[139] observed that single-mode operation on the 514.5 nm laser line could be obtained in the presence of a simultaneous laser oscillation at 488.0 nm. Bass, de Mars, and Statz,[140] Sedelnikov, Sinichkin, and Tuchin,[141] Berndt and Klose,[142] and Gorog and Spong[167] were able to extend stable mode operation to higher excitation levels by applying an axial magnetic field. Yarborough and Hobert[143] and Ramsey[144] investigated the output spectrum of a commercial Ar^+ laser in the presence of an axial magnetic field. In Ref. 143 it was found that for the 488 nm laser line, at an output power of 2 W and at optimum magnetic field intensity, the emission spectrum consisted of one strong mode and two weak side bands at about 1 GHz (10 $c/2L$) from the center frequency. Similar emission spectra were also observed for the 514.5 and 496.5 nm laser lines. Similar results were obtained by Ramsey[144] for the 514.5 nm Ar^+ laser lines in a large model (Spectra-Physics 171-08, with $c/2L = 82$ MHz). Also other lines (454 nm and UV) exhibited similar behaviors. In an elegant experiment Lebedeva, Odintsov, and Salimov[145] investigated the output spectrum of the 488 nm laser line with a maximum temporal resolution of 10^{-7} s. Both stable and unstable operation was observed, as in Ref. 136. In this case, however, it was observed that the laser was operating in a single-frequency regime also in the unstable operation, the instability resulting from jumps between the longitudinal cavity modes, with typical time intervals τ_{av} of a few μs. It was found that τ_{av} becomes shorter with increasing discharge currents and with decreasing resonator losses, and becomes longer with increasing resonator length. The results of Lebedeva et al.[145] were confirmed by Butkevich, Privalov, and Skvortsova, who investigated the noise spectra of the Ar^+ and Kr^+ laser lines.[146] These noise spectra showed a large and broad peak around 200 kHz when the laser was operating in the unstable regime. This peak corresponds to the switching time between modes observed in Ref. 145.

In conclusion, the operation of the noble-gas ion lasers is very close to single-frequency laser emission, with the NLHE gain increase not significantly reduced by saturation dip competition as in the multimode He-Ne laser. However, the size of the NLHE is limited by the $\gamma/\Delta\nu_D$ ratio of the order of 0.1-0.15.

Electron impact is the main excitation mechanism in the gas discharge laser, therefore the gain of the active medium and the output power depend on both the electron temperature T_{el} and the electron density N_{el}. The efficiency of a gas discharge laser is usually very low because a large fraction of the pump power is lost into several channels, including the losses on the laser tube walls. This last loss channel can be reduced by the presence of an axial magnetic field strong enough to reduce the coefficient of transverse ambipolar diffusion $D_\perp$ of the plasma electrons toward the tube walls. The expression for $D_\perp$ is[44,147]

$$D_\perp = \frac{D_0}{1+\omega_e\omega_i/\nu_{en}\nu_{in}} = \frac{D_0}{1+\beta B^2/N^2} \qquad (95)$$

where D_0 is the diffusion coefficient at zero field, ω_e and ω_i are the cyclotron frequencies for the electrons and ions, respectively, ν_{en} and ν_{in} are the collision frequencies between electrons and neutrals, and ions and neutrals, respectively, B is the axial magnetic field intensity, and β is a constant. The magnetic field can substantially reduce the transverse diffusion only if $\omega_e \omega_i > \nu_{en} \nu_{in}$ (i.e., at low pressure and above a given magnetic field intensity), and if the Larmor radius of the electrons is smaller than the diameter of the discharge tube. Essentially, the longitudinal magnetic field reduces the charged particle loss to the tube walls, thus increasing the electron density N_{el} and, consequently, the excitation rate. Experimental investigations by several authors on the Ar plasma gave results in good agreement with this interpretation.[131,148-150] In particular, the excitation rate increase is expected to affect all the atomic levels involved in a similar way. This was verified by measuring the magnetic-field dependence of the spontaneous emission at constant current,[151,152] as illustrated in Figure 48. A spontaneous emission increase by approximately a factor two was observed for all the ion lines. For each line, the laser output follows this spontaneous emission increase with the magnetic field up to an optimum field intensity of 0.1 T, denoted as B_{OPT} in Figure 48. Then, the output power decreases at higher field intensities. On the other hand, the spontaneous emission from neutral atoms decreases with the magnetic field, because the populations of the levels of the neutral atoms are saturated, and their intensities depend on the electron temperature rather than on the

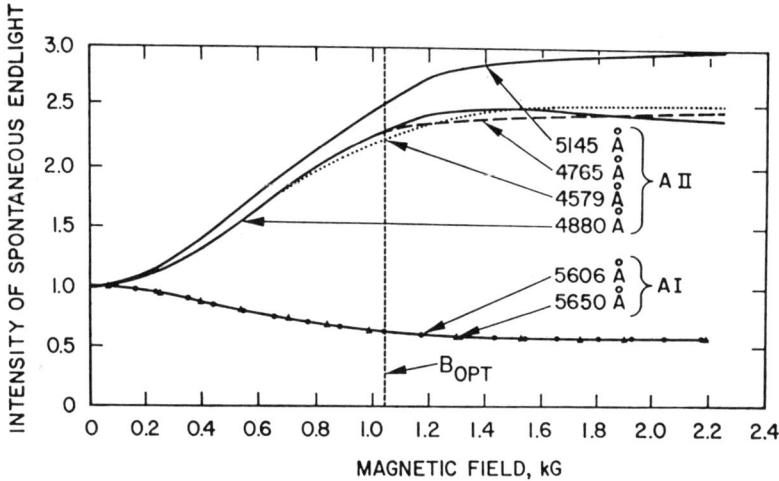

FIGURE 48. Spontaneous emission intensity vs. the axial magnetic field for a number of argon ion (ArII) and neutral argon (ArI) lines. It can be assumed that the variation represents also the gain increase in the corresponding laser line of ArII by the magnetoplasma effect (from Refs. 151 and 152).

electron density. Similar results were reported by Labuda, Gordon, and Miller[45,153] and by Kitaeva, Ostrovskaya, and Sobolev,[149] even if the spontaneous emission increase was found somewhat smaller.

It is not surprising that the effect of a magnetic field on the output power of the noble-gas ion laser has been investigated in so many papers, because of its practical relevance. Unfortunately, most of these investigations were restricted to the 488 nm laser line, and very few of them report quantitative results for other laser lines. Moreover, most papers report PE measurements without providing numerical data and other important parameters (like the small signal gain), so that only qualitative comparisons with the theoretical predictions are possible. The first evidence of a PE due to a magnetic field in an Ar ion laser was reported by Labuda, Gordon, Miller, and Webb.[45,153] They observed a PE of about a factor two for the 488 nm line and of a factor three for the 514 nm line at an optimum magnetic field of about 0.1 T. They also found a practically equal PE in laser tubes with and without polarizing Brewster windows. Sinclair[128] reported similar results for the 488 nm line and observed a PE of about 10 at 0.1 T in a resonator with or without Brewster windows. In the same experiment also the dependence on the magnetic field of the Faraday rotation of the output beam, and of the amount of power reflected by the intracavity Brewster windows, was investigated. The experiment proved that the presence of Brewster windows does not decrease the output power, and that the Faraday rotation is small (a few degrees). Analogous results were obtained by Fodiati and Fridrikhov,[154] even if in this case the PE was only about 1.5 at an optimum magnetic field of 0.08 T. Power enhancements between 2 and 4 have also been reported by several other authors.[147,151,155-158] More detailed measurements on the ArII laser were reported by Gorog and Spong[159,160] for the 488 nm line, and by Bridges and Halsted[152] for several other lines. The data of Gorog and Spong[160] are of particular interest because the ArII laser was operated on the 488 nm line in a stable single-frequency mode by using a cavity with a Fox-Smith interferometer replacing one of the end mirrors. The PE observed under these conditions was of the same order as that observed for a conventional two-mirror cavity (between 2.5 and 6). This is further evidence that the Ar^+ laser is always operating in a single-mode regime on a short time scale.

Bridges and Halsted[161] report the PE for several ArII, KrII, and XeII laser lines; the magnetic fields at which maximum PE is observed range from 0.05 to 0.1 T. Bridges and Mercer,[133] Paananen,[162] Fendley,[163,173] and Latimer[164] report PE effects of the UV laser lines of NeII, ArIII, and KrIII. The optimum field is about 0.12 T for these lines, and a PE larger than 3 has been observed for the KrII 350.7 nm line. Leonard, Yoffee, and Billman[165] observed a different PE for each visible laser line of KrII; analogous results were obtained for ArII by Tae-Soo Kim and Ung Kim.[166]

Different authors have investigated the effect of a magnetic field on pulsed noble-gas ion lasers in the cases of $Ar^{(168-170,172)}$ and $Xe.^{(171)}$ The PE effect is observed at intensities of the pulsed discharge currents which are of the same order of the steady currents at which the effect is observed in CW laser operation. At higher currents the excitation is saturated and the PE becomes negligible (a few percent), as in the case of the He-Ne laser.

The influence of an axial magnetic field has been investigated also on CW ion lasers with large bore tubes (internal diameter of the order of 1 cm). In this case the reduction in the transverse ambipolar diffusion by the axial magnetic field is expected to be effective at lower field intensities.[177] However, in a large-bore tube the excitation rate is less sensitive to the reduction in the transverse diffusion because of the larger distance between the walls and the region where the gas interacts with the laser field. In addition, the CW current density that can be injected before plasma instability is reached is larger than in the case of the commercial small-bore (internal diameters of the order of 3 mm) ion lasers. Thus, in normal operating conditions, the excitation rate in a large-bore tube is near saturation, and the magnetoplasma effect is expected to be small. This is particularly the case of the ArII and KrII laser lines.

In fact, no substantial PE effect was reported in the early investigations on large-bore tube ArII and KrII lasers.[174-176] However Banse, Lüthi, and Seelig were able to observe a PE effect also in large-bore tubes,[177] at least at a limited current density. At higher current densities a very small PE was observed for the 488 nm line[177] and can probably be explained by a NLHE in the upper laser levels, since the effect was maximum at about 5 mT. The gain saturation is expected to be smaller for the ArIII and KrIII UV laser lines. Lüthi et al.[178,179] observed a PE of the order of a few percent for a weak axial magnetic field (between 2.5 and 12.5 mT). Similar results were also reported for the 332.4 nm ($J' = \frac{3}{2} \to J'' = \frac{3}{2}$) NeII laser line.[180] In this case a PE of 33% was observed at a field of 11.5 mT and at a discharge current of 400 A. It is probable that this PE, occurring at a rather weak magnetic field, can be correctly interpreted as a pure NLHE accompanied by no significant magnetoplasma effect.

Much better evidence for a NLHE in the upper laser level of the ArII 488 nm line has been recently obtained by using a magnetic field orthogonal to the laser tube.[181,182] The laser tube, with Brewster windows, could be rotated in order to select the σ ($\Delta M = \pm 1$) or π ($\Delta M = 0$) transitions. The voltage drop across the tube showed a sharp increase, between 30% and 40%, for magnetic fields above 0.045 T.[181] It was estimated that at this critical magnetic field the cyclotron radius of the electrons is of the order of the laser tube radius. Thus also an orthogonal field becomes effective in reducing the electron and ion diffusion toward the walls. The laser output power P showed different behaviors for the σ and π transitions (Figure

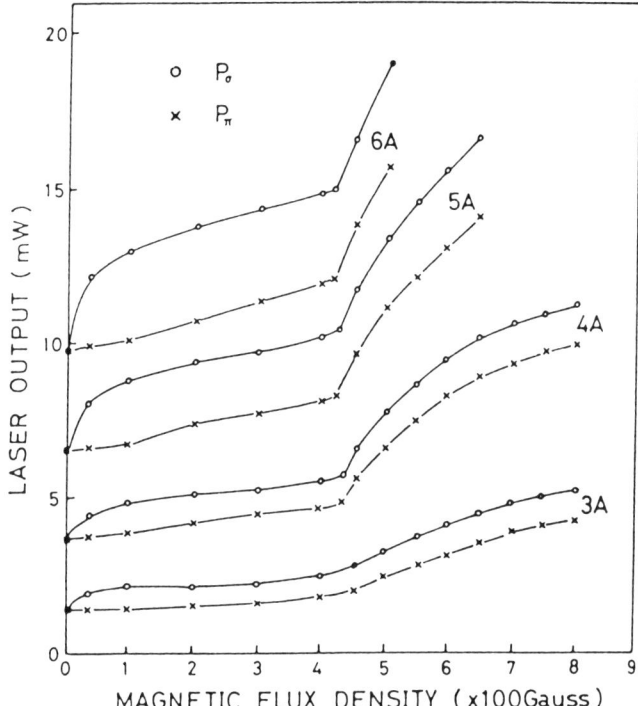

FIGURE 49. Output power of the 488 nm ArII laser line for $\Delta M = 0$ transitions (P_π) and $\Delta M = \pm 1$ transitions (P_σ) vs. the intensity of a magnetic field orthogonal to the laser axis, for different values of the discharge current (from Ref. 182).

49). The value of P_π does not increase substantially up to 0.045 T, with a sudden and large PE above this critical field. A PE of the order of 35% to 60% is observed for P_σ when the field increases from 0 up to about 0.01 T (see also Figure 50), followed by a plateau up to 0.045 T, and a sudden large PE at larger fields, similar to that observed for P_π. The results of Figures 49 and 50 are explained by the occurrence of a NLHE between 0 and 0.01 T affecting P_σ only, while the additional PE observed for both P_σ and P_π above 0.045 T is due to a magnetoplasma effect. In particular, the PEs of Figure 50 are in good quantitative agreement with the predictions of equation (83), as shown in Figure 36. Figure 51 shows P_0, P_π, and P_σ versus the discharge current. The PE observed for P_π is explained as due to the magnetoplasma effect only, and the PE for P_σ as due to both the NLHE and the magnetoplasma effect. These experimental results prove very clearly that the PE effect of a magnetic field on noble-gas ion lasers is due to the combined action of the NLHE and the magnetoplasma effect.

The PE for σ and π transitions can be evaluated from equations (94) and (90). We have fitted the results of Figure 51 by assuming that the zero-field gain is proportional to the square of the discharge current and that the magnetoplasma effect is the same for both polarizations. Figure 52 shows a comparison of PE_σ and PE_π, calculated under the assumption that $F = 2.25$, with the experimental points of Figure 51; the error bars represent our uncertainty in reading Figure 51 and in deriving G from the current intensity. It is remarkable that good agreement has been obtained with only one free parameter (F). This proves that the different behavior of PE_σ and PE_π is entirely due to the NLHE.

Further direct experimental evidence for the NLHE contribution to the PE in ion lasers can be obtained from the measurement of the PE effect on laser lines with different J' and J'' values and observed in the same experimental conditions. In this situation the magnetoplasma effect increases the small signal gain by about the same quantity for all the laser lines, as was observed experimentally (see Figure 48). On the other hand, the NLHE gain increase will be different and, in particular, zero for the $J' = \frac{1}{2} \to J'' = \frac{1}{2}$ transition.

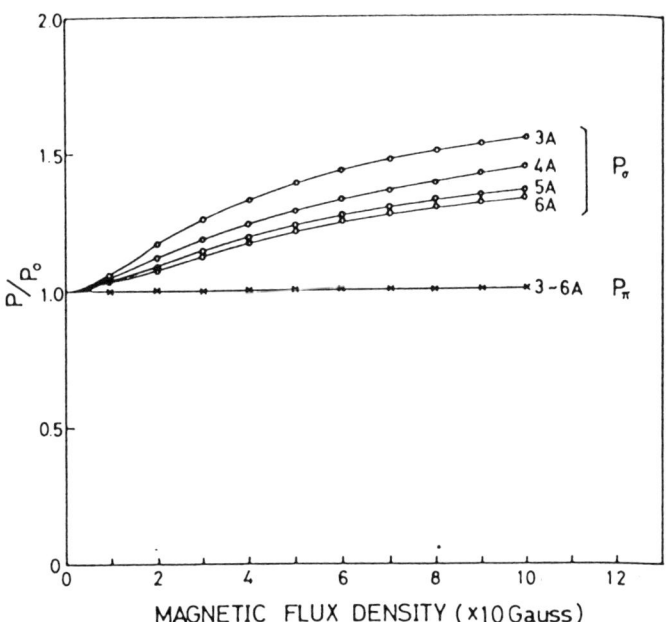

FIGURE 50. Relative output power of the 488 nm laser line for σ and π polarizations in an orthogonal magnetic field. The emitted power remains constant for the π polarization ($\Delta M = 0$ selection rule) (from Ref. 182).

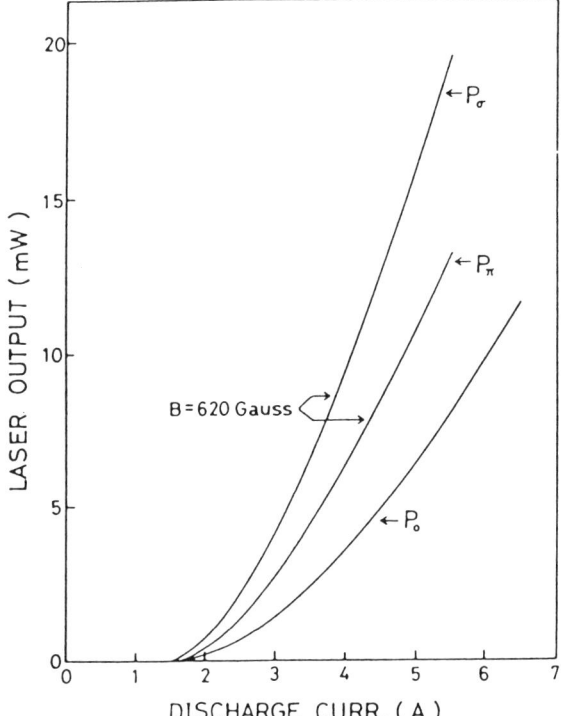

FIGURE 51. Output power for σ and π polarization of the 488 nm laser line vs. the discharge current at constant magnetic field (0.062 T), and at zero field (P_0) (from Ref. 182).

Fortunately, Bridges and Halsted[152] have conducted such an experiment: they measured the output powers of the 488 nm and 514 nm ($J' = \frac{5}{2} \rightarrow J'' = \frac{3}{2}$), 476.5 nm ($J' = \frac{3}{2} \rightarrow J'' = \frac{1}{2}$), and 457.9 nm ($J' = \frac{1}{2} \rightarrow J'' = \frac{1}{2}$) laser lines as functions of the axial magnetic field intensity in the same laser tube with Brewster windows, and at the same pressure and discharge current.

The experimental data of Bridges and Halsted are reported in Figure 53. This figure has been redrawn by normalizing the power of each line to a common zero-field value, in order to compare the PE effects. The maximum PE, observed at about 0.1 T, was 4.18, 3.38, 4.10, and 1.92 for the 514, 488, 476, and 458 nm laser lines, respectively. The last value corresponds to a PE due to the magnetoplasma effect only, since there can be no NLHE on a $J' = \frac{1}{2} \rightarrow J'' = \frac{1}{2}$ transition. We have obtained quantitative theoretical predictions by assuming that the magnetoplasma gain increase is equal to the spontaneous emission increase as given in Figure 48. This assumption is reasonable because all the laser lines show the same multiplet as the lower state.

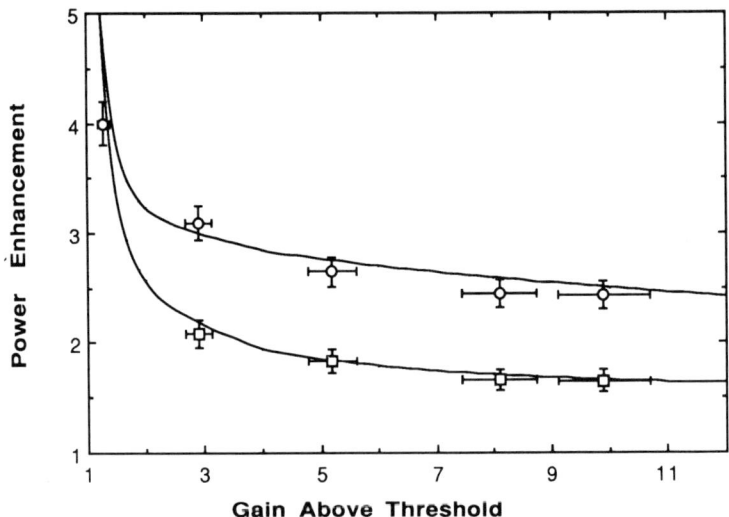

FIGURE 52. Calculated power enhancement (PE) for the 488 nm laser line vs. the excitation parameter G at zero magnetic field. The lower curve corresponds to π polarization and the upper curve to σ polarization. The experimental points have been derived from Figure 51 by assuming that G is proportional to the square of the discharge current in the absence of a magnetic field. The constant F was found to be 2.8 for both P_σ and P_π.

The saturation in the presence of the magnetic field was obtained from equation (94) by substituting $\Delta\nu_F$ with the corresponding Zeeman splitting at 0.1 T, and the zero-field saturation was obtained from equation (90). The corresponding PE as a function of the excitation parameter G at zero field is shown in Figure 54 for each laser line.

The G values corresponding to the experimental data were estimated by introducing into equations (77) and (90) the experimental output powers, the output mirror transmission as given by Bridges and Halsted,[152] and the saturation intensities as given in Refs. 48 and 49. The relatively large horizontal errors of Figure 54 reflect the uncertainty affecting the parameters. Nevertheless, the agreement between theory and experiment is remarkable and could not be obtained by neglecting the NLHE.

In conclusion, the magnetic field PE in noble-gas ion lasers is a result of the combined gain increase induced independently by the NLHE and the magnetoplasma effect. This result is based on the following experimental results: (1) The PE depends on whether the transition selection rule is $\Delta M = \pm 1$ or $\Delta M = 0$, and is smaller in the second case. (2) The PE depends on the quantum numbers of the upper and lower level of the lasing transition even if the small signal gain is about the same. In particular, the smallest PE is observed for the $J' = \frac{1}{2} \to J'' = \frac{1}{2}$ lines. (3) The NLHE gain increase

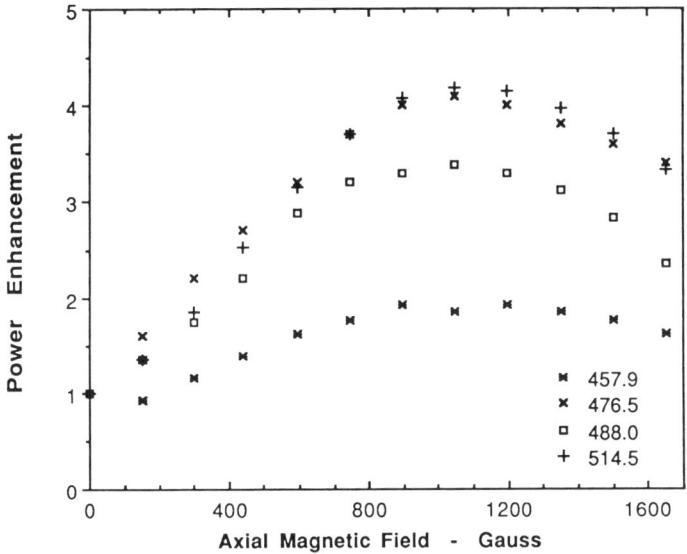

FIGURE 53. Power enhancement of several ArII CW laser lines vs. the intensity of an axial magnetic field. Discharge current: 20 A, laser tube bore: 3 mm (redrawn from Ref. 152).

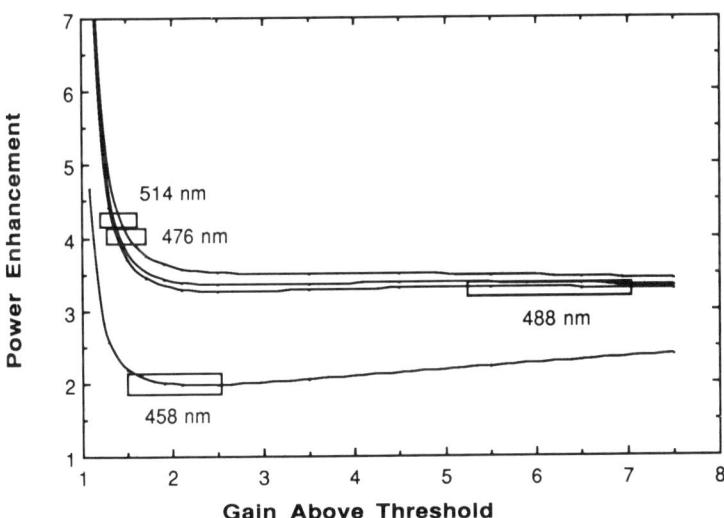

FIGURE 54. Calculated NLHE PE for the ArII laser line of Figure 53: a Gaussian beam and the same magnetoplasma gain increase of Figure 48 are assumed. Starting from the top, the curves correspond to the 476, 514, 488, and 458 nm laser lines. Boxes represent the experimental results deduced from Figure 53.

depends on the saturation with a positive feedback in the output power, thus making the relative contribution of the NLHE and magnetoplasma effect to the PE almost equal.

9.5. Optically Pumped Far-Infrared Lasers

The optically pumped far-infrared (FIR) molecular lasers are a remarkable source of CW coherent radiation in the spectral region between 20 μm and 2 mm, where almost no other source is available. These lasers are described in a large number of recent review papers.[61,203-207,219,233] Here we shall confine ourselves to what is relevant to the NLHE.

The population inversion between a pair of rotational levels of a polar molecule can be achieved by optical pumping, i.e., by absorption of infrared (IR) laser light (typically, a CO_2 laser line in the 9-11 μm region is used). The molecule is thus optically pumped into an excited rotovibrational level manifold. The energy of the excited level is well above kT, and a stationary population inversion between the upper level of the IR transition and other, lower-lying, rotational sublevels of the excited vibrational state is obtained, provided that the vibrational relaxation rate Γ toward the ground vibrational state is sufficiently fast with respect to the rotational relaxation rate γ (see Figure 55). In all molecules $\gamma > \Gamma$, so that there is a bottleneck in the laser cycle. In these conditions, the additional relaxation due to the collisions against the walls of the laser tube allows a stable population inversion only at a sufficiently low density. Thus, the typical operating pressure of an

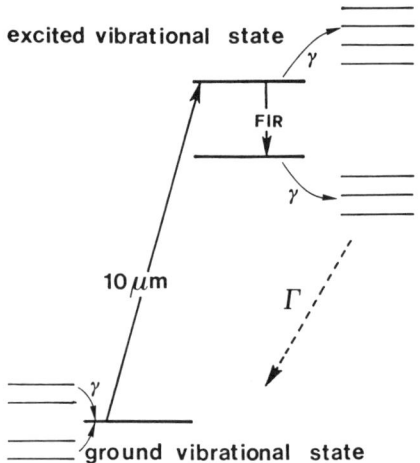

FIGURE 55. Level scheme of the optically pumped far-infrared molecular lasers. Since $\Gamma < \gamma$, CW laser operation is possible only at low gas pressure (a few pascals) when the additional relaxation due to the diffusion toward the laser walls is present.

optically pumped FIR laser is in the range 5–25 Pa. In this density range the absorption of the active medium is very small and the CO_2 pump radiation travels many times up and down the FIR laser cavity. As a consequence, the IR pump transition of the active medium is well saturated.

Laser emission in the FIR region can be obtained on any allowed rotational transition starting from the upper level of a pump transition if the molecule has a permanent electric dipole moment. In fact, thousands of laser lines have been discovered.[207] If the lasing molecule is a symmetric top, or a *special* asymmetric top like CH_3OH, the M sublevels can be shifted in an external electric field by a linear, or almost linear, Stark effect. A few tens of V/cm are, in general, sufficient for separating the sublevels by an amount larger than the homogeneous linewidth. In fact, the pressure self-broadening of both the vibrational and rotational transitions of polar molecules is of the order of 0.15–0.25 MHz/Pa, thus giving a homogeneous halfwidth γ between 0.5 and 3 MHz for the FIR laser-active medium. For comparison, the Doppler width $\Delta \nu_D$ [half-width at $1/e$ as defined in equation (86)] is about 40 MHz at 10 μm (pump transition) and between 1 and 4 MHz in the FIR region. A first important consequence is that the $\gamma/\Delta \nu_D$ ratio in the FIR is of the order of 0.5–1.5, while it is only of the order of 0.05, or smaller, for the IR pump transition. Second, the gain linewidth is smaller than the separation of the longitudinal mode of the FIR cavity [$c/(2L)$ is of the order of 100 MHz]. As a consequence a FIR laser is always operating single-frequency when a cavity mode is tuned in coincidence with a laser line, otherwise the laser action is forbidden because $G < 1$.

The best practical way to generate a constant and homogeneous electric field for the Stark effect on the laser medium is to use a hybrid waveguide terminated by flat mirrors as FIR resonator.[61,208,217] The hybrid waveguides have a rectangular cross section, the two opposite larger walls being conducting (metal plates) and the two smaller walls dielectric. Typical dimensions are 3×0.5 cm. The FIR oscillating EM field can be transmitted in these waveguides only if the electric vector $\mathbf{E}$ is parallel to the conducting walls. On the other hand, also an EM field with $\mathbf{E}$ perpendicular to the conducting walls is almost equally well transmitted at 10 μm.[208] The static electric field is orthogonal to the metal walls of the hybrid waveguide, and the FIR laser is always operating with the selection rule $\Delta M = \pm 1$, while the polarization of the pump radiation can be rotated in order to induce both $\Delta M = 0$ and $\Delta M = \pm 1$ transitions.

In symmetric-top molecules the Stark effect is linear, and the energy shift of the M sublevels in an electric field E is, to first order,

$$\Delta W = -\mu E \frac{KM}{J(J+1)} \qquad (96)$$

where K is the component of J along the symmetry axis of the molecule and μ is the permanent electric dipole moment.

In an asymmetric-top molecule the Stark effect is quadratic, and the energy shift much smaller. However, in a few asymmetric-top molecules with internal rotation, the Stark effect can also be linear and described by equation (96). This is, in particular, the case of the most important source of FIR laser lines, the methyl alcohol (or methanol) molecule CH_3OH. Methanol is a slightly asymmetric-top molecule, but the OH group can rotate around the symmetry axis of the CH_3 group subject to a threefold hindering potential with a barrier height of about 400 cm^{-1}. It follows that the internal rotation is much faster than the overall rotation of the molecule which, as a consequence, has rotational levels similar to those of a symmetric-top molecule, even if new quantum numbers are needed for labeling the states, and the selection rules for the rotational transitions are more relaxed. In particular, the selection rule for the quantum number K is not only $\Delta K = 0$, as for the symmetric top, but also $\Delta K = \pm 1$. Incidentally, this is one of the reasons why CH_3OH is such an important laser source.

We shall now restrict ourselves to the most common Stark multiplets; a more complete description can be found elsewhere.[61] We shall assume that the pump transitions obey the selection rule $\Delta K = 0$, and we shall denote by J and K the quantum numbers of the common upper level of the pump and laser transitions, so that the quantum numbers of the lower level of the pump transition will be K and J or $J \pm 1$, the quantum numbers of the lower level of the laser transitions will be K or $K - 1$, and J or $J \pm 1$. The Stark shift for the pump line is thus:

1. $J - 1 \to J$ (^qR branch lines)

$$\Delta \nu = \frac{EK}{J(J^2 - 1)} \{[\mu''(J+1) - \mu'(J-1)]M - \mu''(J+1)\Delta M\} \quad (97)$$

2. $J \to J$ (^qQ branch lines)

$$\Delta \nu = \frac{EK}{J(J+1)} [(\mu'' - \mu')M - \mu''\Delta M] \quad (98)$$

3. $J + 1 \to J$ (^qP branch lines)

$$\Delta \nu = \frac{EK}{J(J+1)(J+2)} \{[\mu''J - \mu'(J+2)]M - \mu''J\Delta M\} \quad (99)$$

where μ' and μ'' are the static electric dipole moments of the upper and lower level, respectively, and ΔM equals 0 and ± 1 for π and σ transitions, respectively. The selection rules and the Stark behaviors of the FIR laser lines of particular interest are the following (we have assumed $\mu' = \mu''$ for transitions within the same excited vibrational state):

1. $\Delta K = 0$, $J \to J - 1$

$$\Delta \nu = \frac{\mu E K}{J(J^2 - 1)} [2M - (J + 1)\Delta M] \quad (100)$$

2. $K \to K - 1$, $J \to J$

$$\Delta \nu = \frac{-\mu E K}{J(J + 1)} [M + (K - 1)\Delta M] \quad (101)$$

3. $K \to K - 1$, $J \to J - 1$

$$\Delta \nu = \frac{\mu E K}{J(J - 1)} \left[\left(\frac{2K}{J + 1} - 1 \right) M - (K - 1) \Delta M \right] \quad (102)$$

where (1) would be the only possible transition for a true symmetric top, while all cases are possible for CH_3OH. The frequency spread of equations (101) and (102) is larger than that of the pump transition, however in case (3) the first term within the brackets becomes negligible when $2K/(J + 1) \approx 1$ and all the M components collapse into only two narrow packets, corresponding to the $\Delta M = \pm 1$ transitions. In this case, the FIR lasing line is seen to split into only two lines when a large Stark field is applied to the active medium. This is, in fact, the case of the most powerful laser lines of CH_3OH and its isotopic species.[53,61]

When an electric field is applied, the laser gain G is quickly split into a large number of resolved Stark components in cases (1) and (2). Quantity $G(M)$ then becomes easily smaller than 1 and the laser oscillation dies out. On the other hand, G is decreased only by a factor 2 in the case of equation (102) and $K \approx J/2$, so that laser oscillation can be maintained also in correspondence with the two Stark components, as in the case of the 3.39 μm He-Ne laser. The degeneracy between the M sublevels of the pump transition is removed when

$$\gamma \approx \frac{\mu E K}{J(J + 1)} \quad (103)$$

thus well before the energy splitting becomes of the order of $\Delta \nu_D$.

Thus, the rate-equation model must take into account the following key points for describing the NLHE on the optically pumped FIR lasers:

1. The collisional broadening is almost the same for all the molecular levels, with the coherence relaxation term γ_{ab} much smaller than $\gamma_a \approx \gamma_b$.
2. The $\gamma/\Delta\nu_D$ ratio in the FIR laser lines is sufficiently large (≥ 0.3) for each molecule to interact with the FIR power traveling in both directions inside the laser cavity. In addition, the NLHE gain increase in the FIR transition is strongly reduced, as shown in Figure 40.
3. The $\gamma/\Delta\nu_D$ ratio in the pump IR transition is, on the contrary, $\ll 1$, and the transition is well saturated. The pump radiation can be assumed to be uniformly distributed over the inner volume of the FIR cavity, therefore the excitation increase due to the NLHE is directly given by equations (61), (63), and (65) without further corrections.
4. In a hybrid waveguide the selection rule for the laser transition is $\Delta M = \pm 1$, while the pump selection rule can be either $\Delta M = \pm 1$ or $\Delta M = 0$, or any combination of the two.
5. J' and J'' are usually large numbers (typically >10), so that $R(S)$ is large for $\Delta J = 0$ and small for $\Delta J = \pm 1$ (see Figures 17 and 18).
6. The Stark effect can split the FIR transition into a large number of components, thus strongly reducing the gain. Only for $\Delta K = \Delta J = \pm 1$ transitions does the Stark splitting result in two lines only. However, the corresponding Stark splitting of the pump transition is smaller than $\Delta\nu_D$ in any case.

Equations (94) can thus be rewritten more conveniently for the case $\Delta K = \Delta J = -1$ as:

High field and $\Delta M = 0$ pump selection rule

$$\sqrt{1 + 2S} = G(0,0) \frac{\mathcal{R}W(\delta\nu/\Delta\nu_D + i\gamma^\pi\sqrt{1+2S}/\Delta\nu_D)}{\mathcal{R}W(0 + i\gamma/\Delta\nu_D)} \quad (104)$$

High field and $\Delta M = \pm 1$ selection rule

$$\sqrt{1 + 2S} = G(0,0) R(S_{\text{FIR}}) R(S_{\text{IR}}) \frac{\mathcal{R}W(\Delta\nu_F/\Delta\nu_D + i\gamma^\sigma\sqrt{1+S}/\Delta\nu_D)}{\mathcal{R}W(0 + i\gamma/\Delta\nu_D)} \quad (105)$$

where $R(S_{\text{IR}})$ is the gain increase due to the NLHE in the pump transition,

γ^π and γ^σ may be larger than γ if the Stark effect is taken into account, $\delta\nu$ is a possible small detuning of the laser frequency induced by the frequency pulling of the FIR laser cavity, and $\Delta\nu_F$ is the Stark splitting of the FIR line due to the last term in equation (102).

The first power enhancements on optically pumped FIR lasers were observed by Tobin and Jensen on CH_3OH[51] and CH_3F[52] lasers. Laser oscillation on single M components was also observed in CH_3F by further increasing the intensity of the electric field.[52,83,220] PE and splitting of the laser line into two Stark tunable components was observed successively on many CH_3OH laser lines.[32,53,61,208-212,218,225-227,229] In all these papers power enhancements up to a factor five were observed when the pump transition obeyed the selection rule $\Delta M = \pm 1$. On the other hand, no PE was observed for $\Delta M = 0$ pump transitions. In general, however, a power decrease is observed when the spacing between the M components of the laser transition becomes of the order of γ. As a general rule, a large PE was obtained for laser lines with $\Delta K = \Delta J = \pm 1$ (see Figure 5). For the other selection rules, both power enhancement and power decrease could be observed, depending on the competition between the splitting of the FIR transition into isolated M components and the NLHE on the pump transition.

PE was observed for the 496 μm CH_3F laser line ($\Delta K = 0$, $\Delta J = -1$)[51,53,228] and for the 165 μm laser line of CH_3OH ($\Delta K = 0$, $\Delta J = -1$).[32] For laser lines which display only a quadratic and very small Stark shift, the power is unaffected by the presence of an electric field up to the electrical breakdown of the gas. This is the case of several laser lines of CH_3OH, for instance, the 570 and 164 μm lines pumped by the $9R(10)$ CO_2 laser line,[212] the 163 and 251 μm lines pumped by $10R(38)$, and the 145 and 249 μm lines pumped by $10R(42)$.[212,221] The quadratic Stark shift of these lines is due to their large K splitting. PE was also observed on other symmetric-top molecules like CH_3Cl, CH_3Br,[214] and CH_3I.[61] On the other hand, no PE was observed in asymmetric-top molecules like $HCOOH$[213] and CH_2F_2.[222]

More recently large PE effects have also been observed on many strong laser lines of $^{13}CH_3OH$,[223] $^{13}CD_3OH$,[224] and CD_3OH[215,216,226] (see Figure 56). As a general rule, a large PE effect is followed by a splitting into two components if the electric field is further increased. In this case the laser oscillation dies out if the length of the FIR cavity is kept fixed, while laser action can continue up to large fields if the laser cavity is constantly tuned to the Stark-shifted frequencies (Figure 57).

A better insight into the NLHE power enhancement for the FIR lasers can be gained by considering in detail the 118 and 170 μm laser lines of CH_3OH, pumped by the $9P(36)$ CO_2 laser line. The 118 μm line is the most powerful laser line in the FIR spectral region: an output power up to 1 W has been recently observed. Also the 170 μm line is strong, so that a gain well above threshold can be easily obtained. The two lines share a common

Q-branch pump transition, with $J = 16$ and $K = 8$. The condition $K \approx J/2$ is thus satisfied. The 118 μm line corresponds to a $J' = 16$, $K' = 8 \to J'' = 15$, $K'' = 7$ transition [equation (102)], while the 170 μm line corresponds to a $J' = 16$, $K' = 8 \to J'' = 16$, $K'' = 7$ transition [equation (101)]. The small-signal gain Stark patterns for the $\Delta M = \pm 1$ components are quite different for the two laser lines, as shown in Figure 58. Because of the quantum numbers involved in the transition, the 118 μm line is expected to split into only two Stark components (the observed splitting is 26.4 MHz/(kV cm^{-1})$^{(53)}$), while the 170 μm line is expected to display a rapid and strong gain decrease with the electric field. The NLHE absorption increase in the common pump transition is shown in Figure 25, and its effect on the laser emissions is shown in Figures 59 and 60. In the case of the 118 μm line a NLHE power increase is observed with increasing fields up to about 200 V/cm, where the laser emission begins to be split into two components. The output power for one of these components is shown in Figure 59 as a function of the electric field. The FIR gain is halved, thus counterbalancing almost exactly the gain increase due to the NLHE on the pump transition, which reaches its maximum at about 500 V/cm, as shown in Figure 25. For electric fields larger than 2 kV/cm, the Stark splitting of the pump transition becomes larger than the Doppler width and the pumping becomes less and less efficient. The corresponding changes in the pump

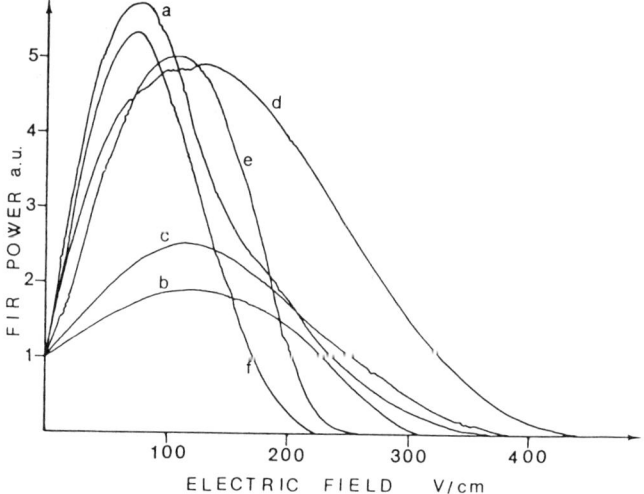

FIGURE 56. NLHE power enhancement for six laser lines of CD$_3$OH pumped by the 10R(34) CO$_2$ laser line: (a) 42.5 μm; (b) 66.4 μm; (c) 76.9 μm; (d) 128.9 μm; (e) 138.4 μm; (f) 181.2 μm. The relative FIR powers are normalized to the same zero-field intensity. The laser lines are excited by four different pump transitions, all belonging to the Q branch ($\Delta J = 0$ selection rule), thus with a large NLHE gain increase (from Ref. 215).

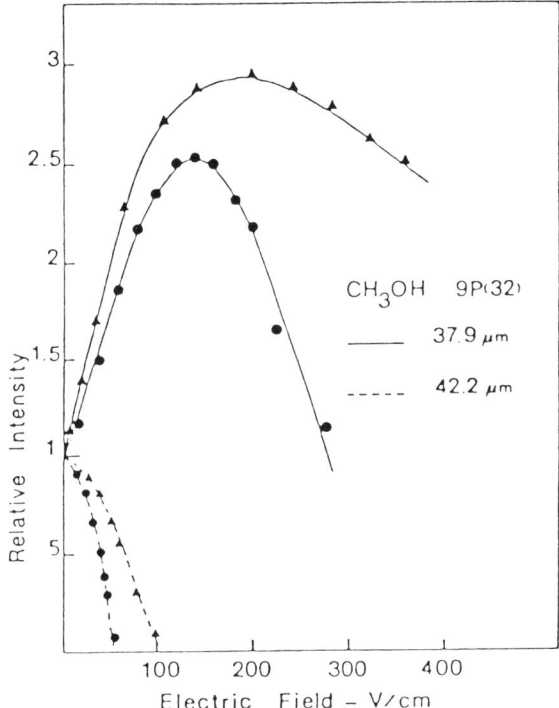

FIGURE 57. NLHE for two FIR laser lines of CH_3OH pumped by the $9P(32)$ CO_2 laser line vs. the applied electric field and with the $\Delta M = \pm 1$ pump selection rule. The triangles correspond to the FIR laser cavity length constantly tuned to maximum output; the dots correspond to the cavity length tuned to maximum output power at zero electric field (from Ref. 212).

transition can be observed by recording the FIR laser power vs. the frequency of the CO_2 pump laser. Typical results are shown in Figure 61. Figure 61a compares the zero-field behavior to the behavior in the presence of a 100 V/cm field. PE is observed at any CO_2 frequency. The central saturation dip is the usual Lamb dip observed as a transferred Lamb dip on the FIR laser output power.[35] The presence of this dip proves that the pump transition is saturated. In Figure 61b the electric field is increased to 700 V/cm. In this case two separate Lamb dips are present, corresponding to the $\Delta M = +1$ and $\Delta M = -1$ transitions. The Lamb dip for $\Delta M = -1$ is hardly observable because it occurs very close to the frequency tuning limit of the CO_2 laser; however, the crossover Lamb dip at the same frequency of the unperturbed line is very clear.

For the 170 μm line, the NLHE gain increase is completely counterbalanced by the large Stark shift, so that only a power decrease in the FIR

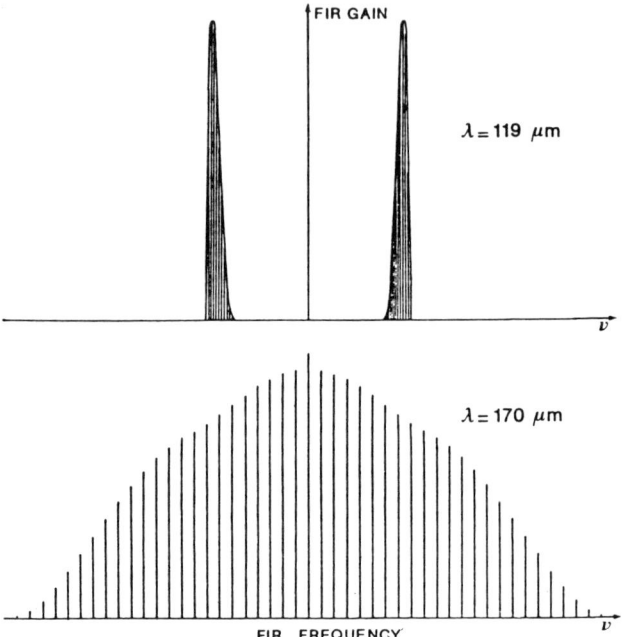

FIGURE 58. Theoretical small-signal gain Stark patterns of the 118 and 170 μm FIR laser lines of CH_3OH, assuming the pump selection rule $\Delta M = \pm 1$ and an electric field of 1 kV/cm. A large number of resolved components can be computed for the 170 μm line, while for the 118 μm line there is a near collapse of the Stark pattern into two components (from Ref. 32).

laser power output is observed when the electric field is increased. Nevertheless, the NLHE can be observed in Figure 60 as a smaller power decrease for the pump selection rule $\Delta M = \pm 1$.

The power enhancement for a $\Delta J = \Delta K = \pm 1$ transition of a FIR laser can be calculated from equations (104) and (105). These equations are formally equivalent to equation (94), where the factor F is now replaced by the NLHE absorption increase in the pump transition $R(S_{IR})$. This factor is about $\sqrt{2}$ times smaller than that due to the magnetoplasma effect in a noble-gas ion laser. However, this last effect reaches its maximum at a magnetic field intensity much larger than the intensity needed for the NLHE full gain increase (see Figures 48 and 40), thus introducing also a gain decrease due to a too large Zeeman splitting $\Delta \nu_F$. On the other hand, $R(S_{IR})$ and $R(S_{FIR})$ reach their asymptotic values at the same Stark field intensity.

In conclusion, the expected power enhancement is about the same as that observed for the noble-gas ion lasers (as shown in Figures 52 and 54), even if for a FIR laser it is only due to a pure NLHE.

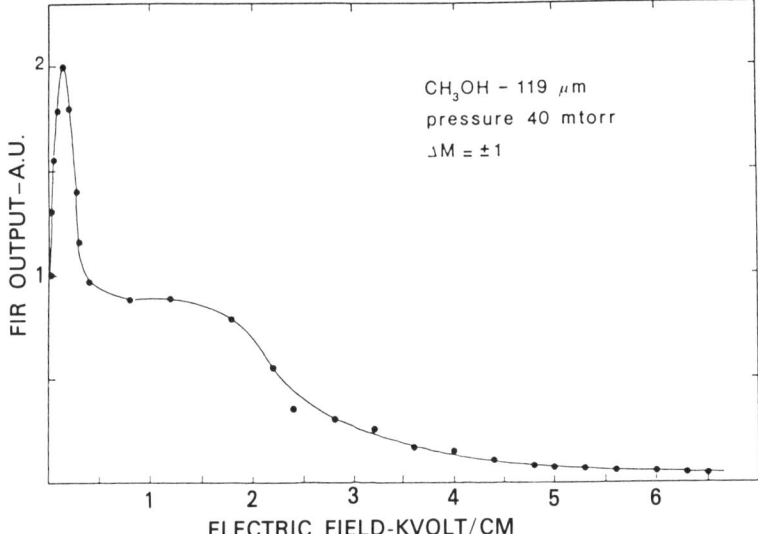

FIGURE 59. Output power of the CH_3OH 118 μm laser line vs. the electric field, for fields up to 6.5 kV/cm. The CO_2 pump power was kept constant throughout and the FIR cavity length was tuned continuously for the maximum output power (from Ref. 208).

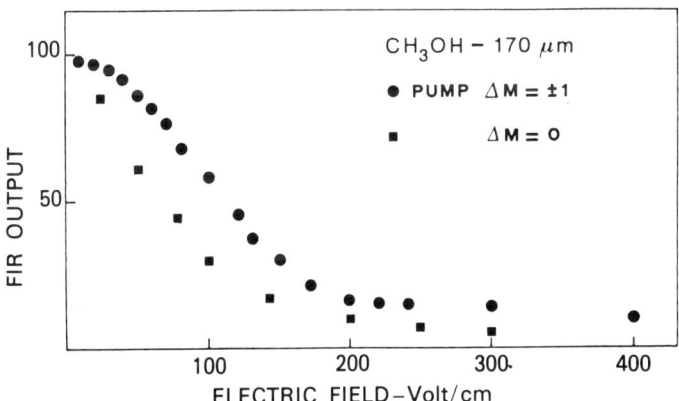

FIGURE 60. Relative output power for the CH_3OH 170 μm laser line vs. the electric field. No PE was observed, but the power was always higher for the pump selection rule $\Delta M = \pm 1$ (from Ref. 32).

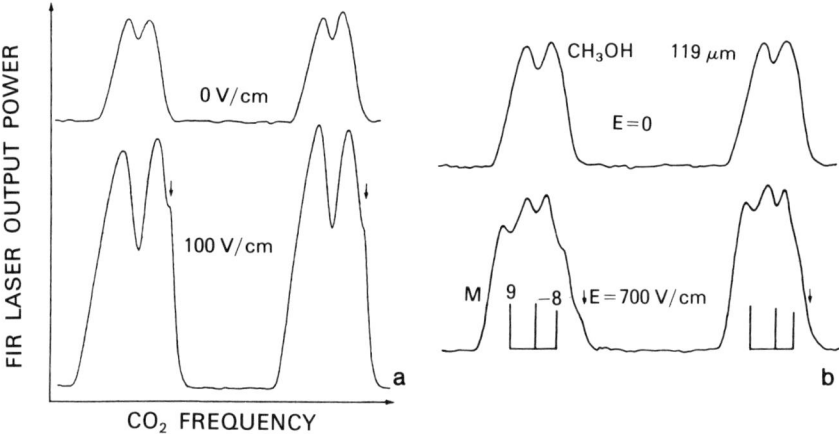

FIGURE 61. Transferred saturation dips observed in the emission of the 118 μm FIR laser line vs. the CO_2 frequency. Two orders are scanned and arrows indicate the position of the CO_2 mode change. Pump selection rule, $\Delta M = \pm 1$. The vertical scale is the same for the upper and bottom traces. The PE observed for a Stark field of 100 V/cm is monitored for any frequency setting of the CO_2 laser. Resolved saturation dips for the $\Delta M = +1$ and $\Delta M = -1$ transitions and the crossover dip are observed for a field of 700 V/cm (from Refs. 32 and 35).

9.6. Other Lasers

The effect of an axial magnetic field has been investigated also on other miscellaneous lasers. Feld et al.[3] observed a PE on the 844.6 nm oxygen laser line $(3p^3P \rightarrow 3p^3S)$ in CW operation. However, no PE was observed for the 5.54 μm Ca laser line $(^1P_1 - ^1D_2)$.[183] This result is explained by the large $\gamma/\Delta\nu_D$ ratio for the infrared transitions, as it was for the Xe laser.

A PE in the presence of an axial magnetic field was observed in the case of the pulsed self-terminated Cu laser for the 510.6 and 578.2 nm lines.[184] The laser tube contained Brewster windows and a contribution of the NLHE to the PE is possible, although the operating conditions, especially for the laser spectral output, were not given in Ref. 184.

9.7. Conclusions

We have seen in this last section that the output power of the gas lasers can be improved by the NLHE if the active medium is located in an external electric or magnetic field which removes the degeneracy of the M sublevels. A gain increase can be achieved if the polarization of the laser radiation and the static field are oriented so that $\Delta M = +1$ and $\Delta M = -1$ are simultaneously excited. On the contrary, no gain increase is observed for $\Delta M = 0$ transitions.

If the laser medium is excited by an electric discharge, the laser output power may be enhanced also by the magnetoplasma effect, which increases the electron density and the excitation rate by magnetic confinement. However, this effect is important only when the excitation current is far from saturation and the gas density is below a given threshold. The two effects, the NLHE and magnetoplasma effect, can be combined to give a larger power enhancement, up to a factor four for some lines of the commercial ion lasers. A rate-equation approach for the NLHE, and for the laser power enhancement, has been developed. This treatment is valid for any quantum number J involved in the transitions. All the available data on He-Ne, He-Cd, noble-gas ion, and FIR molecular lasers can be explained correctly by this model. An interesting result is that the magnetic field PE observed in noble-gas ion lasers can be explained only as a combination of both the nonlinear Hanle effect and magnetoplasma effect.

REFERENCES

1. A. JAVAN, *Bull. Am. Phys. Soc.* **9**, 489 (1964); R. H. CORDOVER, A. SZÖKE, AND A. JAVAN, *Bull. Am. Phys. Soc.* **9**, 490 (1964).
2. M. S. FELD, J. W. PARKS, H. R. SCHLOSSBERG, AND A. JAVAN, *Spectroscopy with Gas Lasers*, in *Physics of Quantum Electronics*, edited by P. L Kelley, B. Lax, and P. E. Tannenwald (McGraw-Hill, New York, 1966), pp. 567-580.
3. M. S. FELD, A. SANCHEZ, A. JAVAN, AND B. J. FELDMAN, Proc. Aussois Conf. Colloques Int. du CNRS, no. 217 (1974), pp. 87-104.
4. M. S. FELD, in *Fundamental and Applied Laser Physics*, edited by M. S. Feld, A. Javan, and N. A. Kermitt (Wiley, New York 1973), pp. 369-420.
5. R. L. FORK, L. E. HARGROVE, AND M. A. POLLACK, *Phys. Rev. Lett.* **12**, 705 (1964).
6. H. R. SCHLOSSBERG AND A. JAVAN, *Phys. Rev. Lett.* **17**, 1242 (1966).
7. T. F. JOHNSTON JR. AND G. J. WOLGA, *Phys. Lett.* **27A**, 639 (1968).
8. M. TSUKAKOSHI AND K. SHIODA, *J. Phys. Soc. Jpn.* **26**, 758 (1969).
9. W. FLYNN, M. S. FELD, AND B. J. FELDMAN, *Bull. Am. Phys. Soc.* **12**, 669 (1967).
10. S. LEVINE, P. A. BONCZYK, AND A. JAVAN, *Phys. Rev. Lett.* **22**, 267 (1969).
11. G. HERMANN AND A. SCHARMANN, *Z. Phys.* **254**, 46 (1972).
12. C. LUNTZ, R. G. BREWER, K. L. FOSTER, AND J. D. SWALEN, *Phys. Rev. Lett.* **23**, 951 (1969).
13. C. LUNTZ AND R. G. BREWER, *J. Chem. Phys.* **53**, 3380 (1970).
14. R. G. BREWER, in *Fundamental and Applied Laser Physics*, edited by M. S. Feld, A. Javan, and N. A. Kermitt (Wiley, New York, 1973), pp. 421-436.
15. H. R. SCHLOSSBERG AND A. JAVAN, *Phys. Rev.* **150**, 267 (1966).
16. M. S. FELD AND A. JAVAN, *Phys. Rev.* **177**, 540 (1969).
17. B. J. FELDMAN AND M. S. FELD, *Phys. Rev. A* **5**, 899 (1971).
18. K. SHIMODA, *Jpn. J. Appl. Phys.* **11**, 564 (1972).
19. M. DUCLOY, *Opt. Commun.* **3**, 205 (1971).
20. B. DECOMPS, M. DUMONT, AND M. DUCLOY, in *Laser Spectroscopy of Atoms and Molecules*, edited by H. Walther (Springer-Verlag, Berlin, 1976), pp. 283-347.

21. M. Ducloy, *Phys. Rev. A* **8**, 1844 (1973); *Phys. Rev. A* **9**, 1319 (1974).
22. S. M. Freund, J. W. C. Johns, A. R. W. McKellar, and T. Oka, *J. Chem. Phys.* **59**, 3445 (1973).
23. B. J. Orr and T. Oka, *Appl. Phys. Lett.* **30**, 468 (1977).
24. B. J. Orr and T. Oka, *J. Mol. Spectrosc.* **66**, 302 (1977).
25. K. Shimoda, in *Laser Spectroscopy of Atoms and Molecules*, edited by H. Walther (Springer-Verlag, Berlin, 1976), pp. 197-252.
26. R. M. Percival, D. Devoy, G. Duxbury, and H. Kato, *J. Opt. Soc. Am.* **4B**, 1188 (1987) and references cited therein.
27. G. Duxbury and J. McCombie, *J. Opt. Soc. Am.* **4B**, 1197 (1987).
28. R. P. Hackel and Sh. Ezekiel, *Phys. Rev. Lett.* **42**, 1736 (1979).
29. P. R. Hemmer, Sh. Ezekiel, and C. C. Leiby Jr., *Opt. Lett.* **8**, 440 (1983).
30. P. R. Hemmer, G. P. Outai, and Sh. Ezekiel, *J. Opt. Soc. Am.* **3B**, 219 (1986).
31. K. Ernst and M. Inguscio, *Riv. Nuovo Cimento* **11**, no. 2, 1 (1988).
32. M. Inguscio, A. Moretti, and F. Strumia, *IEEE J. Quantum Electron.* **QE-16**, 955 (1980).
33. F. Strumia, M. Inguscio, and A. Moretti, in *Laser Spectroscopy V*, edited by A. R. McKellar, T. Oka, and B. P. Stoicheff (Springer-Verlag, Berlin, Heidelberg, 1981), pp. 255-259.
34. M. Inguscio, A. Moretti, and F. Strumia, *Appl. Phys. B* **28**, 89 (1982).
35. M. Inguscio, A. Moretti, and F. Strumia, *Opt. Commun.* **30**, 355 (1979).
36. N. Beverini and M. Inguscio, *Lett. Nuovo Cimento* **29**, 10 (1980).
37. P. Hannaford and G. W. Series, *J. Phys. B* **14**, L661 (1981).
38. B. Barbieri, N. Beverini, G. Bionducci, M. Galli, M. Inguscio, and F. Strumia, *Proc. 3rd Natl Conf. Quantum Electron. and Plasma*, edited by V. De Giorgio (Como, 1982), pp. 337-340.
39. P. Hannaford and G. W. Series, in *Laser Spectroscopy V*, edited by A. R. Mckellar, T. Oka, and B. P. Stoicheff (Springer-Verlag, Berlin, Heidelberg, 1981), p. 94.
40. N. Beverini, K. Ernst, M. Inguscio, and F. Strumia, *Appl. Phys. B* **37**, 17 (1985).
41. N. Sokabe, Y. Tamura, K. Matsushima, and A. Murai, *Int. J. IR and MM Waves* **8**, 1145 (1987).
42. P. Hannaford and G. W. Series, *Phys. Rev. Lett.* **48**, 1326 (1982).
43. R. G. Buser, J. Kainz, and J. Sullivan, *Appl. Opt.* **2**, 86 (1963).
44. M. I. Dyakonov and S. A. Fridrikhov, *Sov. Phys. Usp.* **9**, 837 (1967).
45. E. F. Labuda, E. I. Gordon, and R. C. Miller, *IEEE J. Quantum Electron.* **QE-1**, 273 (1975).
46. M. Sargent III, M. O. Scully, and W. E. Lamb Jr., *Laser Physics* (Addison-Wesley, Reading, Mass., 1974).
47. H. Haken, *Laser Theory*, in *Encyclopedia of Physics*, Vol. XXV/2C (Springer-Verlag, Berlin, 1970).
48. C. C. Davis and T. A. King, in *Advances in Quantum Electronics*, Vol. 3, edited by D. W. Goodwin (Academic Press, New York, 1975), pp. 169-454.
49. M. H. Dunn and J. N. Ross, in *Progress in Quantum Electronics*, Vol. 4, edited by J. H. Sandars and S. Stenholm (Pergamon Press, Oxford, New York, 1977), pp. 233-264.
50. F. Strumia, *J. Phys. (Paris), Colloq.* **44**, C7 (1983).
51. M. S. Tobin and R. E. Jensen, *Appl. Opt.* **15**, 2023 (1976).
52. M. S. Tobin and R. E. Jensen, *IEEE J. Quantum Electron.* **QE-13**, 481 (1977).
53. M. Inguscio, P. Minguzzi, A. Moretti, F. Strumia, and M. Tonelli, *Appl. Phys.* **18**, 261 (1979).
54. C. Cohen-Tannoudji, in *Atomic Physics* 4 (Les Houches, 1985), p. 589.
55. A. Corney, *Atomic and Laser Spectroscopy* (Clarendon Press, Oxford, 1977).
56. M. Ducloy, M. P. Gorza, and B. Decomps, *Opt. Commun.* **8**, 21 (1973).

57. M. Ducloy, *J. Phys. B* **9**, 357 (1976).
58. J. C. Lehmann, in *Frontiers in Laser Spectroscopy*, edited by R. Balian, S. Haroche, and S. Liberman (North-Holland, Amsterdam, 1977), pp. 475–526.
59. M. Broyer, G. Guedard, J. C. Lehmann, and J. Vigué, *Adv. At. Mol. Phys.* **12**, 165 (1976).
60. M. Broyer, F. W. Dalby, J. Vigué, and J. C. Lehmann, *Can. J. Phys.* **51**, 226 (1973).
61. F. Strumia and M. Inguscio, in *Infrared and Millimeter Waves*, edited by K. J. Button, Vol. 5 (Academic Press, New York, 1982), pp. 129–213 and reference cited therein.
62. M. Inguscio, F. Strumia, K. Evenson, D. A. Jennings, A. Scalabrin, and S. Stein, *Opt. Lett.* **4**, 9 (1979).
63. P. Hannaford, D. S. Gough, and G. W. Series, *J. Phys. (Paris)* **44**, C7 (1983).
64. P. Hannaford, D. S. Gough, and G. W. Series, in *Laser Spectroscopy VI*, edited by H. P. Weber and W. Lüthy (Springer-Verlag, Berlin, 1983), p. 95.
65. K. Ernst, M. Grabinska, and M. Inguscio, *Opt. Commun.* **73**, 43 (1989).
66. J. L. Piqué, *J. Phys. B* **11**, L59 (1978).
67. J. A. O'Neill and G. Smith, *Astronom. Astrophys.* **81**, 100 (1978).
68. R. J. McLean, D. S. Gough, and P. Hannaford, in *Laser Spectroscopy VII*, edited by T. W. Hänsch and Y. R. Shen (Springer-Verlag, Berlin, 1985), p. 220.
69. G. Bertuccelli, N. Beverini, M. Galli, M. Inguscio, F. Strumia, and G. Giusfredi, *Opt. Lett.* **10**, 270 (1985).
70. G. Bertuccelli, N. Beverini, M. Galli, M. Inguscio, and F. Strumia, *Opt. Lett.* **11**, 351 (1986).
71. N. Beverini, M. Galli, M. Inguscio, and F. Strumia, in *Laser Spectroscopy VII*, edited by T. W. Hänsch and Y. R. Shen (Springer-Verlag, Berlin, 1985), p. 285.
72. J. E. Thomas, P. R. Hemmer, Sh. Ezekiel, C. C. Leiby Jr., R. H. Picard, and C. R. Willis, *Phys. Rev. Lett.* **48**, 867 (1982).
73. J. Mlynek, R. Grimm, E. Buhr, and V. Jordan, *Appl. Phys.* **B45**, 77 (1988).
74. K. Arnett, R. Ryan, T. Bergeman, H. J. Metcalf, M. Hamilton, J. Brandenberger, and S. Smith, *CLEO 1989 Conference Digest*, paper TU9910; K. Arnett, S. J. Smith, R. E. Ryan, T. Bergeman, H. Metcalf, M. W. Hamilton, and J. R. Brandenberger, *Phys. Rev. A* **41**, 2580 (1990).
75. W. Gawlik, J. Kowalski, R. Neumann, and F. Träger, *Phys. Lett.* **48A**, 283 (1974).
76. S. Giraud-Cotton, V. P. Kaftandjan, and L. Klein, *Phys. Lett.* **88A**, 453 (1982).
77. S. Giraud-Cotton, V. P. Kaftandjan, and L. Klein, *Phys. Rev. A* **32**, 2223 (1985).
78. W. Lange, K. H. Drake, and J. Mlynek, in *Laser Spectroscopy VIII*, edited by W. Persson and S. Svanberg (Springer-Verlag, Berlin, 1987), p. 300.
79. P. Jungner, B. Ståhlberg, T. Fellmann, and M. Lindberg, in *Laser Spectroscopy IX*, edited by M. S. Feld (Academic Press, New York, 1989), p. 92.
80. P. Jungner, T. Fellmann, B. Ståhlberg, and M. Lindberg, *Opt. Commun.* **73**, 38 (1989).
81. P. Jungner, B. Ståhlberg, and M. Lindberg, *Phys. Scr.* **38**, 550 (1988).
82. B. Barbieri, N. Beverini, G. Bionducci, M. Galli, M. Inguscio, and F. Strumia, in *Laser Spectroscopy VI*, edited by H. P. Weber and W. Lüthi (Springer-Verlag, Berlin, 1983), pp. 133–135.
83. F. Strumia, in *Coherence in Spectroscopy and Modern Physics*, edited by F. T. Arecchi, R. Bonifacio, and M. O. Scully (Plenum Press, New York, 1978), p. 381.
84. P. Hannaford and G. W. Series, *Opt. Commun.* **41**, 427 (1982).
85. R. J. McLean, R. J. Ballagh, and D. M. Warrington, *J. Phys. B* **18**, 2371 (1985).
86. R. J. McLean, R. J. Ballagh, and D. M. Warrington, *J. Phys. B* **19**, 3477 (1986).
87. P. Minguzzi, S. Profeti, M. Tonelli, and A. Di Lieto, *J. Opt. Soc. Am. B* **3**, 1075 (1986).
88. B. Ståhlberg and K. Weckstrom, *Phys. Scr.* **22**, 483 (1970).

89. B. Ståhlberg, M. Lindberg, and P. Junger, J. Phys. B **18**, 627 (1985).
90. G. Hermann and A. Scharmann, Phys. Lett. **24A**, 606 (1967).
91. G. Hermann and A. Scharmann, Z. Phys. **208**, 367 (1968).
92. G. Hermann and A. Scharmann, Phys. Lett. **40A**, 293 (1972).
93. F. Graubner, G. Hermann, and A. Scharmann, Z. Phys. **269**, 79 (1974).
94. G. Hermann, G. Lasnitschka, and A. Scharmann, Phys. Lett. **61A**, 99 (1977).
95. G. Hermann, G. Lasnitschka, and A. Scharmann, Z. Phys. A **282**, 253 (1977).
96. F. Graubner and G. Hermann, Z. Phys. A **289**, 21 (1978).
97. F. Graubner and G. Hermann, Z. Phys. A **289**, 31 (1978).
98. G. Hermann, K. H. Abt, G. Lasnitschka, and A. Scharmann, Phys. Lett. **69A**, 103 (1978).
99. G. Hermann, K. H. Abt, and G. Lasnitschka, Z. Phys. A **288**, 113 (1978).
100. K. H. Abt, G. Hermann, G. Lasnitschka, A. Scharmann, and M. Schleicher, Phys. Lett. **74A**, 55 (1979).
101. R. L. Fork and C. K. N. Patel, Appl. Phys. Lett. **2**, 180 (1963).
102. I. Gorog and F. W. Spong, IEEE J. Quantum Electron. **QE3**, 691 (1967).
103. M. Abramowitz and I. A. Stegun, Handbook of Mathematical Functions (N.B.S., Washington, D.C., 1964), pp. 297-330.
104. P. Zory, IEEE J. Quantum Electron. **QE-3**, 390 (1967).
105. P. W. Smith, J. Appl. Phys. **37**, 2089 (1966).
106. P. W. Smith, IEEE J. Quantum Electron. **QE-2**, 62 (1966).
107. P. W. Smith, IEEE J. Quantum Electron. **QE-2**, 77 (1966).
108. R. L. Field, Rev. Sci. Instrum. **38**, 1720 (1967).
109. E. G. Saprikin, R. N. Yudin, and S. N. Atutov, Opt. Spectrosc. (USSR) **34**, 435 (1973).
110. W. Culshaw and J. Kannelaud, Phys. Rev. A **133**, 691 (1964).
111. W. Culshaw and J. Kannelaud, in Quantum Electronics, edited by P. Grivet and N. B. Bloembergen (Dumont and Columbia Press, 1964), p. 523.
112. A. E. Fodiati and S. A. Fridrikhov, Sov. Phys. Tech. Phys. **11**, 416 (1966).
113. D. K. Terekhin and S. A. Fridrikhov, Sov. Phys. Tech. Phys. **11**, 288 (1966).
114. A. N. Alekseeva and D. V. Gordeev, Opt. Spectrosc. (USSR) **23**, 520 (1967).
115. W. E. Bell and A. L. Bloom, Appl. Opt. **3**, 413 (1964).
116. G. Durand, IEEE J. Quantum Electron. **QE-2**, 448 (1968).
117. S. A. Ahmed, R. C. Kocher, and H. J. Gerritsen, Proc. IEEE **52**, 1356 (1964).
118. St. St. Cartaleva, S. V. Gateva, and G. V. Kolarov, Appl. Phys. B **40**, 153 (1986).
119. W. B. Bridges, in Methods of Experimental Physics, Vol. 15A, edited by C. L. Tang (Academic Press, New York, 1979), p. 31.
120. R. L. Fork and C. K. N. Patel, Phys. Rev. **129**, 2577 (1963).
121. J. Kannelaud and W. Culshaw, Appl. Phys. Lett. **9**, 120 (1966).
122. K. G. Hernquist, J. Appl. Phys. **40**, 5399 (1969).
123. T. Hara, M. Hamagaki, K. Matsunaga, and T. Dote, Phys. Lett. **72A**, 349 (1970).
124. A. Dienes and T. P. Sosnowski, Appl. Phys. Lett. **16**, 512 (1970).
125. A. Taszner, A. Kowalski, and J. Heldt, Appl. Phys. **11**, 203 (1976).
126. C. P. Cristescu, A. I. Lupescu, I. M. Popescu, and A. M. Preda, Can. J. Phys. **56**, 1071 (1978).
127. W. Sasaki, H. Ueda, and T. Ohta, IEEE J. Quantum Electron. **QE-19**, 1259 (1983).
128. D. C. Sinclair, J. Opt. Soc. Am. **56**, 1727 (1966).
129. R. T. Menzies, A. Dienes, and N. George, IEEE J. Quantum Electron. **QE-6**, 117 (1970).
130. V. F. Kitaeva, A. N. Odintsov, and N. N. Sobolev, Sov. Phys. Usp. **12**, 699 (1970).
131. M. Sargent III, W. E. Lamb Jr., and R. L. Fork, Phys. Rev. **164**, 436 (1967).
132. W. B. Bridges and A. N. Chester, in CRC Handbook of Lasers, edited by R. J. Pressley (CRC Press, Boca Raton, Florida, 1971), pp. 242-297.

133. W. B. BRIDGES AND G. N. MERCER, *IEEE J. Quantum Electron.* **QE-5**, 476 (1969).
134. R. C. SZE, Y. T. ANTROPOV, AND W. BENNET JR., *Appl. Opt.* **11**, 197 (1972).
135. R. C. SZE AND W. BENNET JR., *Phys. Rev. A* **5**, 837 (1972).
136. T. J. BRIDGES AND W. W. RIGROD, *IEEE J. Quantum Electron.* **QE-1**, 303 (1965).
137. J. M. FORSYTH, *Appl. Phys. Lett.* **11**, 391 (1967).
138. J. M. FORSYTH, *J. Appl. Phys.* **40**, 3049 (1969).
139. V. M. GRIMBLATOV, E. P. OSTAPCHENKO, AND V. V. V. TESELKIN, *Sov. J. Quantum Electron.* **1**, 378 (1972).
140. M. BASS, G. DE MARS, AND H. STATZ, *APpl. Phys. Lett.* **12**, 17 (1968).
141. V. A. SEDELNIKOV, YU. P. SINICHKIN, AND V. V. TUCHIN, *Opt. Spectrosc. (USSR)* **31**, 408 (1971).
142. K. BERNDT AND E. KLOSE, *Opt. Commun.* **35**, 417 (1980).
143. J. M. YARBOROUGH AND J. L. HOBERT, *Appl. Phys. Lett.* **13**, 305 (1968).
144. J. M. RAMSEY, *J. Appl. Phys.* **53**, 1381 (1982).
145. V. V. LEBEDEVA, A. I. ODINTSOV, AND V. M. SALIMOV, *Sov. Phys. Tech. Phys.* **13**, 1122 (1969).
146. V. I. BUTKEVICH, V. E. PRIVALOV, AND G. V. SKVORTSOVA, *Opt. Spectrosc. (USSR)* **61**, 691 (1986).
147. M. ARMAND, *Ann. Radioelectr.* **22**, 191 (1967).
148. V. E. GOLANT, M. V. KRIVOSHEEV, AND V. E. PRIVALOV, *Sov. Phys. Tech. Phys.* **9**, 737 (1964).
149. V. F. KITAEVA, L. YA. OSTROVSKAYA, AND N. N. SOBOLEV, *Sov. J. Quantum Electron.* **1**, 341 (1972).
150. M. S. BORISOVA, YE. F. ISHCHENKO, M. V. LADYGIN, M. A. MOLCHASKIN, YE. F. NASEDKIN, AND G. S. RAMAZANOVA, *Radio Eng. Electron. Phys.* **4**, 526 (1967).
151. W. B. BRIDGES, A. N. CHESTER, A. S. HALSTED, AND J. V. PARKER, *Proc. IEEE* **59**, 724 (1971).
152. W. B. BRIDGES AND A. S. HALSTED, *Gaseous Ion Laser Research*, Final Tech. Rep., AFAL-TR-67-89 (Hughes Research Laboratories, Malibu, CA, 1967), unpublished.
153. E. I. GORDON, E. F. LABUDA, R. C. MILLER, AND C. E. WEBB, in *Physics of Quantum Electronics*, edited by P. L. Kelley, B. Lax, and P. E. Tannenwald (McGraw-Hill, New York, 1966), pp. 664–673.
154. A. E. FODIATI AND S. A. FRIDRIKHOV, *Sov. Phys. Tech. Phys.* **14**, 1292 (1970).
155. I. GOROG AND F. W. SPONG, *Appl. Phys. Lett.* **9**, 61 (1966).
156. J. P. GOLDSBOROUGH, E. B. HODGES, AND W. E. BELL, *Appl. Phys. Lett.* **8**, 137 (1966).
157. I. D. KONKOV, R. E. ROVINSKY, A. G. ROZANOV, AND N. V. CHEBURKIN, *Radio Eng. Electron. Phys. (USSR)* **13**, 2008 (1968).
158. S. A. AHMED, A. CAMPILLO, AND R. P. CODY, *IEEE J. Quantum Electron.* **QE-5**, 267 (1969).
159. I. GOROG AND F. W. SPONG, *RCA Rev.* **28**, 38 (1967).
160. I. GOROG AND F. W. SPONG, *RCA Rev.* **30**, 277 (1969).
161. W. B. BRIDGES AND A. S. HALSTED, *IEEE J. Quantum Electron.* **QE-2**, 84 (1966).
162. R. PAANANEN, *Appl. Phys. Lett.* **9**, 34 (1966).
163. J. R. FENDLEY JR., *IEEE J. Quantum Electron.* **QE-4**, 355 (1968).
164. I. D. LATIMER, *Appl. Phys. Lett.* **13**, 333 (1968).
165. E. T. LEONARD, M. A. YOFFEE, AND K. W. BILLMAN, *Appl. Opt.* **9**, 1209 (1970).
166. TAE-SOO KIM AND UNG KIM, *J. Korean Phys. Soc.* **18**, 206 (1985).
167. A. LE FLOCH AND P. FRÈRE, *Rev. Phys. Appl.* **7**, 409 (1972).
168. K. TOYODA AND C. YAMANAKA, *Tech. Rep. Osaka University (Japan)* **17**, 407 (1967).
169. S. M. GARRETT AND G. C. BARKER, *J. Appl. Phys.* **39**, 4845 (1968).
170. M. BIRNBAUM, *Appl. Phys. Lett.* **12**, 86 (1968).
171. H. S. AMES, *IEEE J. Quantum Electron.* **QE-8**, 808 (1972).

172. S. A. AHMED, T. J. FAITH, AND G. W. HOFFMANN, *Proc. IEEE* **55**, 691 (1967).
173. J. R. FENDLEY, *IEEE J. Quantum Electron.* **QE-4**, 627 (1968).
174. G. HERZIGER AND W. SEELIG, *Z. Phys.* **215**, 437 (1968).
175. G. HERZIGER AND W. SEELIG, *Z. Phys.* **219**, 5 (1969).
176. H. BOERSCH, J. BOSCHER, D. HODER, AND G. SCHÄFER, *Phys. Lett.* **31A**, 188 (1970).
177. K. BANSE, H. R. LÜTHI, AND W. H. SEELIG, *Appl. Phys.* **4**, 141 (1974).
178. H. R. LÜTHI, W. SEELIG, J. STEINGER, AND W. LOBSIGER, *IEEE J. Quantum Electron.* **QE-13**, 404 (1977).
179. H. R. LÜTHI, W. SEELIG, AND J. STEINGER, *Appl. Phys. Lett.* **31**, 670 (1977).
180. H. R. LÜTHI AND J. STEINGER, *Opt. Commun.* **27**, 435 (1978).
181. S. KOBAYASHI, T. KAMIYA, T. HAYASHI, AND T. GOTO, *Appl. Phys. Lett.* **46**, 925 (1985).
182. S. KOBAYASHI, T. KAMIYA, AND K. UJIHARA, *IEEE J. Quantum Electron.* **QE-23**, 633 (1987).
183. V. M. KLIMKIN AND P. D. KOLBYCHEVA, *Sov. J. Quantum Electron.* **7**, 1037 (1977).
184. L. A. CROSS, R. S. JENKINS, AND M. CEM GOKAY, *J. Appl. Phys.* **48**, 453 (1978).
185. W. R. BENNET JR. AND R. C. SZE, *IEEE J. Quantum Electron.* **QE-10**, 908 (1974).
186. M. V. LADYGIN AND B. A. TSARKOV, *Opt. Spectrosc. (USSR)* **30**, 69 (1971).
187. D. K. TEREKHIN, S. A. FRIDRIKHOV, AND C. G. ANTONOV, *Opt. Spectrosc. (USSR)* **26**, 358 (1969).
188. V. G. GUDELEV AND V. M. YASINSKII, *Sov. J. Quantum Electron.* **12**, 904 (1982).
189. D. L. LENSTRA, *Appl. Phys.* **17**, 257 (1978).
190. K. KURODA, Y. KAWASA, AND I. OGURA, *J. Quant. Spectrosc. Radiat. Transfer* **31**, 259 (1984).
191. S. WATANABE, M. CHIHARA, AND I. OGURA, *Jpn. J. Appl. Phys.* **13**, 164 (1974).
192. P. BRUN A. LE FLOCH, AND J. P. TACHE, *C. R. Acad. Sci. Paris, Ser. B* **265**, 993 (1967).
193. P. BRUN, A. LE FLOCH, AND J. P. TACHE, *C. R. Acad. Sci. Paris, Ser. B* **266**, 189 (1968).
194. A. LE FLOCH AND P. BRUN, *C. R. Acad. Sci. Paris, Ser. B* **269**, 23 (1969).
195. A. LE FLOCH AND P. BRUN, *Appl. Phys. Lett.* **17**, 40 (1970).
196. A. LE FLOCH, R. LE NAOUR, G. STEPHAN, AND P. BRUN, *Appl. Opt.* **15**, 2673 (1976).
197. M. I. DYAKONOV AND V. L. PEREL, *Sov. Phys. JETP* **23**, 298 (1967).
198. J. R. FENDLEY JR., I. GOROG, K. G. HERNQVIST, AND C. SUN, *RCA Rev.* **30**, 422 (1969).
199. D. C. BROWN AND T. M. SWIFT, *IEEE J. Quantum Electron.* **QE-10**, 94 (1974).
200. K. DEKI, Y. TAKENAKA, S. MATSUURA, AND T. OHTA, *Jpn. J. Appl. Phys.* **17**, 1593 (1978).
201. M. QIU AND Z. ZHOU, *Opt. Commun.* **43**, 207 (1982).
202. D. L. PERRY, *IEEE J. Quantum Electron.* **QE-7**, 102 (1971).
203. F. STRUMIA, in *Advances in Laser Spectroscopy*, edited by F. T. Arecchi, F. Strumia, and H. Walther (Plenum Press, New York, 1983), p. 267.
204. T. A. DE TEMPLE AND E. J. DANIELEWICZ, in *Infrared and Millimeter Waves*, Vol. 7, edited by K. J. Button (Academic Press, New York, 1983), p. 1.
205. F. STRUMIA, N. IOLI, AND A. MORETTI, in *Physics of New Laser Sources*, edited by N. B. Abraham, F. T. Arecchi, A. Moorodian, and A. Sona (Plenum Press, New York, 1985), p. 217.
206. M. S. TOBIN, *Proc. IEEE* **73**, 61 (1985).
207. K. J. BUTTON, M. INGUSCIO, AND F. STRUMIA (eds.), *Optically Pumped FIR Lasers* (Plenum Press, New York, 1984).
208. G. BIONDUCCI, M. INGUSCIO, A. MORETTI, AND F. STRUMIA, *Infrared Phys.* **19**, 297 (1979).
209. K. P. KOO AND P. C. CLASPY, *Appl. Opt.* **18**, 1314 (1979).
210. J. O. HENNINGSEN, *J. Mol. Spectrosc.* **83**, 70 (1980).
211. M. INGUSCIO, N. IOLI, A. MORETTI, G. MORUZZI, AND F. STRUMIA, *Opt. Commun.* **37**, 211 (1981).

212. M. INGUSCIO, A. MORETTI, G. MORUZZI, AND F. STRUMIA, *Int. J. Infrared Millimeter Waves* **2**, 943 (1981).
213. S. SOKABE, K. HORIKAWA, N. ZUMOTO, AND A. MURAI, *Int. J. Infrared Millimeter Waves* **6**, 893 (1985).
214. P. JANSSEN AND H. VANDERSTRAETEN, *Int. J. Infrared Millimeter Waves* **8**, 415 (1987).
215. N. IOLI, A. MORETTI, AND D. PEREIRA, in *Interaction of Radiation with Matter*, volume in honor of A. Gozzini (Scuola Normale Superiore, Pisa, 1987), p. 387.
216. G. CARELLI, N. IOLI, A. MORETTI, D. PEREIRA, AND F. STRUMIA, *Appl. Phys. B* **44**, 111 (1987).
217. F. K. KNEUBÜHL, *J. Opt. Soc. Am.* **677**, 959 (1977).
218. M. INGUSCIO, F. STRUMIA, AND J. O. HENNINGSEN, in *Optically Pumped FIR Lasers*, edited by K. J. Button, M. Inguscio, and F. Strumia (Plenum Press, New York, 1984), p. 105.
219. K. WALZER, in *Infrared and Millimeter Waves*, Vol. 7, edited by K. J. Button (Academic Press, New York, 1983), p. 119.
220. M. INGUSCIO, F. STRUMIA, K. M. EVENSON, D. A JENNINGS, A. SCALABRIN, AND S. R. STEIN, *Opt. Lett.* **4**, 9 (1979).
221. N. IOLI, A. MORETTI, G. MORUZZI, P. ROSSELLI, AND F. STRUMIA, *J. Mol. Spectrosc.* **105**, 284 (1984).
222. N. IOLI, A. MORETTI, G. MORUZZI, F. STRUMIA, AND F. D'AMATO, *Int. J. Infrared Millimeter Waves* **6**, 1017 (1985).
223. N. IOLI, A. MORETTI, F. STRUMIA, AND F. D'AMATO, *Int. J. Infrared Millimeter Waves* **7**, 459 (1986).
224. N. IOLI, A. MORETTI, AND F. STRUMIA, *Appl. Phys. B* **48**, 305 (1989).
225. N. SOKABE, A. HIGUCHI, K. IMURA, Y. YASUDA, AND A. MURAI, *Opt. Commun.* **34**, 255 (1980).
226. T. YOSHIDA, M. KOBAYASHI, T. YISHIHARA, K. SAKAI, AND S. FUJITA, *Opt. Commun.* **40**, 45 (1981).
227. J. O. HENNINGSEN, *J. Mol. Spectrosc.* **91**, 430 (1982).
228. M. KAWAMURA AND Y. KOKUBO, *IEEE J. Quantum Electron.* **QE-18**, 903 (1982).
229. N. SOKABE, T. MIYATAKA, Y. NISHI, AND A. MURAI, *J. Phys. B* **16**, 4487 (1983).
230. G. DUXBURY, J. C. PETERSEN, H. KATO, AND M. L. LE LERRE, *J. Mol. Spectrosc.* **107**, 261 (1984).
231. D. J. BEDWELL, G. DUXBURY, H. HERMAN, AND C. A. ORENGO, *Infrared Phys.* **18**, 453 (1978).
232. W. B. BRIDGES, in *CRC Handbook of Laser Science*, Vol. II (CRC Press, Boca Raton, Florida, 1982), p. 171.
233. D. J. E. KNIGHT, in *CRC handbook of Laser Science*, Vol. II (CRC Press, Boca Raton, Florida, 1981), p. 421.
234. V. S. LETOKHOV AND V. P. ZAROV, *Laser Optoacoustic Spectroscopy*, Springer-Verlag, Berlin-Heidelberg, 1986.
235. B. BARBIERI, N. BEVERINI AND A. SASSO, *Rev. Mod. Phys.*, 1990 (in press).

CHAPTER 5

APPLICATIONS OF THE HANLE EFFECT IN SOLAR PHYSICS

JAN OLOF STENFLO

1. INTRODUCTION

In laboratory applications of the Hanle effect the observed polarization phenomena in a known magnetic field are used to obtain information on the atomic structure. In astrophysical applications the problem is generally turned around. It is the magnetic field that is the unknown quantity, to be derived from the observed polarization of the radiation from the celestial object, using our knowledge about the atomic physics involved.

Magnetic fields play a fundamental role everywhere in cosmos, e.g., in active galactic nuclei, for neutron star radiation, for star formation, or for the physics of our nearest star, the sun. The solar atmosphere provides us with a giant physics laboratory, in which the sun itself performs the experiments, and we have to remotely diagnose what is going on. This laboratory is unique, since most of the plasma phenomena cannot be simulated in earth laboratories due to the scaling laws for field strengths and electric currents when the huge spatial dimensions are scaled down. Other stars, on which similar processes occur, cannot be explored in such detail as the sun, since the closest ones are about a million times more distant than the sun.

Magnetic fields were first discovered on the sun in 1908, when Hale[1] found that the absorption lines in the solar spectrum were split and polarized in sunspots, in accordance with the Zeeman effect if the sunspots were seats of strong magnetic fields. Since then the Zeeman effect has been systematically exploited to obtain increasingly detailed information on the structure

JAN OLOF STENFLO • Institute of Astronomy, ETH-Zentrum, CH-8092 Zürich, Switzerland.

and evolution of the magnetic field in the solar photosphere, the layer of the solar atmosphere where the visible spectrum originates.

Although the Hanle effect was discovered by Hanle already in 1923,[2] its potential for magnetic-field diagnostics on the sun has been realized fairly recently, which has led to a large number of theoretical and observational papers on the subject during the last decade. As will be explained in more detail below, the Hanle effect responds to magnetic fields in a different parameter regime as compared with the Zeeman effect, and therefore provides an important extension of the diagnostic possibilities for solving some fundamental problems in solar physics, which cannot be adequately dealt with by the Zeeman effect alone. Various applications of the Hanle effect in astrophysics have been reviewed by Leroy[3] and Kazantsev.[4]

2. BRIEF REVIEW OF THE PROPERTIES OF SOLAR MAGNETIC FIELDS

Solar activity, like sunspots, flares, or the 11 yr activity cycle, is the result of an interaction between magnetic fields, turbulent motions in the sun's convection zone, and the rotation of the sun. The overall theoretical explanation for this is generally sought within the framework of the so-called dynamo theories, which try to explain the existence of large-scale magnetic fields in cosmos, like galactic magnetic fields or the earth's magnetic field, through processes in turbulent, rotating media.[5-8] The role of rotation is to break the symmetry between left and right in the turbulent motions via the Coriolis force, while convection induces differential rotation (gradients in the angular velocity). The sun provides a testing ground for such dynamo theories. It represents the special case of an oscillating dynamo, with a period of 22 yr (since the magnetic polarities reverse every 11 yr, the full magnetic cycle is 2×11 yr).

The circular polarization in the wings of spectral lines sensitive to the Zeeman effect can be scanned across the solar disk using a solar optical telescope in combination with a spectrograph, electrooptical polarization modulation, and photoelectric detection.[9,10] The circular polarization is to a first approximation proportional to the line-of-sight component of the magnetic flux through the spatial resolution element on the solar disk, in the atmospheric layers where the chosen spectral line is formed (longitudinal Zeeman effect). Such a map of the circular polarization across the solar disk is therefore normally called a magnetogram.

Full-disk magnetograms have been recorded daily at the Mount Wilson Observatory (where Hale made his first discovery of magnetic fields on the sun) since 1959, and at the Kitt Peak National Observatory since 1976. In addition, numerous studies of the small-scale structure of the field have been conducted, in an on-going quest for ever increasing spatial resolution. In solar physics like in high-energy physics, it appears that the fundamental physical processes occur at the smallest scales.

The observations show that the sun's magnetic field is extremely intermittent.[11-14] More than 90% of the total magnetic flux that we see as circular polarization in magnetograms is in fact due to strong fields with field strengths of 1–2 kG, in spite of the average, spatially smeared flux density being typically only a few G. The strong-field flux is confined to discrete elements with a magnetic filling factor (fractional photospheric volume occupied by the strong field) of generally a fraction of one percent only. Most of the basic flux elements or "fluxtubes" are smaller than can be resolved by existing telescopes. Nevertheless it has been possible to overcome the limitations in resolution and deduce the actual field strengths in the unresolved structures by making diagnostic use of the nonlinear relation between flux density and circular polarization amplitude for intermediately strong fields. Since this effect is different in different, simultaneously recorded spectral lines, it was possible to separate the flux effects from field-strength effects and deduce the kG nature of the small-scale fields.[11]

New magnetic flux emerges all the time all over the sun, but the largest flux concentrations occur in the activity zones in the form of sunspot groups. Sunspots represent large, resolved fluxtubes, which are darker than the surroundings since the magnetic field inhibits the convective energy transport, and the fluxtube interior is thermally insulated from the surroundings by being optically thick for radiative exchange with the hot fluxtube walls. Large fluxtubes like sunspots decay by fragmentation, whereby the kG field strengths are retained in the flux fragments, which however get scattered over an increasing area by being randomly carried around by the turbulent motions. These flux remnants together with new flux from the solar interior organize themselves into large-scale patterns, which exhibit remarkable regularities in their cyclic evolution.[15]

The evolution of the large-scale pattern can be explored using the continuous synoptic data set of magnetograms extending back to 1959. Decomposition of the magnetic maps in their spherical harmonics coefficients, and power spectrum analysis of the time series of the coefficients, have revealed an underlying resonant modal structure of the fields. The rotationally symmetric modes, described by spherical harmonic degree l, are governed by a strict parity selection rule. Modes with odd values of l (representing patterns anti-symmetric with respect to reflections in the equatorial plane) are dominated by the 22 yr resonance. The even modes (symmetric around the equator) show no trace of the 22 or 11 yr cycle, but exhibit another dispersion relation, for which the modal frequency increases with the *l* value.[16,17]

Inside the sun, in the convection zone, the magnetic field is dominated by hydrodynamic forces due to the high mass density, and can be effectively tangled up and amplified by the turbulent motions. In the photosphere the strong fluxtubes can withstand the dynamic forces, but are still controlled by gas pressure effects. Higher up, however, in the chromosphere and corona, the atmosphere becomes magnetically controlled due to the exponential drop in gas density, and a "force-free" situation is approached. Here electric currents (corresponding to twisted fields) and various magnetohydrodynamic waves play important roles, and appear to be responsible for the heating of the corona to over a million degrees.[18]

In these higher layers where the coronal heating takes place the Zeeman effect is inadequate for diagnosing the magnetic field for reasons explained in Section 3, while the Hanle effect has promising potential not yet exploited (cf. Section 8). Another area in which the Hanle effect has distinct advantages is for diagnosing the magnetic fields in solar prominences, arch-shaped, cloud-like condensations of denser and cooler matter in the solar corona. This is the area in which the Hanle effect has most systematically been put to use in solar physics (cf. Section 5).

A key question for the understanding of the nature of the sun's magnetic field and the operation of the solar dynamo is what magnetoturbulent processes occur on the smallest scales. Between the kG photospheric fluxtubes that contribute to the longitudinal Zeeman effect observed in the magnetograms we expect a tangled, "turbulent" magnetic field of mixed polarities, such that no net circular-polarization signal is seen due to cancellation of the signals of opposite signs within the resolution element. The Hanle effect can provide crucial information for solving this problem (cf. Section 7).

3. OVERVIEW OF THE DIAGNOSTIC POSSIBILITIES AND LIMITATIONS OF THE HANLE EFFECT

The Hanle effect is a coherence phenomenon fundamentally due to quantum interference, and occurs in the solar spectrum only when coherent scattering contributes to the spectral line formation. The ordinary Zeeman-effect polarization that has been exploited in the past for mapping the fields and making magnetograms has nothing to do with coherence effects. In the absence of magnetic fields and coherent scattering the polarization is zero, but it is produced in the presence of a magnetic field, and increases with field strength.

In coherent scattering, however, maximum linear polarization occurs in the absence of magnetic fields. When a magnetic field is introduced, the polarization is modified via the Hanle effect. The magnetic field influences the polarization primarily in two ways: (1) rotation of the plane of linear polarization; (2) depolarization. The Hanle effect is significant only in the linear polarization, while the ordinary Zeeman effect is most sensitive and useful in the circular polarization (longitudinal Zeeman effect). The polarization signature of the Hanle effect (polarization profile across the spectral line) is very different from the signature of the transverse Zeeman effect. Also, while the transverse Zeeman effect can occur for almost all lines in the spectrum, the Hanle effect only occurs in the rather few lines that are partially formed by coherent scattering.

In comparison with the ordinary Zeeman effect, the Hanle effect has both advantages and disadvantages. We begin with an overview of the main advantages.

The Hanle effect is sensitive to weaker fields than the ordinary Zeeman effect. While the magnitude of the polarization signal from the Zeeman effect scales with the ratio between the Zeeman splitting and the width of the spectral line (which is normally dominated by Doppler broadening), the Hanle effect scales with the ratio between the Zeeman splitting and the damping width of the transition (inverse life time of the upper atomic level). As the damping width is generally much smaller than the Doppler width, the Zeeman and Hanle effects are sensitive in different field-strength regimes.[19]

The advantages of the Hanle effect are particularly pronounced above the solar photosphere, in the chromosphere and the transition zone between the chromosphere and corona (where the temperature rises by two orders of magnitude and the processes heating the corona occur), as well as in prominences. There are several reasons why these regions are preferred: (1) The density decreases exponentially with height, and therefore also the rate of depolarizing collisions, which wipe out the phase coherence in the scattering processes. From being predominantly incoherent in the photosphere (although the coherent contribution is well observable, see Section 6), the scattering becomes almost entirely coherent (scattering atom undisturbed by collisions when it is in its excited state) in the upper atmosphere. (2) The spectral lines arising in the chromosphere-corona transition region appear almost exclusively in the vacuum ultraviolet, as resonant emission lines below about 1700 Å. Due to the λ^2 dependence of the Zeeman effect and the height decrease of the magnetic field strength, the Zeeman effect gives an almost negligible polarization signal for these lines, while this is the right regime for the Hanle effect. (3) The lines in the visible spectrum from the chromosphere or prominences are very broad (largely Stark-broadened), in particular the Balmer lines of hydrogen and the helium D_3

line. Because of their large widths, their Zeeman-effect polarization is small. (4) Due to the low density in the higher layers, many spectral lines are optically thin (in contrast to the photosphere), which means that radiative-transfer effects can be neglected in the interpretation, and only the single-scattering case needs to be considered.

A small-scale turbulent magnetic field in the solar photosphere does not contribute to any Zeeman-effect circular-polarization signal if the opposite magnetic polarities are well mixed within each spatial resolution element, and such a field is therefore not "seen" in magnetograms. The Hanle effect however responds to this apparently invisible field by depolarizing the scattering polarization (since this depolarization has the same "sign" for both magnetic polarities). The observed depolarization can be used to constrain the properties of the turbulent field on the smallest, unresolved spatial scales (cf. Section 7).

A practical advantage of the Hanle effect over the Zeeman effect is that it is possible to work with low spectral resolution to use the polarization effects as integrated over the line profile. The circular-polarization profile of the longitudinal Zeeman effect is nearly antisymmetric with respect to the line center, so that the polarization integrated over the entire line is close to zero. The polarization related to the Hanle effect, on the other hand, has the same sign across the line profile, which makes spectral integration feasible. This relaxes the demands on spectral resolution, such that the photon statistics can be enhanced.

Let us now turn to the main disadvantages of the Hanle effect. One difficulty is that the full magnetic field vector, having three unknown vector components, cannot be determined unambiguously from the Hanle effect in a single spectral line, since the Hanle effect gives effectively only two observables (angle of rotation of the plane of linear polarization, and depolarization). This problem can generally be overcome by making simultaneous recordings in two lines with different sensitivities to the Hanle effect, or by using supplementary information, e.g., constraints from Maxwell's equations, theoretical models, extrapolations from fields observed in the photosphere via the Zeeman effect, etc.

To derive the magnetic-field effects from the observed scattering polarization, we have to know what the polarization would be if there were no magnetic fields. This requires knowledge of the scattering geometry. For optically thin prominences this knowledge is based on the good model assumption that the illuminating radiation is described by the limb-darkening law of the underlying solar disk at the wavelength considered. In the photosphere, for the diagnostics of turbulent magnetic fields, we eventually need to derive the effects of the scattering geometry using radiative-transfer calculations for a model atmosphere (giving the temperature-density stratification) that has been established by other diagnostic means (analysis of the unpolarized spectrum). For the chromosphere–corona transition region and lines in the vacuum ultraviolet it may in principle be possible to eliminate the effects of scattering geometry by making simultaneous polarization observations in two or more lines with different sensitivities to the Hanle effect (which is also needed to deduce the vector field).

Another disadvantage is the difficulty of obtaining unique answers due to symmetries in the scattering matrix. Further, the Hanle effect vanishes when the incident radiation field is symmetrical around the magnetic field vector (which is normally the case when the field is vertical in the solar atmosphere).

The last main disadvantage of the Hanle effect, which is already apparent from the above discussion, is that the interpretation of the observations is very complex, generally involving the combination of quantum interference, radiative transfer with non-LTE calculations of the level populations, and scattering polarization in complicated geometries. The theory to handle all this is still in a rather early state of development. This is the reason why the Hanle effect has not been more exploited in the past. However, as we learn how to overcome these obstacles, we expect the Hanle effect to take on an increasingly important role in solar diagnostics.

4. BASIC THEORETICAL CONCEPTS FOR APPLICATIONS IN ASTROPHYSICS

Magnetic fields in the atmospheres of the sun and stars can be diagnosed by observing and interpreting their influence on spectral lines via the Zeeman and Hanle effects, provided that we have a theory for the formation of polarized spectral lines in stellar atmospheres in the presence of these effects. This problem becomes particularly complex in the case of the Hanle effect, since the polarization that arises through coherent scattering depends on the detailed scattering geometry, the population and polarization of the atomic levels, multiple scattering (optically thick medium), the relative contribution of nonscattering transitions, and depolarizing collisions. In this section we will try to give a brief introductory overview of the theoretical concepts needed to solve this problem, and which will be used in the following sections.

Let us first see how quantum-mechanical coherences can be incorporated into a convenient and general description of polarized radiation. We begin by considering a general quantum-mechanical system, which can be described as a coherent superposition of orthogonal states:

$$\psi = \sum_{n=1}^{N} c_n \psi_n \qquad (1)$$

The given state, including its coherences, is fully described by the density matrix

$$\rho_{\mu\nu} = c_\mu c_\nu^* \qquad (2)$$

The expectation value of an operator **A** is obtained through the trace operation

$$\langle \mathbf{A} \rangle = \mathrm{Tr}(\boldsymbol{\rho}\mathbf{A}) \qquad (3)$$

These general concepts can also be applied to photons by letting $N = 2$, describing any polarized photon as a superposition of two states of orthogonal polarizations (e.g., left and right circular polarization; two orthogonal linear polarizations; more generally two orthogonal elliptical polarizations). The photon is described by a two-component vector, with each component being characterized by its amplitude *and* phase.

Such a two-component vector (often called a Jones vector) can adequately describe the state of polarization of individual photons, which are 100% elliptically polarized, but it cannot describe a beam of light that is unpolarized or partially polarized. An unpolarized beam can be considered to consist of an *incoherent* superposition of an ensemble of photons with different polarizations. However, such a beam can be fully described by the four components of the density matrix, which thus provides a more general description of a natural light beam consisting of many photons. Another, equivalent description of a beam of natural light in terms of four parameters is by the Stokes vector

$$\mathbf{I} = \begin{pmatrix} I \\ Q \\ U \\ V \end{pmatrix} \qquad (4)$$

which has been found most suitable for practical applications, since the four Stokes parameters lend themselves to a straightforward operational interpretation. In equation (4) I is the ordinary intensity of the beam, Q is the difference between the beam intensities recorded with an ideal (without internal absorption) linear polarizer set at position angles 0° and 90°, respectively, for the direction of the electric vector, U is the difference intensity obtained with the linear polarizer set at position angles 45° and 135°, and V is the difference between the intensities recorded with two ideal circular polarizers, one for right-handed and the other for left-handed circular polarization.

As both the density matrix and the Stokes parameters describe exhaustively the polarization properties of a light beam, one description can be transformed into the other. Thus the Stokes parameters can be obtained from the components of the density matrix through[20,21]

$$I = \text{Tr}(\boldsymbol{\rho}\boldsymbol{\sigma}_0) = \rho_{11} + \rho_{22}$$

$$Q = \text{Tr}(\boldsymbol{\rho}\boldsymbol{\sigma}_1) = \rho_{11} - \rho_{22}$$

$$U = \text{Tr}(\boldsymbol{\rho}\boldsymbol{\sigma}_2) = \rho_{12} + \rho_{21}$$

$$V = \text{Tr}(\boldsymbol{\rho}\boldsymbol{\sigma}_3) = i(\rho_{12} - \rho_{21})$$

(5)

σ_i, $i = 0, 1, 2, 3$, are the standard Pauli spin matrices.

The influence on the radiation by a medium can be described in the case of individual photons by the 2×2 matrix **w**:

$$\begin{pmatrix} \psi_1 \\ \psi_2 \end{pmatrix} = \mathbf{w} \begin{pmatrix} \psi'_1 \\ \psi'_2 \end{pmatrix} \quad (6)$$

or by the 4×4 matrix **M** (often called the Mueller matrix):

$$\mathbf{I} = \mathbf{M}\mathbf{I}' \quad (7)$$

As noted above, **M** is more general than **w**, since it can handle partially polarized light as well. However, obtaining **w** for individual quantum processes, we can always transform it to **M**, and then perform a statistical averaging over many processes, to obtain a description of an observed macroscopic system. If we let **W** denote the direct tensor product

$$\mathbf{W} = \mathbf{w} \otimes \mathbf{w}^* \quad (8)$$

where $*$ denotes complex conjugation, the Mueller matrix becomes[21]

$$\mathbf{M} = \mathbf{T}\mathbf{W}\mathbf{T}^{-1} \quad (9)$$

where

$$\mathbf{T} = \begin{pmatrix} 1 & 0 & 0 & 1 \\ 1 & 0 & 0 & -1 \\ 0 & 1 & 1 & 0 \\ 0 & i & -i & 0 \end{pmatrix} \quad \text{and} \quad \mathbf{T}^{-1} = \frac{1}{2} \begin{pmatrix} 1 & 1 & 0 & 0 \\ 0 & 0 & 1 & -i \\ 0 & 0 & 1 & i \\ 1 & -1 & 0 & 0 \end{pmatrix} \quad (10)$$

The most complete quantum-mechanical treatment of the interaction between the atomic system and the radiation field that leads to atomic level polarization and coherences in the emitted radiation is in terms of the density matrix formalism.[22-25] Two types of coupled systems of equations are needed: (1) The statistical equilibrium equations for the density matrix, describing the level populations and polarizations. (2) The radiative transfer equation, which is usually formulated in terms of the Stokes vector, but which determines the density matrix for the radiation field used in the statistical equilibrium equations, and which needs the density matrix for the atomic system obtained from the solution of the statistical equilibrium equations.

Since the full problem is intractable in its generality, we need to use idealized models to obtain practical results. One very direct way of seeing the basic physics involved, including the interferences between the atomic sublevels of different M and J quantum numbers, the frequency variation of the polarization effects throughout the spectral line, etc., is to use the Kramers-Heisenberg dispersion formula[26,27] for the scattering amplitude $w_{\alpha'\alpha}$ describing coherent scattering in the dipole approximation, where α' symbolizes the polarization angle of the incident photon, α the scattered photon. If we omit a constant of proportionality, which is unimportant for the consideration of the polarization effects, take into account that the dominating contribution to the scattering comes near the resonance, and disregard the generally unimportant nonresonant term representing emission followed by absorption, then

$$w_{\alpha'\alpha} = \sum_{bJ''} \sum_{M''} \left[\frac{\langle aJM|\varepsilon^{\alpha*}\mathbf{D}|bJ''M''\rangle\langle bJ''M''|\varepsilon^{\alpha'}\mathbf{D}|a'J'M'\rangle}{\omega - \omega_{JJ''} - \Delta\omega_H + i\Gamma/2} \right] \quad (11)$$

Primed quantities refer to the incident, unprimed to the scattered radiation; Γ is the natural width of the transition, and $\mathbf{D}$ is the dipole moment operator. The coherences between levels of different M'' and J'' quantum numbers are manifest by the sum over all intermediate levels.[28-30] Quantity $\omega_{JJ''}$ is the resonant frequency between the unsplit upper and lower levels. The Zeeman splitting $\Delta\omega_H$ is determined by the Landé factors $g_{a,b}$ and the magnetic quantum numbers M and M'' of the final and intermediate levels through

$$\Delta\omega_H = (g_b M'' - g_a M)\omega_L \quad (12)$$

where ω_L is the Larmor frequency. If the field strength B is expressed in G, then

$$\omega_L = 8.803 \times 10^6 B \text{ s}^{-1} \quad (13)$$

It is convenient to introduce the quantization axis along the magnetic field, and describe the polarization vectors $\boldsymbol{\varepsilon}$ in terms of a linear polarization basis, with $\alpha = 2$ if the electric vector is in the meridian plane, $\alpha = 1$ if it is perpendicular to this plane. The matrix elements in expression (11) can then be expanded in the spherical tensor components of $\mathbf{D}$ and $\boldsymbol{\varepsilon}$ and be evaluated in a standard way, conveniently making use of 3-j symbols.

The phase matrix $\mathbf{M}$ for scattering of the Stokes vector can then be obtained from relations (8) and (9), with the tensor product $\mathbf{W}$ containing a summation over all the initial and final states. Omitting unimportant factors of proportionality and interferences between the sublevels of the initial state (the off-diagonal elements of the density matrix for the initial state), the elements of the $\mathbf{W}$ matrix can be expressed as

$$\sum_{J'M'} N_{J'M'} \sum_{JM} w_{i'k} w_{j'l}^* \tag{14}$$

where the w-indices assume values of 1 and 2.$^{(30,31)}$ Quantity $N_{J'M'}$ is the relative population of the sublevels of the initial state (the diagonal elements of the density matrix), and can be set equal to unity if the initial state is not polarized. Otherwise it has to be obtained by solving the statistical equilibrium equations for the density matrix. Usually it is a good approximation to consider the initial state to be unpolarized (but for the He I D_3 line, coherences in the lower level are significant, as will be seen in Figure 4 below).

So far we have only considered coherent scattering processes in the rest frame of the scattering atom. To obtain the correct frequency redistribution we have to convolve $\mathbf{M}$ with a distribution of Doppler velocities; $\mathbf{M}$ is a function both of the scattering geometry (directions $\mathbf{n}'$ and $\mathbf{n}$ of the incident and scattered photons) and of the incoming and outgoing frequencies ν' and ν. In the limit of weak magnetic fields (Zeeman splitting much smaller than the Doppler linewidth), the frequency redistribution part becomes to a good approximation identical for all the matrix elements, so that one can write

$$\mathbf{M} = F(\nu', \mathbf{n}'; \nu, \mathbf{n}) \mathbf{P}(\mathbf{n}', \mathbf{n}) \tag{15}$$

where $\mathbf{P}$ is the 4×4 phase matrix usually normalized such that the first component P_{11} becomes unity when averaged over all directions; F is the frequency redistribution function.

It is illuminating to explicitly consider the case of a normal Zeeman triplet with an unsplit lower level, i.e., with $J' = J = 0$ and $J'' = 1$. Let us introduce the convenient notation

$$b_q = \frac{e^{iq(\phi - \phi')}}{\omega - \omega_0 - q g_b \omega_L + i\Gamma/2} \tag{16}$$

where ω_0 is the unsplit resonance frequency, and $q = M'' - M = 0, \pm 1$. Omitting a common proportionality factor, we find

$$w_{1'1} = -(b_+ + b_-)$$

$$w_{2'2} = (b_+ + b_-) \cos \theta' \cos \theta + 2b_0 \sin \theta' \sin \theta$$

$$w_{1'2} = i(b_+ - b_-) \cos \theta$$

$$w_{2'1} = -i(b_+ - b_-) \cos \theta' \qquad (17)$$

These fairly simple expressions contain all the dependences we need on frequency, field strength (through ω_L), relative azimuth $\phi - \phi'$, and colatitudes θ' and θ. The mathematical complexity arises when we want to derive the scattering matrix **M**, integrate over a distribution of Doppler velocities, and transform to a coordinate system in which the magnetic field vector has an arbitrary direction (until now the magnetic field has defined the quantization axis).

Generally, the expressions obtained when evaluating the various components of the tensor product **W** are lengthy and not very transparent due to the complicated frequency dependence. If, however, the Zeeman splitting $\Delta\omega_H$ is small as compared with the Doppler width, we may integrate over all frequencies, resulting in a frequency-averaged matrix **P** that is representative for the Doppler core of the line. Let us, as an example, write down the explicit expression in our Zeeman triplet case for the frequency-averaged first component of the **W** matrix:

$$\langle W_{11} \rangle = \langle |b_+ + b_-|^2 \rangle \sim 1 + \cos \alpha_2 \cos [2(\phi - \phi') - \alpha_2] \qquad (18)$$

The magnetic-field dependence is contained in the angle α_2, defined by

$$\tan \alpha_2 = \frac{2g_b \omega_l}{\Gamma} = \Omega \qquad (19)$$

where, for convenience, we have introduced and defined the variable Ω to be used later below.

Already in equation (18) we see the two main ingredients of the Hanle effect at work: depolarization occurs through the factor $\cos \alpha_2$, rotation of the plane of polarization through α_2 being subtracted from twice the azimuth angles.

The other terms in the **W** matrix contain these two effects coupled in similar but different ways to the geometrical factors. When the Zeeman splitting goes to zero, the nonmagnetic phase matrix for Rayleigh scattering is retrieved from the general Hanle-effect scattering matrix. In Section 6 below our formalism will be applied to nonmagnetic scattering when there is quantum interference between widely separated states of different total angular momentum J''.

The above frequency-averaged expressions are valid in the Doppler core of the lines when the Zeeman splitting is much smaller than the Doppler width. Let us now consider the dispersion wings of the line, where $|\omega - \omega_0|$ is much larger than the Doppler width (and the Zeeman splitting and damping width). There the influence of $\Delta\omega_H$ and Γ in the denominator of expression (11) vanishes, and the scattering behaves like in the nonmagnetic case. We thus see that the Hanle effect is primarily present in the Doppler core of the spectral lines, but vanishes in the dispersion wings, in contrast to the nonmagnetic scattering polarization, which is generally quite pronounced with polarization maxima in the wings. In Section 6 this behavior will be illustrated by observed Stokes profiles (Figure 7).

The expressions used so far are only valid for the case that the quantization axis is along the magnetic field vector **B**. The phase matrix for an arbitrary direction of **B** with respect to the quantization axis can be obtained from our special case by applying rotation matrices.[32] Explicit expressions for the Hanle phase matrix for the general case have recently been published.[33]

Let us now consider the general weak-field case for an arbitrary atomic transition, including the effect of collisions, but neglecting coherences in the lower atomic level. The Hanle scattering matrix for this general case was derived by Stenflo[32] using results from a redistribution treatment by Omont et al.[34] Let us consider the resulting expressions for the scattered Stokes vector in the case that the incident radiation is unpolarized and in the vertical direction, the scattering angle is 90°, and the field vector is assumed to be in the horizontal plane. The direction of **B** is defined by its azimuth angle χ with respect to the direction of the scattered radiation, which is also in the horizontal plane. Then

$$I = 1 + \tfrac{3}{8}k_c W_2[\tfrac{1}{3}(1 - 3\cos^2\chi) - \sin^2\chi \cos^2\alpha_2]$$

$$Q = \tfrac{3}{8}k_c W_2[\sin^2\chi + (1 + \cos^2\chi)\cos^2\alpha_2]$$

$$U = \tfrac{3}{8}k_c W_2 \cos\chi \sin 2\alpha_2 \qquad (20)$$

As Stokes V is zero in the incident beam, it remains zero in the scattered beam, since V is decoupled from the other Stokes parameters in the scattering phase matrix. The damping constant that determines the value of α_2 via equation (19) is now understood to be given by

$$\Gamma = \Gamma_N + \Gamma_c \qquad (21)$$

Γ_N is the natural damping width or inverse lifetime of the excited level, while Γ_c is the collisional depolarization rate. When adequate atomic physics data are lacking, the collisional depolarization rate may be taken to be the same as the collisional broadening rate; k_c is then given by

$$k_c = \Gamma_N/(\Gamma_N + \Gamma_c) \quad (22)$$

the probability that the atom does not suffer a depolarizing collision while it is in its excited state.

The factor W_2 (the notation has nothing to do with the **W** matrix that we have used) represents "atomic depolarization," and depends only on the J quantum numbers of the upper and lower levels of the transition. It can be viewed as representing the fraction of the scattering processes that occur as polarizing dipole-type scattering. The remaining fraction $1 - W_2$ occurs as unpolarized, isotropic scattering. For a $J = 0$, $J'' = 1$ transition, $W_2 = 1$.

Figure 1, from Stenflo,[32] displays in the two diagrams to the upper and lower left the polarization characteristics of the case described by system (20). The fractional polarizations $p_Q = -Q/I$ and $p_U = -U/I$ are given for various values of the field azimuth χ, as a function of the field-strength parameter Ω defined by equation (19). The diagram to the upper left represents the case that the joint atomic-collisional depolarization factor $k_c W_2 = 1$, while in the diagram to the lower left this factor is 0.5. In the diagrams to the right the magnetic field vector makes an angle $\vartheta_B = 45°$ with the vertical instead of being in the horizontal plane.

Another commonly used way to illustrate the state of polarization of the scattered radiation is shown in Figure 2. The diagrams represent the same case as before, given by equations (20), but now the two "Hanle parameters" p/p_{max}, describing the Hanle depolarization, and β, the angle of rotation of the plane of polarization, are given instead of p_Q and p_U. They are determined by

$$p = \sqrt{Q^2 + U^2}/I \quad (23)$$

and

$$\tan 2\beta = U/Q \quad (24)$$

p_{max} is the degree of linear polarization for zero magnetic field.

On inserting equations (20) in the above expressions, we obtain

$$p/p_{max} = \frac{[1 + \Omega^2 + (\tfrac{1}{2}\Omega^2 \sin^2 \chi)^2]^{1/2}}{1 + \Omega^2 + \tfrac{1}{2}p_{max}\Omega^2 \sin^2 \chi} \quad (25)$$

APPLICATIONS IN SOLAR PHYSICS

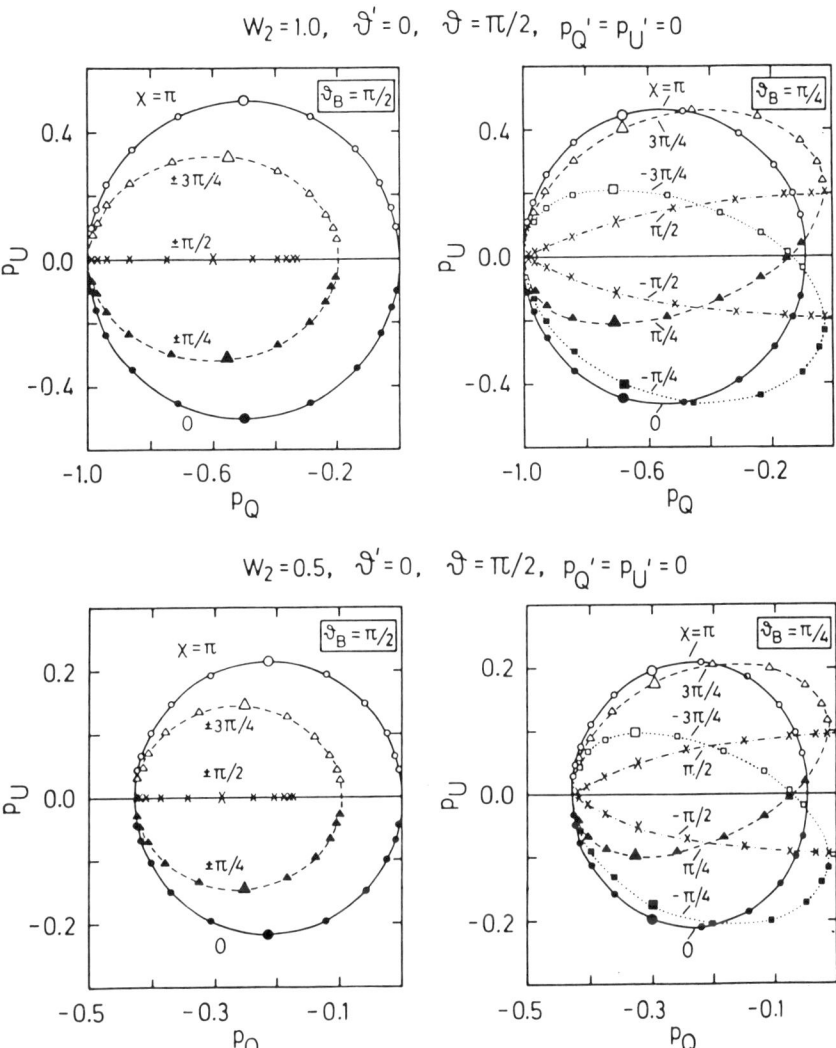

FIGURE 1. Polarization $p_Q = -Q/I$ and $p_U = -U/I$ of the scattered radiation as a function of magnetic field strength and field azimuth χ. The incident radiation is unpolarized and parallel to the vertical axis ($\vartheta' = 0$), while the scattered radiation is in the horizontal plane ($\vartheta = 90°$). ϑ_B is the angle between the magnetic field vector and the vertical axis. The symbols along the various curves correspond to different values of the field-strength parameter Ω, defined in equation (19), increasing in steps of 0.2 in log Ω, from left to right. The larger symbols correspond to $\Omega = 1$ (from Stenflo[32]).

and

$$\tan 2\beta = \frac{\Omega \cos \chi}{1 + \frac{1}{2}\Omega^2 \sin^2 \chi} \tag{26}$$

The upper diagram in Figure 2 represents the case that $p_{\max} = 1$, the lower diagram the limiting case that $p_{\max} \to 0$ [the difference between the two cases being caused by the last term in the denominator of equation (25)].

Although it may be appealing to use the representation of Figure 2 when displaying data, since it directly illustrates the two "Hanle parameters," the representation of Figure 1 is more adequate for the analysis

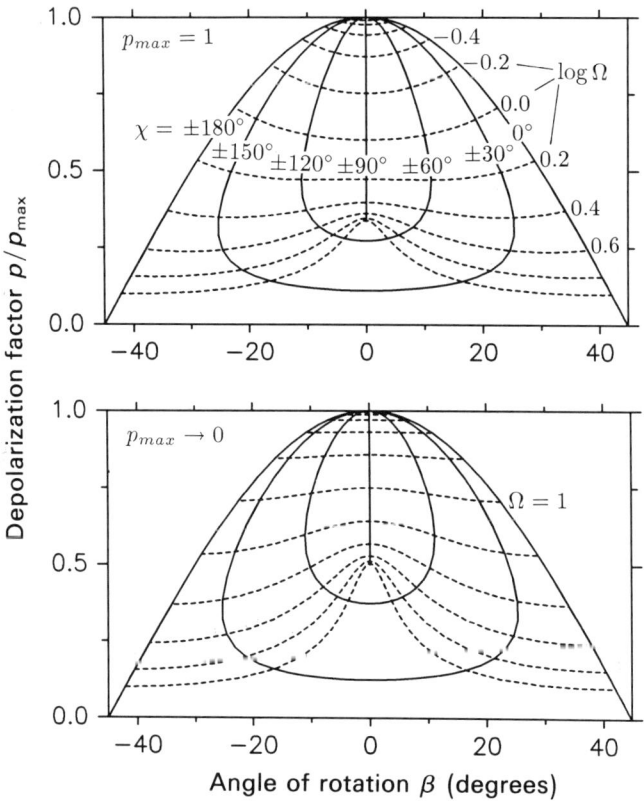

FIGURE 2. Hanle depolarization factor and angle of rotation of the plane of linear polarization, as a function of field-strength paremeter Ω and field azimuth χ, for the same scattering geometry as the two diagrams to the left in Figure 1. The upper diagram represents the same case as the upper left diagram of Figure 1, while the lower diagram represents the limiting case when the nonmagnetic polarization $p_{\max} \to 0$ (which would occur, e.g., through $W_2 k_c \to 0$).

of real observational data. The reason is that Q and U are the directly observed parameters, and they are affected by observational errors. The error boxes in the p_Q-p_U diagram are rectangles (usually squares, as is the case when Q and U are treated on an equal basis in the observations), with a size that is almost independent of position in the diagram (the small variation being due to the Ω and χ dependence of Stokes I). This is not the case in the β-p diagram, however, since the error in β increases to infinity when p goes to zero.

In most astrophysical situations, except for some special cases (notably solar prominences, see Section 5 below), we have to do with optically thick media and accordingly with multiple scattering processes. To treat this general case, we need to consider the transport of the Stokes vector through the medium. The transfer equation can be written as[32]

$$\frac{d\mathbf{I}_\nu}{ds} = -\kappa_c \mathbf{I}_\nu - \boldsymbol{\kappa}_L \mathbf{I}_\nu + \mathbf{j}_c + \mathbf{j}_L \quad (27)$$

Index c symbolizes continuum processes, index L line processes, while index ν is the frequency ($=\omega/2\pi$) of the Stokes vector $\mathbf{I}$. The absorption coefficient for continuum radiation is a scalar, while the emission vector $\mathbf{j}_c$ generally contains two contributions, one unpolarized part proportional to the unity vector $\mathbf{1}$ (first position unity, remaining positions zero), the second part representing nonmagnetic dipole-type scattering due to Thomson scattering at free electrons and Rayleigh scattering at neutral hydrogen atoms. This second part is obtained through angular averaging over the Rayleigh phase matrix acting on the incident Stokes vector, and multiplication with the scattering cross section.

Atomic polarization may affect both the 4×4 matrix $\boldsymbol{\kappa}_L$ and the line emission vector $\mathbf{j}_L$, but under normally occurring circumstances the coherence effects have a much larger influence on the Stokes vector emerging from the atmosphere via the scattering term in $\mathbf{j}_L$ than via the absorption matrix $\boldsymbol{\kappa}_L$ (through lower-level coherences), although there are situations for which such an assumption is not valid.[35,36] If we neglect atomic polarization effects in the absorption term, we may write

$$\boldsymbol{\kappa}_L = (N_i B_{ij} - N_j B_{ji}) \frac{h\nu}{4\pi} \mathbf{K} \quad (28)$$

where $\mathbf{K}$ is a normalized 4×4 matrix containing the Zeeman splitting and magnetooptical effects (anomalous dispersion), but without coherence effects.[37-41] Quantities N_i and N_j represent the populations of the lower and upper levels of the transition considered, while B_{ij} and B_{ji} are the Einstein coefficients for absorption and stimulated emission.

It is convenient to consider the line emission vector as consisting of two parts,

$$\mathbf{j}_L = \mathbf{j}_{sc} + \mathbf{j}_{th} \tag{29}$$

where $\mathbf{j}_{sc}$ contains all the scattering transitions that we have taken into account with the Kramers-Heisenberg dispersion formula, and $\mathbf{j}_{th}$ stands for all the other processes that lead to a spontaneous transition from level j to level i, normally without polarization effects, and therefore may be considered as "thermal" emission. In this case $\mathbf{j}_{th}$ is proportional to **K1**.

The scattering term can be written in the form

$$\mathbf{j}_{sc} = \sigma \int \frac{d\omega'}{4\pi} \int d\nu' \, F(\nu', \mathbf{n}'; \nu, \mathbf{n}) \mathbf{P}(\mathbf{n}', \mathbf{n}) \mathbf{I}_{\nu'}(\mathbf{n}') \tag{30}$$

where σ is the scattering cross section, **P** the 4×4 phase matrix, and F the frequency redistribution function as described in connection with equation (15). The product $F\mathbf{P}$ is obtained by expressions (8), (9), and (15), but the so-derived phase matrix **P** has to be transformed via rotation matrices to a coordinate system fixed with respect to the solar atmosphere, valid for an arbitrary direction of the magnetic field, before it is inserted into equation (30).

5. DIAGNOSTICS OF MAGNETIC FIELDS IN SOLAR PROMINENCES

Solar prominences are clouds or condensations in the solar corona, which form around an apparently arch-like magnetic structure connecting footpoints of opposite polarities down in the photosphere (where the visible solar spectrum originates). They usually extend to considerable heights in the corona. In comparison with the surrounding one million degree corona, in which they are embedded, the prominences are much cooler (typically 10,000–50,000 K) and denser formations, radiating mainly in chromospheric lines like the hydrogen Balmer and Lyman lines, He and Na lines, etc. Some prominences are quiescent, lasting for weeks, and unrelated to active regions and sunspots. Others are eruptive and directly associated with active regions. When observed with high spatial resolution, the prominence emission exhibits a pronounced, threadlike fine structure.

Although many attempts have been made to explain these peculiar plasma condensations, lack of observational guidance has not yet allowed us to select between the various, often diverging models and to establish a consensus view on how the prominences are formed and evolve. The reason is the difficulty in diagnosing the key physical parameter in the problem, the magnetic field, via the normally used Zeeman effect. This situation is changing with the introduction of the Hanle effect and our improved understanding how to apply it.

As a matter of fact the prominences have served as a testing ground for the application of the Hanle effect on the sun, and are the objects for which the Hanle effect has been predominantly used so far. The efforts were first concentrated on the interpretation of polarimetric observations made in the He I D_3 line at a wavelength of 5876 Å,[42-51,24,25] but has later been extended to the Balmer lines Hα and Hβ,[52,53] as well as to Na I D_2 at 5890 Å.[54] The main reasons for the particular interest in prominences, and the initial emphasis on the He D_3 line are the following: (1) Prominences are optically thin objects in the He D_3 line. This means that radiative transfer effects can be disregarded, and we have to consider a single scattering process only. (2) Due to the low number density of He atoms, collisional depolarization can be neglected. (3) The radiation field incident on the scattering atoms in the prominence is highly anisotropic, which is a necessary requirement for appreciable amounts of polarization in the scattered radiation. Moreover, the geometry of the incident radiation is known, since the illumination of the prominence atoms comes from the underlying solar disk, which covers a solid angle simply determined by the height of the scattering atoms above the solar photosphere, and which has a known limb-darkening function (intensity variation from disk center to limb at the wavelength of the incident radiation). (4) The prominence magnetic fields are difficult to diagnose by other means. The only real alternative, the use of the longitudinal Zeeman effect, gives a circular-polarization signal that is about an order of magnitude smaller than the Hanle effect in the linear polarization signal. (5) The field strengths in the prominences happen to fall in the range where the Hanle effect has its greatest sensitivity for the most useful spectral lines.

The line most used in applications of the Hanle effect in prominences is the He D_3 5876 Å line ($3d\ ^3D_{3,2,1} \to 2p\ ^3P_{2,1,0}$), which is optically thin in the prominences. Figure 3, from Bommier,[25] shows the energies of the magnetic sublevels of the three fine structure components $^3D_{3,2,1}$ of the upper level, as a function of the magnetic field strength B. We note that level crossings between the $J'' = 2$ and 3 levels occur already for field strengths of about 10 G, while level crossing with the $J'' = 1$ level requires fields stronger than about 200 G. For the lower ^3P level the separations are larger, requiring fields >500 and 5000 G for level crossings. The ordinary Hanle effect, on the other hand, depends on the ratio between the Zeeman splitting and the natural width (inverse lifetime) of the sublevels. In the case of the $^3D_1$, $^3D_2$, and $^3D_3$ levels, the Zeeman splitting equals the natural width for field strengths of 16.0, 6.9, and 6.0 G, respectively. For the lower

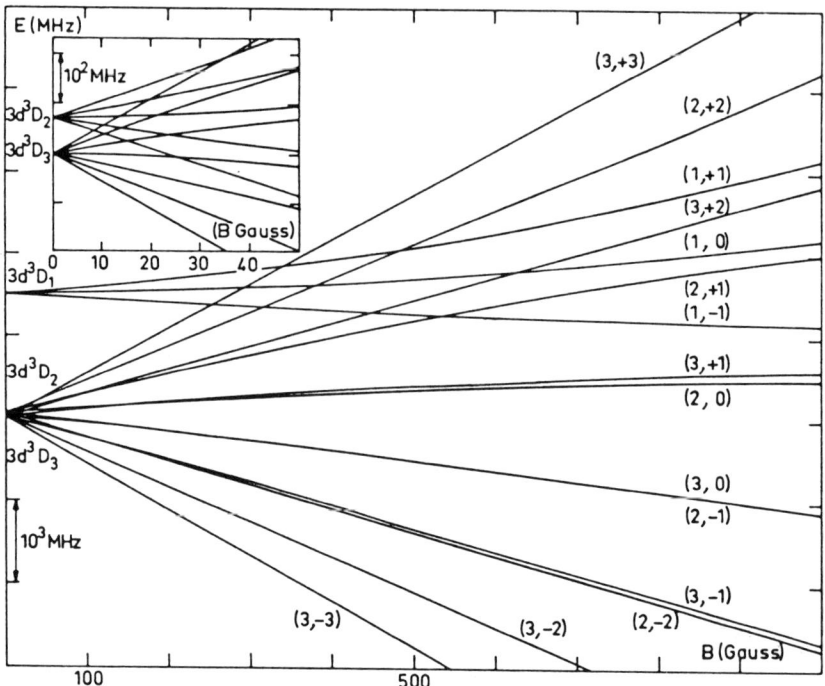

FIGURE 3. Energies of the excited sublevels of the He I D_3 transition, as functions of the field strength B. Each line is marked by its (J, M) quantum numbers (from Bommier[25]).

levels, due to their longer lifetimes, the corresponding values are less than 1 G.

We thus see that the level crossing between the $^3D_2$ and $^3D_3$ fine structure components occurs partly within the range that represents the optimum response of the Hanle effect, while the other level crossings occur at much higher field strengths and therefore have no significant effect on the Hanle diagnostics.

The polarization properties of the emitted radiation is obtained by solving the statistical equilibrium equations for the density matrix, which takes into account the coherence transfer between the various levels, and then using the density matrices for the excited, upper levels, to derive the emissivity Stokes vector.[25,47,22] The incident radiation field is determined by the limb-darkening function of the underlying solar disk.

Some results of such calculations are shown in Figure 4, from Bommier,[25] giving the magnetic depolarization factor p/p_{max} as a function of the rotation angle β of the plane of linear polarization, as was done in our Figure 2 of Section 4 above; p_{max} is the polarization that would result if the magnetic field were zero.

For p_{max} to differ from zero, the incident radiation field has to be anisotropic. If the height h of the scattering atom above the solar disk (photosphere) were zero and there were no limb darkening, the illumination would be uniform over a half sphere, and the scattered radiation would be unpolarized. In the extreme case that the height h goes to infinity, or the limb darkening function is extremely peaked around the disk center, we get the case of 90° scattering. In reality we have a situation intermediate between these two extremes, but normally the degree of anisotropy is quite small, resulting in large "geometrical depolarization" due to the angular averaging over the incident radiation field.

Our understanding of the effect of the "geometrical depolarization" is greatly helped by the circumstance that the unpolarized, incident radiation

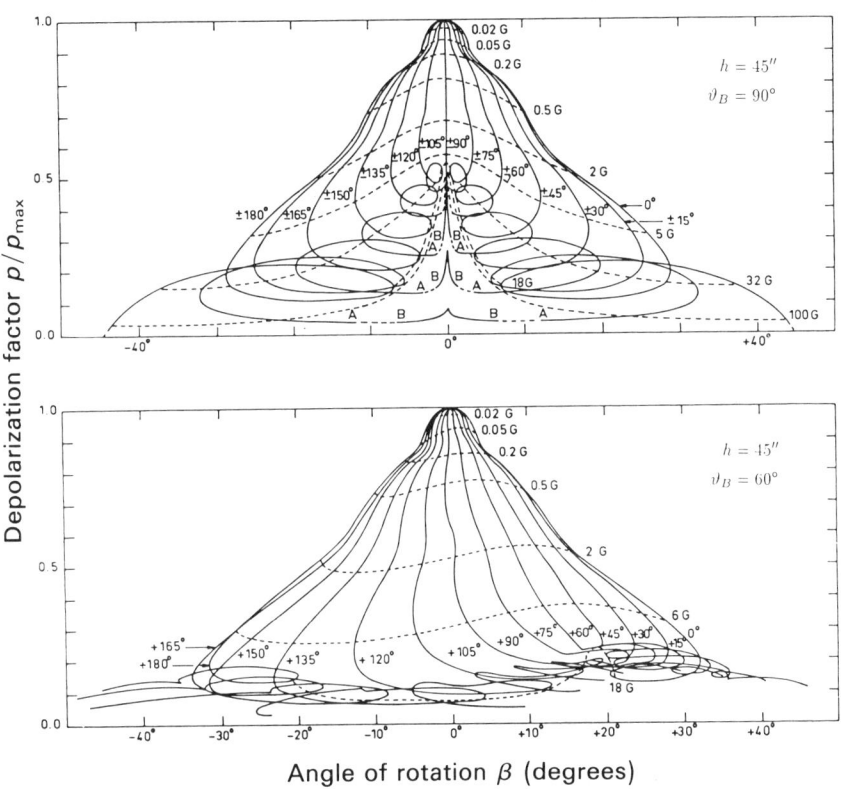

FIGURE 4. Polarization diagrams as in Figure 2, but for scattering by a He I atom 45 s of arc above the solar limb. The loops arise through the level crossings between the $^3D_2$ and $^3D_3$ levels shown in Figure 3. The undulations of the outer contours are due to coherences in the lower sublevels. In the bottom diagram the symmetry is broken by the field vector not being in the horizontal plane as in the upper diagram (from Bommier[25]).

field can be described as a sum of two components: For the first component the distribution of the electric vectors at the scattering atom is isotropic, for the second component the electric vectors lie in the horizontal plane only (the plane perpendicular to the direction of the solar radius).[53] Only the second component can be the source of polarization in the scattered radiation. It acts as if it were coming from the disk center only and thus represents the 90° scattering case. The first component only produces unpolaized scattered photons, diluting the polarized ones from the first component. The amount of dilution is determined by the relative magnitude of the two components.

Let us denote by k_G the ratio between the magnitude of the second, anisotropic component and the total incident intensity. For 90° scattering it would be unity. The effect of k_G on the scattered radiation is the same as that of the collisional depolarization factor k_c and the atomic depolarization factor W_2 in equations (20). We can therefore call it the geometrical depolarization factor. It is included by replacing $k_c W_2$ in expressions like (20) by $k_G k_c W_2$. This causes a corresponding scaling of the value of p_{max}. Through the normalization of the amount of polarization p by p_{max} as in Figure 4, we effectively reduce the diagram to the case of 90° scattering. However, it is not only a matter of scaling, which can be seen from the case described by equation (25), in which p_{max} occurs in the denominator on the right-hand side. With more angular averaging, p_{max} is reduced toward zero, with the consequence that the situation described by the bottom diagram of Figure 2 is approached. This diagram is thus much more representative of the real sun than the upper diagram, since the anisotropy of the incident radiation field is generally quite small.

If there were no limb darkening, the geometrical depolarization factor would be determined alone by the height h of the scattering atom:

$$k_G = \tfrac{1}{2}(1 + \cos \gamma) \cos \gamma \qquad (31)$$

where γ is defined by

$$\sin \gamma = R_\odot/(R_\odot + h) \qquad (32)$$

$R_\odot$ being the solar radius. When $h \to 0$, $k_G \to 0$ as expected. With limb darkening, $k_G \neq 0$ even when $h = 0$. This is the case that will be considered in Section 6 below, as the scattering atoms are down in the photosphere itself.

The polarization diagrams of Figure 4 provide a good overview of the various coherence effects. Let us first consider the loops that occur toward the bottom of the diagram. They are due to the level crossings between the $^3D_2$ and $^3D_3$ components (cf. Figure 3), and therefore only show up in the bottom part of the polarization diagrams, where the field strength exceeds about 20 G. As the loops cause multiple-valuedness in the relation between polarization and magnetic field, this part of the diagrams is less useful for diagnostic purposes.

The undulations of the solid contour lines for field strengths below 1 G is due to coherences in the fine structure components of the *lower* level. The Zeeman splitting equals the natural width for a field strength of 0.075 G for the $^3S_1$ level, 0.83 G for the $^3P_{0,1,2}$ levels. The different field-strength scalings existing in the system results in the undulations seen.

The observations provide the two parameters p and β, which can be entered as points in theoretical diagrams like those of Figure 4 to read off the field strength B and field azimuth χ. This, however, cannot be accomplished in a unique way from one spectral line alone, like the He D_3 line. Disregarding the problem with the loops, which only occur within a certain field-strength range, we must deal not only with one polarization diagram, but with a family of diagrams like those of Figure 4, one for each value of the angle ϑ_B between the magnetic field vector and the vertical direction. There is no way to uniquely determine three unknowns (the three spatial components of **B**) from two observables (p and β). Another ambiguity is caused by symmetries in the problem, as is most clearly seen by the top diagram of Figure 4 (valid for horizontal fields), which is identical for positive and negative values of the field azimuth χ (with respect to the line of sight).

These various ambiguities can be resolved by using different types of supplementary information. One possibility is to record the polarization simultaneously in two or more spectral lines with different sensitivities to the Hanle effect. The two lines must be formed in the same part of the prominence, and the magnetic field must not be vertical (since the Hanle effect vanishes for all lines when the incident radiation field is symmetric around the magnetic field vector). One useful combination that has been tested[55] is to separately record the two adjacent, spectrally resolved components $3d\,^3D_{3,2,1} \to 2p\,^3P_{2,1}$ and $3d\,^3D_1 \to 2p\,^3P_0$ of the He D_3 line. Another method is to simultaneously record the circular polarization (Stokes V) and obtain the line-of-sight component of the field directly via the longitudinal Zeeman effect.[55,51]

One ambiguity can be directly removed simply by postulating that the prominence magnetic field is horizontal ($\vartheta_B = 90°$), which is well justified (except in the prominence "legs") for reasons of stability, since the condensed plasma must be supported in an equilibrium situation by the magnetic field against the force of gravity. Then only two unknown spatial components of the field have to be determined.

The observational errors of $p_{Q,U}$ in the prominences are typically 0.1%. Inversions with the methods outlined above[55,47] are claimed to give (when $B \approx 10$ G) accuracies in the field strength of about 0.5 G, in the field azimuth of about 5°, and in the inclination angle ϑ_B of about 10°, although noisy data can also result in fictitious values of ϑ_B. The deduced values of the field strength and azimuth are, however, insensitive to errors in the inclination.[3]

The polarization diagrams of Figure 4 are based on the idealized assumption that the only source of depolarization is the Hanle effect (apart from the geometrical effect of angular averaging). Thereby three other effects have been neglected: (1) multiple scattering, (2) collisional depolarization, (3) inhomogeneous magnetic-field structure.

The He D_3 line is free from multiple scattering and collisional effects, since the optical thickness and helium number density are low. If, however, the magnetic-field strength and direction vary within the spatial resolution element of the instrument or along the line of sight within the prominence, the effect will primarily be to decrease the angle of rotation β as compared with the case of a homogeneous magnetic field that carries the same average magnetic flux.

Accordingly, observational points can fall near the outer contours of the polarization diagram only if the magnetic field is *both* oriented almost along the line of sight, *and* the field is rather homogeneous within the effective resolution element (which typically corresponds to 5000–10,000 km on the sun). For prominences observed edge-on (line of sight almost parallel to the length axis of the prominence), the direction of the magnetic field is expected to be nearly parallel to the line of sight. The observations show that about $\frac{1}{4}$ of all recordings made in such prominences give points that fall near the outer contours in a polarization diagram like the top one of Figure 4, in spite of the long line-of-sight integration occurring when the prominence is viewed edge-on.[3]

This result gives strong evidence that small-scale inhomogeneities in the prominence magnetic field are not very pronounced, in contrast to the extremely intermittent photospheric magnetic field, and that therefore the assumption of a homogeneous field is justified for the diagnostics of the prominences, although it should be applied with care. This conclusion is rather remarkable, considering the threadlike fine structure of the prominence emission, showing that the temperature and density structure is highly inhomogeneous on scales much smaller than what is normally resolved in polarimetric observations. It apparently implies that the small-scale fluctuations in temperature and density are largely uncoupled to the magnetic-field variations.

Multiple-scattering effects need to be taken into account in the hydrogen lines, and have been considered in detail for the Balmer line $H\alpha$.[53] Although the interpretation is greatly complicated by the introduction of these effects, they bring a significant diagnostic advantage. As we have shown in Figure 4 for the optically thin case and a horizontal magnetic field, there is an ambiguity in the field direction, since the same polarization signature occurs when the field vector is reflected around the line of sight. This symmetry, however, gets broken by multiple scattering in the prominence geometry.[53] Thereby it becomes possible in principle to at least partially eliminate this ambiguity. Another way to suppress this ambiguity is by statistical means, by making observations in prominences with different angles between their length axes and the line of sight, and considering the deduced distribution of the angle between the field vector and the prominence length axis.[56]

Quasi-simultaneous polarization measurements in the He D_3 and $H\beta$ lines have shown[52] that the $H\beta$ line is also subject to collisional depolarization in addition to the Hanle effect that influences both lines. As the collisional depolarization rate is proportional to the electron density, polarization observations in these two spectral lines allow the electron density in prominences to be diagnosed, in addition to the magnetic field.

In Figure 5, from Bommier et al.,[52] the effect of collisional depolarization in the $H\beta$ line is illustrated for a statistical sample. The histograms show for the D_3 and $H\beta$ lines how the observed amounts of polarization p are distributed for a sample of 100 measurements in 22 prominences. The

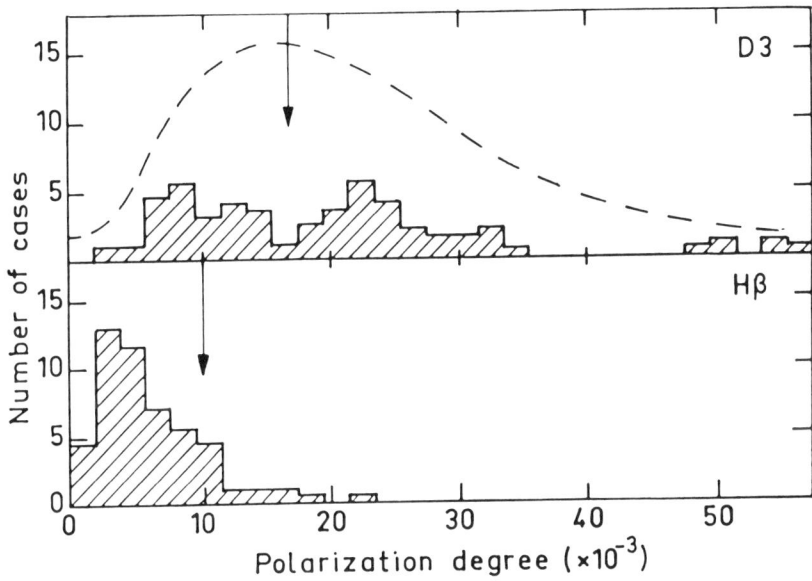

FIGURE 5. Histograms of observed values (100 measurements in 22 prominences) of the degree of linear polarization, for the He I D_3 line (top) and the hydrogen $H\beta$ line (bottom), from Bommier et al.[52] The dashed curve outlines the overall behavior of all the 3000 measurements since 1974.[42] The arrows mark the polarization expected for an 8 G horizontal magnetic field with an azimuth angle of 30° with respect to the line of sight, without accounting for collisional depolarization. The systematically low polarization values observed in $H\beta$ indicate that collisional depolarization is important for this line, information that can be used for the diagnostics of the electron density in prominences.

dashed curve gives the distribution that is representative for the total of about 3000 observations in the D_3 line since 1974. The arrows show the expected amount of polarization for an 8 G horizontal field with azimuth $\chi = 30°$, assuming no collisional depolarization. While for the D_3 line the arrow is centered on the histogram, it is located too much to the right of it in the case of the $H\beta$ line. This discrepancy is explained by collisional depolarization in $H\beta$.

Simultaneous observations in the He D_3 and hydrogen $H\beta$ lines give us four observed parameters: depolarization p/p_{max} and angle of rotation β of the plane of polarization, for each of the lines. As the two lines have different sensitivities to the Hanle effect, four unknowns may in principle be determined: the three vector components of the magnetic field, as well as the electron density.[52]

A prime objective of the magnetic-field determinations in the solar prominences is to be able to decide between competing theoretical models describing the origin and structure of these enigmatic objects. The two main competing models have been the older, more widely used Kippenhahn-Schlüter model,[57] and the Kuperus-Raadu model.[58] They make opposite predictions concerning the direction of the magnetic field in the prominence. Thus, in the Kuperus-Raadu model, the field direction in the prominence is predicted to be opposite to the field direction in the underlying photosphere, while the Kippenhahn-Schlüter model predicts the direction in the photosphere and prominence to be the same.

Observations based on the application of the Hanle effect show[56,3] that for all prominences appearing at high heliographic latitudes, and for prominences extending higher than 30,000 km into the corona, the measured field direction is consistent with that predicted from the Kuperus-Raadu model. However, in a considerable number of cases for prominences lower than about 30,000 km appearing at low heliographic latitudes, the observed field direction is consistent with the Kippenhahn-Schlüter model. It is clear that more detailed observations and interpretations are needed to clarify this somewhat confused situation. The Hanle effect is our prime tool to achieve this.

6. SURVEY OF SCATTERING POLARIZATION ON THE SOLAR DISK

Although the prominences have served in the important role as a testing ground for the Hanle effect and are quite interesting objects in themselves, they only represent a very special case for the potential use of the Hanle effect on the sun. For more general applications in the visible spectrum, we need to consider how the Hanle effect operates in the solar photosphere, on the solar disk. This, however, leads to a number of considerable complications, since we have to do with an optically thick, rather dense medium, with multiple scattering and large collisional depolarization. The anisotropy in the incident radiation field is not externally given, as it was for the prominences, but arises in the scattering medium itself through the radiative-transfer effects. The spectral lines are no longer emission lines as for the prominences, but absorption lines against a large background of continuum-spectrum photons.

At a first glance it might seem that these severe complications make the problem intractable, but this is not so. Although the theoretical tools are not yet so well developed, the Hanle effect has already provided useful information with considerable future potential, as will be seen below (and in Section 7). The complicated physics involved has, however, greatly delayed a more extensive exploitation of the Hanle effect.

Due to the complex nature of the Hanle effect in photospheric lines, we need, carefully in small steps, to build a foundation for its future general use as a diagnostic tool. Since the Hanle effect is the modification by a magnetic field of the polarization that is caused by coherent scattering in the line-formation process, we must first explore and understand the processes of coherent scattering on the sun in the absence of a magnetic field, to give us the background against which the Hanle effect can be gauged. For the prominence case, this merely amounted to the determination of p_{max}, which was simply obtained from geometrical considerations (height of the scattering atoms above the photosphere, limb darkening function), since the incident radiation field was given externally.

The situation is entirely different in the photosphere. For most lines in the visible spectrum on the solar disk, coherent scattering plays a very minor role in comparison with other competing processes (e.g., collisional excitation and incoherent scattering). The collisional broadening of the excited level is usually an order of magnitude larger than the natural width. For most lines the excited level is significantly coupled through collisional and radiative transitions not only to one lower level, but to several lower and higher levels.

Among this jungle of atomic processes, however, linear polarization measurements in regions where the transverse Zeeman effect is insignificant singles out the generally small coherent scattering part, since this part is the only source of the linear polarization. The linear polarization due to the transverse Zeeman effect is generally very small outside active regions, and it has a line profile signature entirely different from the scattering polarization. Since it does not depend on the way the upper level has been populated, it occurs with comparable strength in most spectral lines, while the scattering polarization singles out the smaller number of lines for which coherent scattering is important. Thus the coherence effects in the solar spectrum can easily be separated in a clean way from all the other effects by recording the linear polarization across the solar spectrum.

The theoretical basis for treating the coherence effects in the photosphere has been developed in various investigations.[59-68,32,30] Observationally, the coherence effects have been explored throughout the whole visible solar spectrum using instruments of the National Solar Observatory, including the HAO Stokesmeter at the NSO Sacramento Peak Observatory for selected lines in the UV spectrum,[69] the vertical grating spectrometer at the NSO Kitt Peak McMath telescope to survey the linear polarization in the UV range 3165-4230 Å,[70] and the Fourier transform spectrometer of the NSO McMath telescope to survey the polarization in the range 4200-9950 Å.[71] All recordings were made at a disk position 10" of arc inside the limb near either the south or the north heliographic pole, to eliminate the influence of strong magnetic fluxes in an effort to survey the nonmagnetic scattering polarization. This undertaking also marked the first application of a Fourier transform spectrometer as a polarimeter.

As is generally the case when entering into an almost completely unexplored territory, entirely unexpected effects were encountered, the most striking of which was the polarization signature found around the Ca II H and K lines, which are very strong and broad, and have their central wavelengths at 3968 and 3933 Å, respectively. These two lines have a common lower level, with $J = \frac{1}{2}$. The upper level has $J'' = \frac{3}{2}$ for the K line, $\frac{1}{2}$ for the H line. These J numbers determine the atomic depolarization factor W_2 in equations (20), which becomes $\frac{1}{2}$ for the K line, zero for the H line, if the two lines are considered independently of each other. It means, in particular, that the H line should be unpolarizable. The observations, given over a 100 Å range by the solid line in Figure 6, show however a striking "double S" profile, with the electric vector parallel to the solar limb on the short wavelength side of the K line and the long wavelength side of the H line, but with the electric vector *perpendicular* to the limb in a broad region between the K and H lines.

Shortly after its discovery,[69,30] this remarkable spectral feature could be explained in terms of quantum-mechanical interference between the $J'' = \frac{3}{2}$ and $\frac{1}{2}$ states of the upper level.[30] Although the central wavelengths of the H and K lines are separated by as much as 35 Å, many orders of magnitude larger than the damping width of each of the two excited levels, scattering in the two lines does not occur independently in each line, but they have to be treated as a single, indivisible quantum system. Although the damping cores of the levels are so widely separated, the dispersion wings overlap coherently, leading to an interference term in the scattering amplitude that is of the same order of magnitude as the incoherent terms, and that causes the sign reversal and the "double S" shape of the observed polarization curve.

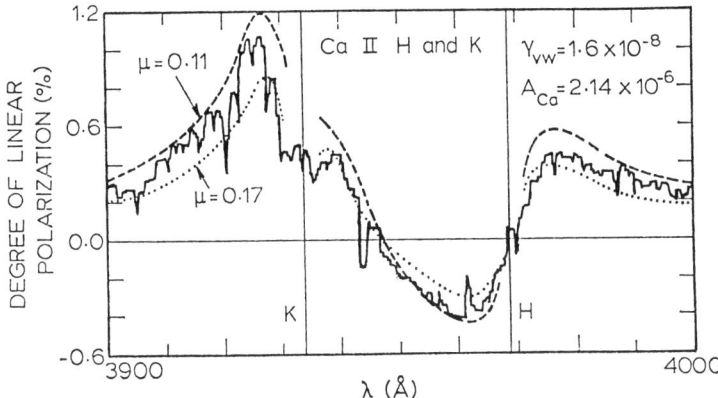

FIGURE 6. Linear polarization in the Ca II H and K lines. The solid curve represents the observations 10″ of arc inside the solar limb ($\mu \approx 0.14$), after a crude removal of the effect of the depolarizing blend lines.[30] The dashed and dotted curves are based on radiative-transfer calculations using values for the van der Waals collisional damping constant γ_{VW} and calcium abundance relative to hydrogen A_{Ca} as indicated. The two vertical lines mark the locations of the central wavelengths of the K and H lines (from Auer et al.[63]).

Under laboratory circumstances one would not expect to see effects so far away from the damping cores, since the probability of scattering there is vanishingly small. However, it is not zero. Due to the enormous number of Ca II ions in the sun, the line opacity remains large in the dispersion wings over distances on the order of 50 Å from line center. This means that of all the photons that we receive from the sun so far from the line centers, a substantial fraction of them are associated with this scattering transition. As we move further away from the line centers, the line photons get more and more diluted with photons from the continuous spectrum. This example demonstrates the usefulness of the solar atmosphere as an atomic physics laboratory.

As in the case of the Hanle effect, the effect of the coherent superposition of the dispersion wings of the two excited levels can be calculated from the Kramers–Heisenberg dispersion formula (11), but as we are so far away from the resonance, the magnetic-field effects in $\Delta\omega_H$ and the damping width Γ can be disregarded. The end result of these calculations[30] can most conveniently be given in terms of the atomic depolarization factor W_2 that we introduced in equations (20). Instead of having the discrete values $\frac{1}{2}$ and 0 for the K and H lines, W_2 becomes frequency dependent, as approximately given by

$$W_2 = \frac{(\nu - \nu_K)^{-2} + 2(\nu - \nu_K)^{-1}(\nu - \nu_H)^{-1}}{2(\nu - \nu_K)^{-2} + (\nu - \nu_H)^{-2}} \qquad (33)$$

where $\nu_{K,H}$ are the resonance frequencies of the K and H lines. It is the interference term (second term in the nominator) that causes the sign reversal of the polarization between the K and H lines.

To compare with the observed polarization curve in Figure 6, we have to compute the anisotropy of the incident radiation field, the effects of continuum absorption and polarization, collisional depolarization, and multiple scattering effects. A full radiative-transfer treatment using the W_2 of equation (33) has been performed.[63] The results, given by the dashed and dotted curves in Figure 6, show excellent agreement with the observations.

It is, however, also possible to obtain a similarly good agreement by means of a simple analytical model that highlights the basic physical ingredients in the problem (while the radiative-transfer formalism acts more like a "black box"). Thus the observed linear polarization p can be written[30,72]

$$p = \frac{kA + \sigma_c p_c}{B + \sigma_c} \tag{34}$$

where

$$A = \left(\frac{\lambda_H}{\lambda_K}\right)^2 a_K^2 + 2\left(\frac{\lambda_H}{\lambda_K}\right) a_H a_K$$

$$B = a_H^2 + 2\left(\frac{\lambda_H}{\lambda_K}\right)^2 a_K^2 \tag{35}$$

and

$$a_{H,K} = (\lambda_{H,K}^{-2} - \lambda^{-2})^{-1} \tag{36}$$

$\lambda_{H,K}$ are the central wavelengths of the H and K resonances.

The various symbols in expression (34) have the following physical meaning: The ratio A/B equals W_2, and is thus the fraction of the scattering processes that occur as polarizing dipole-type scattering. The second term in the expression for A in equations (35) is responsible for the quantum-mechanical interference effect. The scale factor k includes two effects: (1) the anisotropy of the radiation field, (2) collisional depolarization. Quantity σ_c is the continuum opacity, assumed to be wavelength-independent, and which we will give below in units of the line opacity σ_L halfway between the centers of the H and K lines; p_c is the continuum polarization.

If the continuum opacity were zero, p would simply be proportional to W_2. In reality, the line photons are diluted more and more by continuum photons the further out in the line wings we go; $B \to 0$, $A/B \to 1$, and $p \to p_c$ when $|\nu - \nu_{H,K}| \to \infty$.

This model, with the three free parameters k, σ_c, and p_c, should describe all transitions with $J = \frac{1}{2}$, $J'' = \frac{1}{2}, \frac{3}{2}$, i.e., not only the Ca II H and K lines, but also the Na I D_1 and D_2 lines at 5896 and 5890 Å,[69] as well as the Mg II h and k lines at 2803 and 2796 Å, observed with a polarimeter on the SMM (Solar Maximum Mission) satellite.[72] For the H and K lines, an excellent fit is obtained using $k = 0.0176$, $\sigma_c/\sigma_L = 1.0$, and $p_c = 0.001$. For the D_1 and D_2 lines the continuum has a much more dominant effect, and a $\sigma_c/\sigma_L \approx 8$ is needed. For the Mg II h and k lines the situation is the opposite, and σ_c/σ_L is as small as 0.01.

The survey of the linear polarization throughout the visible solar spectrum[70,71] has revealed many other cases of nonmagnetic quantum-mechanical interference between fine-structure components, also with various other combinations of J'' quantum numbers in different atomic multiplets, e.g., in Fe I, Mg I, and Co I. Another effect of great significance for the observed polarization is fluorescent scattering, e.g., when ultraviolet photons are absorbed in the Ca II H and K transition (multiplet no. 1 of Ca II), and are reradiated in the infrared triplet (multiplet no. 2) of Ca II at 8498, 8542, and 8662 Å. Similarly, in hydrogen, excitations in the Lyman lines may reradiate in the Balmer lines. The observed amount of polarization in the Balmer lines then depends on the anisotropy of the radiation field at the Lyman line wavelengths, not so much on the radiation field in the Balmer lines themselves.

Of all the lines in the visible solar spectrum, Ca I 4227 Å shows the largest amplitude, about 2% at a position 10″ of arc inside the solar limb. Most of the strongly polarizing lines are found in the UV, where the limb darkening is more pronounced (larger anisotropy of the incident radiation field), and resonant scattering plays a larger role in the formation of the lines. Examples of such lines are Ni I 3235 Å, Ti II 3242 Å, Cu I 3248 Å, Fe I 4272 Å, Cr I 4254 Å, Sr I 4607 Å, and Ba II 4554 Å. The CN molecular bands in the range 3840-3884 Å also exhibit polarization, increasing with wavelength within each band until the band head is reached.

The majority of lines in the solar spectrum show, however, only very small or no polarization. Since the continuum is polarized (increasingly with decreasing wavelength) due to Thomson scattering at free electrons and Rayleigh scattering at neutral hydrogen, the effect of unpolarized line radiation is to dilute the polarized continuum photons and thus suppress the continuum level of Stokes Q (which is proportional to the number of linearly polarized photons), giving the Stokes Q spectrum the appearance of an absorption-line spectrum like Stokes I, with the exception of a smaller number of polarizing lines, which then appear as "emission features" in Stokes Q.

Prepared by this overview of the nonmagnetic coherence effects in the solar spectrum, we can now turn to the effect of the magnetic field on the scattering polarization. Figure 7 shows the first recording of the Hanle effect on the solar disk.[73] It was made in 1978 in an active region (but outside

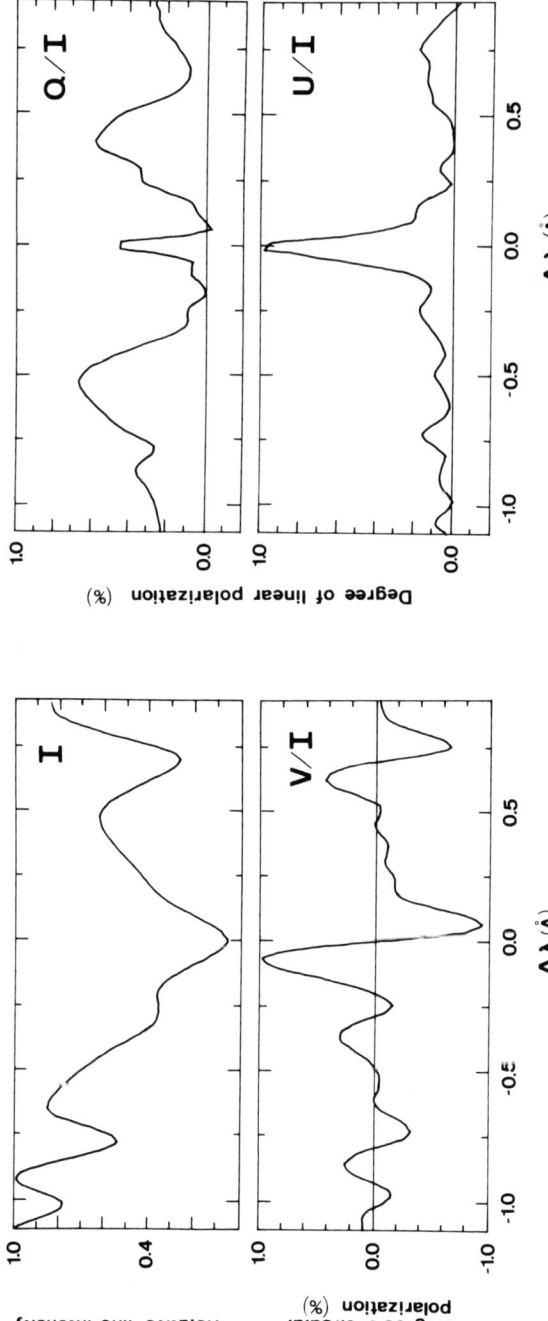

FIGURE 7. Stokes profiles of the Ca I 4227 Å line, recorded in an active region 70″ of arc inside the solar limb. While V/I shows the antisymmetric signatures in the Ca I Doppler core as well as in the surrounding blend lines, characteristic for the longitudinal Zeeman effect, the observed linear polarization is determined by the effects of coherent scattering. The Hanle effect reveals itself by converting the linear polarization in the Doppler core from Stokes Q into Stokes U (rotation of the polarization plane), but the effect vanishes in the line wings, as predicted theoretically (from Stenflo[73]).

sunspots), 70" of arc inside the solar limb, in the Ca I 4227 Å line. The intensity profile (Stokes I) is characterized by the very deep Doppler core, surrounded by wide dispersion wings on which blend lines of other elements are superposed. The fractional circular polarization (V/I) shows the antisymmetric profile signature in the Doppler core, characteristic for the longitudinal Zeeman effect (in a first approximation, Stokes V is proportional to $\partial I/\partial \lambda$).[14] The fractional linear polarization oriented parallel to the nearest solar limb (Q/I) shows the signature found for scattering polarization in nonmagnetic regions, with maxima in the line wings and in the Doppler core, except that the amplitude of the core peak is substantially reduced with respect to the nonmagnetic case. The striking feature is, however, the isolated core peak for U/I, which represents linear polarization at an angle 45° to the limb. In the nonmagnetic case Stokes U is zero. In the magnetic case Q is converted into U via the Hanle effect through rotation of the plane of linear polarization. As explained in Section 4 above, the Hanle effect only operates in the Doppler core, not in the dispersion wings of the line. This is the reason why no significant U polarization appears in the line wings. Figure 7 indicates how the Hanle effect can be used on the solar disk for magnetic-field diagnostics.

Another observation of the Hanle effect on the solar disk has been in the form of maps of the linear polarization in the core of the Ca I 4227 Å line as well as in the adjacent continuous spectrum, across an active region near the limb.[10] Both the continuum and line core showed depolarization effects cospatial with the plage emission features, which mark magnetic-field concentrations. The continuum depolarization must be understood in terms of the different geometry of scattering in the magnetic fluxtubes, which have more of a cylindrical geometry in contrast to the largely plane-parallel stratification of the external atmosphere. Thereby the anisotropy of the incident radiation field becomes different. The depolarization effects in the line core were, however, found to be larger than in the continuum. This additional depolarization has its natural explanation in terms of the Hanle effect. Thus, by ratioing the polarization fluctuations in the line core and continuum, the Hanle depolarization may be separated from the geometrical depolarization.

7. DIAGNOSTICS OF TURBULENT MAGNETIC FIELDS

As described in Section 2 above, the magnetic field in the solar photosphere has an extreme fine structure, with more than 90% of the magnetic flux recorded in magnetograms (using the longitudinal Zeeman effect) in kG form (also far away from active regions). As the average flux density is only a few G, these strong fields occupy only a fraction of one percent of the photospheric volume. Although the magnetograms do not provide information on what goes on in the remaining 99% of the volume, it is clear that it cannot be field-free. The photosphere and deeper regions are in a turbulent state. If there were equipartition between the energies in the magnetic field and the turbulent motions, a tangled field with a strength of a few hundred G would be expected in the photosphere.

For such a tangled or turbulent magnetic field the opposite polarities are likely to be mixed on a very small scale, smaller than the spatial resolution element used in recordings with a solar magnetograph. In such a case it will not contribute to any net circular-polarization signal due to cancellation within the resolution element of contributions of opposite signs. A turbulent field is thus "hidden" to magnetographs based on the longitudinal Zeeman effect.

This "hidden" field may, however, be revealed and diagnosed by the Hanle effect, since the Hanle depolarization only has one "sign" (*reduction of the polarization*) regardless of the orientation of the field vector, and thus is not subject to signal cancellation within the spatial resolution element. If we can separate the Hanle depolarization from the other depolarization effects (geometrical, collisional, etc.), and make some assumption on the distribution of field vectors within the resolution element (e.g., an isotropic distribution), we may determine the strength B_t of the turbulent field.

A complete solution of this problem would require making a detailed radiative-transfer calculation of the nonmagnetic scattering polarization, adjusting the uncertain parameters in the atmospheric model and atomic structure (in particular the collisional effects) until good agreement with the observed polarized profile is achieved in the line wings (where the Hanle effect is absent). Comparison between the observed and theoretical polarization amplitudes in the Doppler core would then reveal the magnitude of the Hanle depolarization. This should be performed in a number of different types of spectral lines, to verify that the results obtained are consistent in the different lines. Such an ambitious program has not yet been executed, due to the complex radiative-transfer physics involved.

Instead, a much more modest version of this program has been carried out, in which the radiative-transfer problem has been bypassed.[77] It allows only crude (and uncertain) estimates of the turbulent field strength B_t, but the approach has the advantage that the physics involved becomes more transparent.

The basis for this approach is that the polarization of the scattered radiation scales with the various, independent depolarization factors, such that one can formally express the observed polarization p as the product

$$p = \alpha W_2 k_G k_c k_H \tag{37}$$

Here k_H represents the Hanle depolarization factor (denoted $p/p_{\max}$ in

Sections 4 and 5 above), W_2 the atomic, and k_c the collisional depolarization factor, as described in the preceding sections. Quantity α can be called a "non-LTE depolarization factor," and represents the fraction of all spontaneous radiative transitions from the upper to the lower level, which have been immediately preceded by radiative excitation from the lower level, i.e., which represent a scattering event; k_G is determined by the anisotropy in the radiation field, and is the factor by which the scattering polarization is reduced due to angular averaging, as compared with the case of 90° scattering when $W_2 = k_c = 1$.

Equation (37) has been deduced by considering a single scattering process, with an unpolarized incident radiation field having an angular distribution that determines k_G. In the real, optically thick solar atmosphere, we of course have multiple scattering, but it is primarily the *last scattering process* that determines the observed polarization. Each scattering event generates some polarization, but as the anisotropy of the radiation field is small, the polarization of the radiation field that is incident on the last scattering particle is small, and generally has no significant effect on the polarization of the light emerging from the last scattering process. The determining factor is the anisotropy of the incident radiation field, not its small degree of polarization, which therefore may be disregarded. Thereby the problem is reduced to the case of a single scattering process, as implied by equation (37).

While the factor W_2 is independent of the conditions in the solar atmosphere, the other factors in equation (37) may vary considerably with height. Even if the last scattering approximation is valid, as indicated above, we have to know at what height in the atmosphere this last scattering event takes place in order to be able to estimate the values of α, k_G, and k_c. While k_c depends on the local physical parameters in the atmosphere, this is not the case for k_G and α, which are coupled to the local radiation field that has nonlocal sources. In general one would need a full non-LTE radiative-transfer treatment to calculate their values. However, in such a full treatment one would not need to go the way via the factorization of equation (37), since the Stokes vector emerging from the atmosphere, and thus p, would be obtained as part of such a computation.

Nevertheless, one may obtain useful estimates of p via equation (37) through the following considerations. Generally, the incident radiation field to be used at the last scattering process in the line depends on the details of the frequency redistribution (due to Doppler motions in combination with the natural probability distribution of the energy levels). Near the central wavelength, however, the general case of partial redistribution is not very far from the special case of frequency coherence (no redistribution), since the dominating source-function contribution comes at the central wavelength. Thus to obtain crude estimates of the expected polarization, we may use the radiation field at the central wavelength at an atmospheric height where the optical depth at this wavelength is about unity. For a position on the solar disk characterized by $\mu = \cos \vartheta$, where ϑ is the heliocentric angle, i.e., the angle between the solar radius and the line of sight, the last scattering process occurs approximately at $\tau_\nu = \mu$, where τ_ν is the optical depth in the vertical direction at core frequency ν. This means that when observing at the extreme limb of the sun, where $\mu \approx 0$, the last scattering particle is on top of the atmosphere. The incident radiation field of the last scattering process is then simply determined by the limb-darkening function of the radiation emerging from the atmosphere at this wavelength. This is a directly observed quantity, and does not depend on any theory.

The observational data set used to estimate the value B_t of the turbulent field strength has been obtained at a position $10''$ of arc inside the solar limb, which corresponds to $\mu \approx 0.14$. This is close to the extreme limb, so the use of the observed limb-darkening function for the calculation of k_G should be a reasonable approximation. It may, however, lead to k_G being somewhat overestimated (since the anisotropy, which produces k_G, decreases with depth in the atmosphere).

Another circumstance that makes equation (37) useful for a quantitative analysis is that a number of strongly polarizing resonance lines can be found (several of which were mentioned in Section 6 above), for which it can be safely assumed that $\alpha = 1$ (line formation entirely by resonant scattering). Then the problem of calculating the level populations and the relative importance of competing types of transitions gets eliminated.

Finally we need to consider the collisonal depolarization factor k_c, which is small in the lower photosphere, but approaches unity asymptotically with height, due to the exponential decrease in gas density. As, near the limb, we receive the radiation from the upper part of the atmosphere, k_c generally does not differ greatly from unity there. In the previous work to determine B_t,[73] all spectral lines used were estimated to have k_c values in the range 0.4–1.0. Because of this natural saturation level for k_c, estimates of its value cannot be wrong by a large factor.

Thus, having estimated all the other factors in equation (37), the Hanle depolarization factor k_H can be derived from the observed polarization p. The next question that arises is how we get from the "observed" value of k_H to the value B_t of the turbulent field. This translation requires some assumption concerning the angular distribution of the field vectors of the unresolved, tangled field inside the spatial resolution element.

In general, we expect that the field vectors are distributed symmetrically around the vertical direction. The integration over all field azimuths leads to zero rotation of the plane of linear polarization (for symmetry reasons), with depolarization (k_H) as the only signature of the Hanle effect. For the inclination angles ϑ_B with the vertical direction, we may explore different possibilities, the most natural and likely one being an isotropic distribution. However, due to the exponential density decrease with height, leading to strong buoyancy forces acting on the flux elements, it is conceivable to have deviations from isotropy in the sense of some preference for the vertical direction.

A general, one-parameter family of ϑ_B distributions has members proportional to $\cos^a \vartheta_B$, with the exponent a as the free parameter. The value $a = 0$ corresponds to an isotropic distribution of field vectors, which gets more and more peaked around the vertical direction with increasing value of a. In this general case the Hanle depolarization k_H is related to B_t through[74]

$$k_H = 1 - \frac{6}{(a+3)(a+5)}[(a+1)\sin^2 \alpha_1 + \sin^2 \alpha_2] \qquad (38)$$

where

$$\sin^{-2} \alpha_q = 1 + \left(\frac{B_0}{qk_c B_t}\right)^2 \qquad (39)$$

and

$$B_0 = \frac{2mc}{e}\frac{\Gamma_b}{g_b} \qquad (40)$$

m and e are the electron mass and charge, c the speed of light, Γ_b and g_b the natural width and Landé factor of the excited level b.

Quantity B_0/k_c represents the characteristic field strength around which the Hanle effect has its greatest sensitivity. The Hanle effect is useful in a field-strength range that is typically 0.1 to 10 times this value. Where this range happens to be in ordinary field-strength units (G) depends on the spectral line used, as seen by equation (40). The relation to the Ω parameter that we introduced in equation (19) and used in Figure 2 is given by

$$\Omega = \frac{2k_c B_t}{B_0} \qquad (41)$$

Figure 8, from Stenflo,[73] shows k_H as a function of $B_0/(k_c B_t) = 2/\Omega$, for an isotropic distribution [$a = 0$ in equation (38)], as well as for a distribution with the field vectors confined to the horizontal plane (which

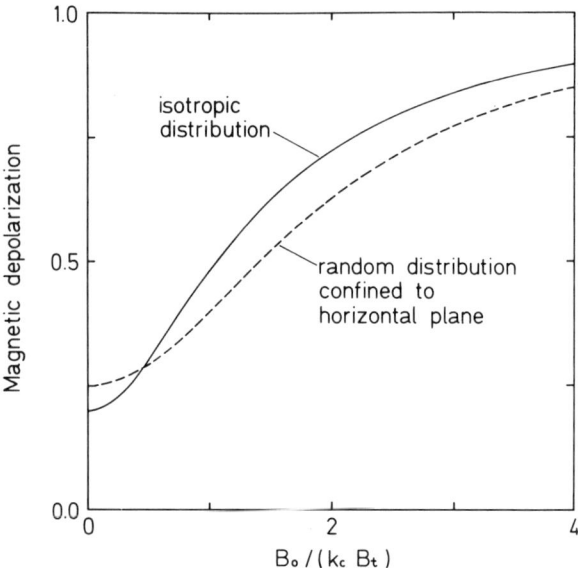

FIGURE 8. Hanle-effect depolarization k_H for an isotropic distribution of field vectors of strength B_t (solid line), as well as for the case that the field vectors are confined to the horizontal plane but are randomly oriented (dashed line). B_0 is the field strength for which the Larmor frequency equals the inverse lifetime of the excited level. k_c is the collisional depolarization factor (from Stenflo[73]).

gives the same k_H as a distribution with $a = -1$). We note in particular that k_H does not go to zero (complete depolarization) when $B_t \to \infty$, but to 0.20 for an isotropic distribution, 0.25 for a horizontal distribution. The reason is that the Hanle depolarization disappears ($k_H = 1$) for a vertically oriented field (when the incident radiation field is symmetrical around the field vector), as well as for horizontal fields that are perpendicular to the line of sight. If all field vectors were parallel to the line of sight, then $k_H \to 0$ as $B_t \to \infty$.

The similarity between the two curves in Figure 8 shows that k_H is not sensitive to the assumed field distribution, as long as it is not peaked around the vertical direction. A recent analysis using symmetry properties of the ordinary transverse Zeeman effect[74] has allowed constraints to be determined for the shape of the field distribution in the "hidden," turbulent magnetic flux on the sun. Accordingly, it is found that the parameter $a \lesssim 1.3$, possibly still much smaller, which implies that the deviations from an isotropic distribution cannot be very large.

Using previous observations of p[69] and assuming an isotropic distribution of field vectors, careful analysis with the methods outlined above, based on equation (37), has resulted in a *lower* limit for the turbulent field strength B_t of about 10 G (the upper limit could not be well constrained).[73] If, on the other hand, B_t were very large, it would contribute significantly to the broadening of the unpolarized line profiles, since the Zeeman broadening is independent of the field polarity and thus does not average out over the spatial resolution element. Searching for a correlation between line broadening and Landé factor, an *upper* limit to B_t of about 100 G could be set.[75]

Combination of the various techniques has thus constrained the turbulent field to have a strength between 10 and 100 G, and to have a nearly isotropic distribution of field vectors.

8. DIAGNOSTICS OF MAGNETIC FIELDS IN THE CHROMOSPHERE-CORONA TRANSITION REGION AND ABOVE

The chromosphere is the region in the solar atmosphere where the temperature inversion starts. Throughout the sun the temperature steadily decreases with distance from the center until it reaches a minimum of around 4000 K at the top of the photosphere. From there on, however, there is an extremely steep temperature increase to the million degrees that are found in the solar corona. The lower, denser part of this temperature inversion layer, where the hydrogen spectrum is formed (below about 40,000 K), is the chromosphere. Then follows the transition zone that brings the temperature up to the coronal levels over a distance of less than 1000 km. The corona is rather isothermal due to the high heat conductivity mediated by the free electrons, and the temperature decreases only slowly when moving away from the sun. For instance, the kinetic temperature of the electrons in the solar wind (the expanding corona) is as high as about 10^5 K at the earth's distance from the sun.

To produce and maintain these high temperatures, the outer atmosphere must be efficiently heated, by waves (developing shocks due to the exponential drop-off in density) or by Joule heating involving electric currents. Whatever the detailed heating mechanisms are, it is clear that the magnetic fields play a central role. The emission structures of the transition region and corona are very closely correlated with the magnetic fields, as observed in the photosphere below, or extrapolated from there into the corona.

Magnetic fields in these layers are, however, not easily accessible to observations, since the radiation from this hot-temperature plasma occurs primarily in the vacuum ultraviolet, below about 1700 Å, and must be observed with instruments outside the earth's atmosphere. Since the Zeeman splitting is proportional to λ^2, the polarization signal from the Zeeman effect is very small at these wavelengths, and it is further reduced as the field strength decreases with height in the atmosphere. Accurate polarimetry in the vacuum ultraviolet is also difficult technologically.

The Hanle effect, on the other hand, has significant advantages for magnetic-field diagnostics in these high atmospheric layers. Resonance scattering plays a dominant role for a large fraction of the emission lines in this spectral range. This scattering is coherent, since depolarizing collisions can be neglected due to the low density. The radiation field is generally highly anisotropic, as evidenced by the high contrast of the emission features at these wavelengths. The continuum opacity is generally negligible in comparison with the line opacity (emission-line spectrum). Therefore the expected degree of linear polarization is relatively large, making it more accessible to the difficult polarimetric observations. The expected magnetic field strengths fall in the range that is optimum for the Hanle effect for most of the spectral lines available. Ambiguities in the interpretation can be removed by making observations simultaneously in two lines with different sensitivities to the Hanle effect.[76]

The first step toward a full exploitation of the Hanle effect is to explore and understand the physics of coherent scattering polarization in the absence of magnetic fields, as was done in Section 6 for the visible spectrum on the solar disk. Thus radiative-transfer calculations have been carried out to investigate the formation of optically partially thick lines formed in the chromosphere–corona transition region, and to show how the observed scattering polarization depends on the optical thickness of this narrow transition zone.[77] Another approach is to avoid the radiative-transfer problem altogether by selecting only the optically thin situations. Thus the diagnostic use of the Hanle effect in the O VI 1032 Å line under optically thin conditions in the solar corona has been examined in detail.[78] Extensive calculations have also been performed to explore the possible use of the H Ly α line for the diagnostics of magnetic fields in the optically thin corona.[79]

A first attempt to measure scattering polarization in the solar vacuum ultraviolet was made in 1976 on the Soviet Intercosmos 16 satellite,[80,81] which achieved an accuracy of about 1% in the H Ly α 1216 Å line, with low spatial resolution, and demonstrated the feasibility of such measurements. A much more powerful and versatile instrument, the UVSP (Ultraviolet Spectrometer Polarimeter), on board the SMM (Solar Maximum Mission) satellite, allowed the Zeeman-effect circular polarization to be recorded above a large sunspot in the C IV 1548 Å line.[82]

The UVSP on the SMM satellite could also be used to record the scattering polarization across a 15 Å wavelength range centered around the strong Mg II h and k resonance lines at 2803 and 2796 Å.[72] These lines are due to transitions with the same quantum numbers as the Ca II H and K lines and the Na I D_1 and D_2 lines discussed in Section 6 above, and must therefore be subject to the same type of quantum-mechanical interference between the upper $J'' = \frac{1}{2}$ and $\frac{3}{2}$ levels, as described by equations (33)-(36). As a matter of fact detailed radiative-transfer calculations[63] predict the polarization amplitude in the Mg II h and k lines to be larger than that of any line in the visible solar spectrum, including Ca I 4227 Å, and that the quantum interference effects should be even more dominant than was the case for the Ca II H and K lines (displayed in Figure 6 above).

The observations[72] confirm the theoretical predictions in the line wings on the short wavelength side of the k line and the long wavelength side of the h line, but do not show the predicted negative polarization between the k and h lines. The analysis suggests that the discrepancy is due to effects of partial frequency redistribution, which may play a significant role for these lines in combination with the quantum interference phenomenon, since the separation between the two lines is only 7 Å (as compared with 35 Å in the case of the Ca II H and K lines). Partial redistribution is a very complex problem of radiative-transfer physics, the effects of which on the scattering polarization have been explored for a number of special cases.[61,64,67,68]

Further observational progress in this area is hampered by lack of instrumentation. No presently existing instrument in space can measure such polarization. It is likely, however, that new generations of space experiments for measuring linear polarization in the solar vacuum ultraviolet will be developed, due to the scientific importance of magnetic-field diagnostics in this portion of the solar spectrum.

9. CONCLUDING REMARKS

We have seen that the Hanle and Zeeman effects have different and only partially overlapping regimes, in which they are most useful for solar diagnostics. They therefore nicely complement each other. The main areas in which the Hanle effect has the most promise are: (1) diagnostics of vector magnetic fields in prominences; (2) diagnostics of magnetic fields in the atmospheric layers, where the solar corona is being heated (requiring spacecraft observations in the vacuum ultraviolet); (3) diagnostics of turbulent magnetic fields in the solar atmosphere, i.e., fields that are tangled up on a scale smaller than can be spatially resolved.

In addition to magnetic-field diagnostics via the Hanle effect, the scattering polarization provides information on macroscopic electric fields,[83] the collision rates in the scattering medium, the electron density, the optical thickness of the line-forming layers, the three-dimensional geometry of the atmospheric structures (contributing to the anisotropies in the radiation field), subtle effects in radiative-transfer physics, including effects of partial frequency redistribution, as well as certain aspects of atomic physics (including fluorescent scattering and level-crossing effects), etc. These examples illustrate the usefulness of accurate recordings of linear polarization, and demonstrate the versatility of the solar atmosphere as a physics laboratory.

To separate the various effects from each other and to determine all three components of the magnetic-field vector via the Hanle effect, one needs to make simultaneous recordings of the linear polarization in two or more spectral lines with different sensitivities to the Hanle effect, but which are otherwise formed in the same atmospheric layers. Such an approach has in the past proven necessary and extremely fruitful also in the application of the Zeeman effect to photospheric spectral lines,[11] since the photospheric magnetic field due to its extreme fine structure that is much smaller than can be spatially resolved, cannot be properly diagnosed from information collected in a single spectral line.

Progress in the application of the Hanle effect has been relatively slow, and we are still only at the beginning of its full exploitation. The theoretical interpretation of the polarization observations is very complicated, involving various aspects of physics and generally complex geometries, but we are beginning to get these problems under control. Progress on the observational side has been hampered by the lack of suitable instrumentation for accurate linear polarization observations, both ground-based and from space. One main obstacle has been that almost all existing major solar telescopes are afflicted by serious instrumental polarization, predominantly linear, arising inside the telescope itself and generally being much larger than the intrinsic solar polarization that we want to measure.

As the key role of magnetic fields for understanding the sun is more and more being appreciated, future major solar telescopes will be much better equipped to make detailed polarization observations of the sun. An example is the French THEMIS project, an intermediate-size (90 cm aperture) "polarization-free" solar telescope to be placed on the island of Tenerife in the 90s.[84] The major observational breakthrough is, however, expected to come with LEST (Large Earth based Solar Telescope), a planned international next-generation-type solar telescope with "polarization-free" design and a 2.4 m diffraction-limited aperture.[85,86] Internal "seeing" will be reduced by filling the telescope with helium, and external seeing (turbulence in the earth's atmosphere) will be compensated for by the use of adaptive optics, such that a spatial resolution of 0.1″ of arc can be achieved. LEST, which should become operational in the second half of the 90s depending on the funding situation, will allow high spatial and spectral resolution to be combined with high polarimetric accuracy for the detailed mapping of Stokes vectors across the solar disk and the solar spectrum. This would bring a new dimension to the exploration of solar magnetic fields, via both the Zeeman and Hanle effects.

REFERENCES

1. G. E. HALE, *Astrophys. J.* **28**, 100 (1908).
2. W. HANLE, *Z. Phys.* **30**, 93 (1924).
3. J.-L. LEROY, in *Measurements of Solar Vector Magnetic Fields*, edited by M. J. Hagyard, NASA Conf. Publ. 2374 (1985), p. 121.
4. S. A. KAZANTSEV, *Sov. Phys. Usp.* **26**, 328 (1983).
5. E. N. PARKER, *Astrophys. J.* **122**, 293 (1955).
6. E. N. PARKER, *Ann. Rev. Astron. Astrophys.* **8**, 1 (1970).
7. M. SCHÜSSLER, in *Solar and Stellar Magnetic Fields: Origins and Coronal Effects*, edited by J. O. Stenflo, IAU Symp. **102**, (1983), p. 213.
8. P. A. GILMAN, in *Solar and Stellar Magnetic Fields: Origins and Coronal Effects*, edited by J. O. Stenflo, IAU Symp. **102**, (1983), p. 247.
9. J. O. STENFLO, *Rep. Prog. Phys.* **41**, 865 (1978).
10. J. W. HARVEY, in *Small Scale Magnetic Flux Concentrations in the Solar Photosphere*, edited by W. Deinzer, M. Knölker, and H. H. Voigt, *Abh. Akad. Wiss. Gött.* No. 38 (1986), p. 25.
11. J. O. STENFLO, *Sol. Phys.* **32**, 41 (1973).
12. J. W. HARVEY, *Highlights Astron.* **4**, 223 (1977).
13. J. O. STENFLO, *Adv. Space Res.* **4**, 5 (1984).
14. J. O. STENFLO, *Sol. Phys.* **100**, 189 (1985).
15. J. O. STENFLO, *Mitt. Astron. Ges.* **65**, 25 (1986).
16. J. O. STENFLO AND M. VOGEL, *Nature* **319**, 285 (1986).
17. J. O. STENFLO, *Astrophys. Space Sci.* **144**, 321 (1988).
18. E. R. PRIEST, *Solar Magnetohydrodynamics*, Geophysics and Astrophysics Monographs (Reidel, Dordrecht, 1982).
19. E. LANDI DEGL'INNOCENTI, *Sol. Phys.* **85**, 33 (1983).
20. L. L. HOUSE, *J. Quant. Spectrosc. Radiat. Transfer* **11**, 367 (1971).
21. A. ROBSON, *The Theory of Polarization Phenomena* (Clarendon Press, Oxford, 1974).
11. E. LANDI DEGL'INNOCENTI, *Sol. Phys.* **85**, 3 (1983).
23. F. K. LAMB AND D. TER HAAR, *Phys. Rep.* **2C**, 253 (1971).
24. V. BOMMIER AND S. SAHAL-BRÉCHOT, *Astron. Astrophys.* **69**, 57 (1978).
25. V. BOMMIER, *Astron. Astrophys.* **87**, 109 (1980).
26. H. A. KRAMERS AND W. HEISENBERG, *Z. Phys.* **31**, 681 (1925).
27. P. A. M. DIRAC, *Quantum Mechanics* (Oxford University Press, London, 1958).
28. L. L. HOUSE, *J. Quant. Spectrosc. Radiat. Transfer* **10**, 909 (1970).
29. L. L. HOUSE, *J. Quant. Spectrosc. Radiat. Transfer* **10**, 1171 (1970).
30. J. O. STENFLO, *Astron. Astrophys.* **84**, 68 (1980).
31. C. M. PENNEY, *J. Opt. Soc. Am.* **59**, 34 (1969).
32. J. O. STENFLO, *Astron. Astrophys.* **66**, 241 (1978).
33. M. LANDI DEGL'INNOCENTI AND E. LANDI DEGL'INNOCENTI, *Astron. Astrophys.* **192**, 374 (1988).
34. A. OMONT, E. W. SMITH, AND J. COOPER, *Astrophys. J.* **182**, 283 (1973).
35. M. LANDOLFI AND E. LANDI DEGL'INNOCENTI, *Astron. Astrophys.* **167**, 200 (1986).
36. D. V. KUPRIANOV, I. M. SOKOLOV, AND S. V. SUBBOTIN, *Zh. Eksp. Teor. Fiz.* **93**, 127 (1987).

37. J. O. STENFLO, in *Solar Magnetic Fields*, edited by R. Howard, IAU Symp. **43** (1971), p. 101.
38. A. WITTMANN, *Sol. Phys.* **35**, 11 (1974).
39. L. L. HOUSE AND R. STEINITZ, *Astrophys. J.* **195**, 235 (1975).
40. L. H. AUER, J. N. HEASLEY, AND L. L. HOUSE, *Astrophys. J.* **216**, 531 (1977).
41. L. G. STENHOLM AND J. O. STENFLO, *Astron. Astrophys.* **67**, 33 (1978).
42. J.-L. LEROY, G. RATIER, AND V. BOMMIER, *Astron. Astrophys.* **54**, 811 (1977).
43. J.-L. LEROY, *Astron. Astrophys.* **60**, 79 (1977).
44. S. SAHAL-BRÉCHOT, V. BOMMIER, AND J.-L. LEROY, *Astron. Astrophys.* **59**, 223 (1977).
45. J.-L. LEROY, *Astron. Astrophys.* **64**, 247 (1978).
46. J.-L. LEROY, *Sol. Phys.* **71**, 285 (1981).
47. E. LANDI DEGL'INNOCENTI, *Sol. Phys.* **79**, 291 (1982).
48. J.-L. LEROY, V. BOMMIER, AND S. SAHAL-BRÉCHOT, *Sol. Phys.* **83**, 135 (1983).
49. R. G. ATHAY, C. W. QUERFELD, R. N. SMARTT, E. LANDI DEGL'INNOCENTI, AND V. BOMMIER, *Sol. Phys.* **89**, 3 (1983).
50. M. B. GORNYI, D. V. KUPRIANOV, AND B. G. MATISOV, *Astron. Zh.* **61**, 1158 (1984).
51. C. W. QUERFELD, R. N. SMARTT, V. BOMMIER, E. LANDI DEGL'INNOCENTI, AND L. L. HOUSE, *Sol. Phys.* **96**, 277 (1985).
52. V. BOMMIER, J.-L. LEROY, AND S. SAHAL-BRÉCHOT, *Astron. Astrophys.* **156**, 79, 90 (1986).
53. E. LANDI DEGL'INNOCENTI, V. BOMMIER, AND S. SAHAL-BRÉCHOT, *Astron. Astrophys.* **186**, 335 (1987).
54. M. LANDOLFI AND E. LANDI DEGL'INNOCENTI, *Sol. Phys.* **98**, 53 (1985).
55. V. BOMMIER, J.-L. LEROY, AND S. SAHAL-BRÉCHOT, *Astron. Astrophys.* **100**, 231 (1981).
56. J.-L. LEROY, V. BOMMIER, AND S. SAHAL-BRÉCHOT, *Astron. Astrophys.* **131**, 33 (1984).
57. R. KIPPENHAHN AND A. SCHLÜTER, *Z. Astrophys.* **43**, 36 (1957).
58. M. KUPERUS AND M. A. RAADU, *Astron. Astrophys.* **43**, 189 (1974).
59. S. DUMONT, A. OMONT, AND J. C. PECKER, *Sol. Phys.* **28**, 271 (1973).
60. J. O. STENFLO, *Astron. Astrophys.* **46**, 61 (1976).
61. S. DUMONT, A. OMONT, J. C. PECKER, AND D. REES, *Astron. Astrophys.* **54**, 675 (1977).
62. D. E. REES, *Publ. Astron. Soc. Jpn.* **30**, 455 (1978).
63. L. H. AUER, D. E. REES, AND J. O. STENFLO, *Astron. Astrophys.* **88**, 302 (1980).
64. D. E. REES AND G. J. SALIBA, *Astron. Astrophys.* **115**, 1 (1982).
65. E. LANDI DEGL'INNOCENTI, *Sol. Phys.* **91**, 1 (1984).
66. E. LANDI DEGL'INNOCENTI, *Sol. Phys.* **102**, 1 (1985).
67. G. J. SALIBA, *Sol. Phys.* **98**, 1 (1985).
68. M. FAUROBERT, *Astron. Astrophys.* **178**, 269 (1987).
69. J. O. STENFLO, T. G. BAUR, AND D. F. ELMORE, *Astron. Astrophys.* **84**, 60 (1980).
70. J. O. STENFLO, D. TWERENBOLD, AND J. W. HARVEY, *Astron. Astrophys., Suppl. Ser.* **52**, 161 (1983).
71. J. O. STENFLO, D. TWERENBOLD, J. W. HARVEY, AND J. W. BRAULT, *Astron. Astrophys., Suppl. Ser.* **54**, 505 (1983).
72. W. HENZE AND J. O. STENFLO, *Sol. Phys.* **111**, 243 (1987).
73. J. O. STENFLO, *Sol. Phys.* **80**, 209 (1982).
74. J. O. STENFLO, *Sol. Phys.* **114**, 1 (1987).
75. J. O. STENFLO AND L. LINDEGREN, *Astron. Astrophys.* **59**, 367 (1977).
76. J. O. STENFLO, in *The Energy Balance and Hydrodynamics of the Solar Chromosphere and Corona*, edited by R. M. Bonnet and P. Delache, IAU Coll. **36** (1977), p. 143.
77. J. O. STENFLO AND L. STENHOLM, *Astron. Astrophys.* **46**, 69 (1976).
78. S. SAHAL-BRÉCHOT, M. MALINOVSKY, AND V. BOMMIER, *Astron. Astrophys.* **168**, 284 (1986).
79. V. BOMMIER AND S. SAHAL-BRÉCHOT, *Sol. Phys.* **78**, 157 (1982).
80. J. O. STENFLO, H. BIVEROT, AND L. STENMARK, *Appl. Opt.* **15**, 1188 (1976).

81. J. O. Stenflo, D. Dravins, N. Wihlborg, Y. Öhman, A. Bruns, V. K. Prokof'ev, I. A. Zhitnik, H. Biverot, and L. Stenmark, *Sol. Phys.* **66**, 13 (1980).
82. W. Henze, Jr., E. Tandberg-Hanssen, M. J. Hagyard, B. E. Woodgate, R. A. Shine, J. M. Beckers, M. Bruner, J. B. Gurman, C. L. Hyder, and E. A. West, *Sol. Phys.* **81**, 231 (1982).
83. B. Favati, E. Landi Degl'Innocenti, and M. Landolfi, *Astron. Astrophys.* **179**, 329 (1987).
84. P. Mein and J. Rayrole, *Vistas Astron.* **28**, 567 (1985).
85. J. O. Stenflo, *Vistas Astron.* **28**, 571 (1985).
86. A. A. Wyller, *LEST, Large Earth-based Solar Telescope—An Overview* (Royal Swedish Academy of Sciences, 1986).

CHAPTER 6

APPLICATIONS OF THE HANLE EFFECT IN SOLID STATE PHYSICS

G. E. PIKUS AND A. N. TITKOV

1. INTRODUCTION

The present volume of review chapters is entitled *The Hanle Effect and Level-Crossing Spectroscopy*. If, however, we were to attempt a comprehensive survey of a variety of effects in solids where level crossing and anticrossing are important, we would have had to consider, in addition to solid state optics which is in itself an unlimited field, also nuclear magnetic resonance (NMR), electron paramagnetic resonance (EPR), and related fields of radiospectroscopy. This is clearly an impossible task for a single review, even one of much greater size. We therefore restrict ourselves to works dealing with the proper Hanle effect, i.e., the influence of an external magnetic field on the polarization of the luminescence from a crystal excited with polarized light.

While the Hanle effect in vapors and gases has a history of over sixty years, the story as regards solids spans a much shorter period. Here the Hanle effect has found wide application in semiconductors following the discovery of the phenomenon of optical orientation of spins of free electrons by Lampel[1] and excitons by Gross *et al.*[2] In both studies the Hanle effect itself has not been used yet: Lampel made use of nonoptical methods to record the electron spin polarization in Si, and the Hanle effect is not possible for A excitons in CdSe investigated by Gross *et al.* The first observation of the Hanle effect on free electrons was performed in GaSb by Parsons,[3] and soon afterward in GaAlAs by Ekimov and Safarov[4] and in GaAs by Zakharchenya *et al.*[5] The Hanle effect on excitons was observed

G. E. PIKUS AND A. N. TITKOV • A. F. Ioffe Physico-Technical Institute, USSR Academy of Sciences, 194021 Leningrad, USSR.

independently in GaSe by Veshchunov et al.,[6] in CdS by Bonnot et al.,[7] and in InP by Weisbuch and Lampel.[8] In all these studies the excitons were generated by circularly polarized light and spin polarization was being observed. A little later, Bir and Pikus[9] showed that excitons could be not only polarized, but also aligned when generated by linearly polarized light. Such alignment and the Hanle effect for aligned excitons have been observed in CdSe by Bonnot et al.[10] and in CdS and CdSe by Permogorov et al.[11]

For impurity centers, the orientation of electron spins during excitation by circularly polarized light was observed as early as 1963 by Karlov et al.[12] at the F center in KBr and then, in 1965, by Anderson et al.[13] on Tu^2 in CaF_2 and Hull et al.[14] on Cr in Rb. However, the Hanle effect for optically oriented electrons on impurity centers came to be studied only in 1975 by Kaplyanskii et al.[15] on Eu^2 in CaF_2 and, as yet, has not found wide application.

This review has two objectives: The first is to show in what respects the Hanle effect in solids differs from that in atoms. The differences result, above all, from the variety of symmetries inherent in a solid, from the influence of the surface, from the possibility of orienting atomic nuclei which leads to the creation of the strongly coupled electron–nuclear spin system, and, finally, from the variety of spin relaxation mechanisms. The second objective is to present the Hanle effect as a very efficient method for disclosing fine details of electronic spectra and those properties pertaining to the interaction of electrons with hole or nuclear spins which do not occur in conventional absorption or luminescence spectra. The Hanle effect is also a simple method for measuring short lifetimes and spin relaxation times. Indeed, the Hanle effect makes it possible in steady-state conditions, using easily attainable magnetic fields up to 10 T, to measure times as small as 10^{-12}–10^{-13} s.

The following sections deal with works which treat predominantly the aforementioned problems. A more comprehensive reference list of papers devoted to the optical orientation of electrons and excitons can be found in reviews by D'yakonov and Perel,[16] Pikus and Titkov,[17] Mirlin,[18] Fleisher and Merkulov,[19] Planel' and Benoit a la Guillaume,[20] and Pikus and Ivchenko.[21]

2. THE HANLE EFFECT ON FREE ELECTRONS

In semiconductors, the Hanle effect was observed during experiments on optical orientation of spins of carriers and on alignment of their momenta. These effects are observable during excitation with, respectively, circularly or linearly polarized light and revealed in the corresponding polarization of luminescence.

2.1. Optical Orientation of Electron Spins

2.1.1. General Principles and Occurrence in Polarized Luminescence. The role played by a magnetic field in experiments on optical orientation of free carriers is considered. These experiments were conducted most extensively on direct gap III-IV compounds. The band structure of these compounds is characterized by the occurrence of the conduction band minimum and of the valence band maximum at the center of the Brillouin zone (see Figure 1). The conduction band Γ_6 is formed by S-type atomic states and is twofold degenerate in spin. The valence band is formed by P-type atomic states and is characterized by an orbital momentum of unity. Spin-orbit interaction splits the valence band into two subbands, an upper (Γ_8) and a lower (Γ_7) one. The Γ_8 subband has total angular momentum $J = \frac{3}{2}$ and is fourfold degenerate at the center of the Brillouin zone, $k = 0$. At wave vector values $k \neq 0$ it splits into the light-hole and heavy-hole bands, each twofold degenerate. The value $\pm\frac{3}{2}$ of the angular momentum projection on the **k** axis corresponds to the heavy-hole band, and the value $\pm\frac{1}{2}$ to the light-hole band. The Γ_7 subband has total angular momentum $J = \frac{1}{2}$ and is twofold degenerate.

In a coordinate system having the z axis directed along **k**, the selection rules for transitions between the valence subbands and the conduction band are the same as for transitions in an atom between P terms with $J = \frac{3}{2}$ or $J = \frac{1}{2}$ and an S state with $J = \frac{1}{2}$: the projection of the angular momentum on the z axis, J_z, is conserved if the polarization of the light wave electric vector **e** is parallel to z, or changes by ± 1 if **e** is perpendicular to z. The relative probabilities of the corresponding transitions are shown on the right in Figure 1, which demonstrates that excitation of electrons by circularly polarized light, σ^+ or σ^-, makes possible a preferential population of one

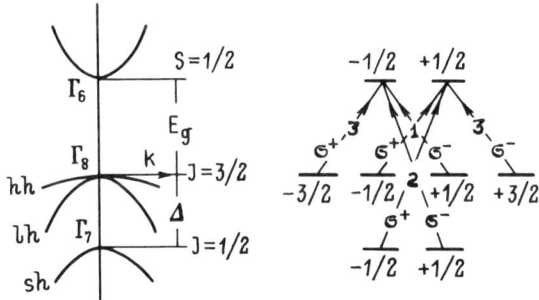

FIGURE 1. Energy band diagram, E vs. k, in III-V cubic crystals near the center of the Brillouin zone showing the energy gap E_g and spin-orbit splitting Δ of the valence subbands. On the right, the allowed optical transitions from the Γ_8 and Γ_7 valence subbands to the Γ_6 conduction band and their relative probabilities for excitation by σ^+ and σ^- light are shown.

of the spin sublevels ($-\frac{1}{2}$ or $+\frac{1}{2}$) in the conduction band. In the laboratory coordinate system the selection rules depend also on the direction of the electron wave vector, given by the unit vector $\mathbf{\nu} = \mathbf{k}/|\mathbf{k}|$, relative to the direction of light propagation $\mathbf{n}$. As a consequence, the average values of spin for electrons with different orientations of wave vector $\mathbf{k}$ will vary. D'yakonov and Perel'[22] have shown that the quantity $\mathbf{S}(\mathbf{\nu})$ for electrons excited in transitions hh-c, lh-c, and sh-c is given by

$$hh\text{-}c \qquad \mathbf{S}(\mathbf{\nu}) = -\frac{\mathbf{\nu}(\mathbf{\nu} \cdot \mathbf{n})}{1 + (\mathbf{\nu} \cdot \mathbf{n})^2} \qquad (1)$$

$$lh\text{-}c \qquad \mathbf{S}(\mathbf{\nu}) = \frac{3\mathbf{\nu}(\mathbf{\nu} \cdot \mathbf{n}) - 2\mathbf{n}}{5 - 3(\mathbf{\nu} \cdot \mathbf{n})^2} \qquad (2)$$

$$sh\text{-}c \qquad \mathbf{S}(\mathbf{\nu}) = \frac{\mathbf{n}}{2} \qquad (3)$$

The average spin of all electrons in the conduction band, $\mathbf{S}_0$, is obtained by averaging $\mathbf{S}(\mathbf{\nu})$ over all wave vector orientations. For the Γ_6 conduction band, the average electron spin at the moment of excitation happens to be the same as for corresponding transitions in atoms: for electrons excited from the Γ_8 valence band one obtains $\mathbf{S}_{01} = -\frac{1}{4}$, and for the Γ_7 valence band $\mathbf{S}_{02} = \frac{1}{2}$, i.e., in these two cases the average spins have opposite orientations. Therefore, as the energy of exciting light is increased up to the value $\hbar\omega \simeq E_g + \Delta$, when electrons can be excited from both valence subbands Γ_8 and Γ_7, the average electron spin in the conduction band decreases. For larger energies of light, the optical orientation of electron spins can disappear. This result is due to the fact that the electric field of a light wave directly affects only the orbital motion of an electron, but not its spin. Spin orientation is achieved through spin-orbit interaction, which is less significant for the valence band states with energies markedly exceeding the magnitude of spin-orbit splitting, Δ.

The value and orientation of the average electron spin S_0 may change significantly when a uniaxial stress is applied to a crystal. Deformation splits the Γ_8 valence band into heavy-hole and light-hole subbands. Under stress the fourfold degeneracy of the Γ_8 band at the point $\mathbf{k} = 0$ is lifted and, in addition, the correlation between the angular momentum and the wave vector for the states of the upper split-off subband will be distorted for the case in which the kinetic energy is smaller than the value of deformational splitting. Such states of the valence subband are characterized by projections of the angular momentum on the deformation axis instead of the wave vector direction. The average spin of the electrons excited from these states will also be determined not by the direction of the wave vector, but by the orientation of the deformation axis relative to the light beam. For stresses along the principal axes [100], [111], and [110] the value and orientation of the average spin, S_0, for excitation from the heavy-hole or light-hole subband, whichever happens to be higher, is defined by expressions derived from equation (1) or (2), respectively, by substituting the unit vector in the direction of the deformation axis **p** for the vector $\boldsymbol{v}$. It follows from these expressions that for an arbitrary angle between the direction of light, **n**, and the deformation axis **p**, the orientation of the average spin of the photoexcited electrons, S_0, may not coincide with either of these directions while remaining in the plane containing both of them. The value of the average electron spin in crystals under stress also changes and may assume values in the range $+\frac{1}{2}$ to $-\frac{1}{2}$. As indicated above, the effect of stress is appreciable only for states of the upper valence subband whose kinetic energy is less than or comparable to the value of deformational splitting at $k = 0$. Therefore, for this effect to occur excitation of electrons must take place only from the upper split-off subband, leaving the lower subband intact. In III–V compounds, a compressive stress pushes the light-hole band up. The heavy-hole band turns out to be uppermost at tensile stress.

The quantity S_0 introduced above is a measure of the initial average electron spin at the moment of excitation. During the lifetime in the conduction band electrons may get involved in various spin relaxation processes which diminish the value of **S**. Accordingly, under continuous excitation a quasi-equilibrium value of the average spin **S** is found to be smaller and dependent on the ratio of the lifetime τ to the spin relaxation time τ_s for electrons,

$$\mathbf{S} = \mathbf{S}_0 \frac{1}{1 + \tau/\tau_s} \qquad (4)$$

In semiconductors, the ratio τ/τ_s varies within the broadest limits depending on temperature and intrinsic properties of the crystal. In the case $\tau \gg \tau_s$ the initial spin orientation is quickly lost and the quasi-equilibrium spin orientation cannot be reached. The optical spin orientation was observed in nearly all direct band III–V compounds as well as in silicon and a number of II–VII and III–VI compounds, as already noted in the Introduction.

Achieved by optical means, the orientation of electron spins in the conduction band can be recorded again by optical methods. Indeed, non-equilibrium population of the spin sublevels in the conduction band causes, as its natural consequence, the appearance of circular polarization of the recombination luminescence. In the following treatment it will be assumed that the polarization of luminescence is due exclusively to electron spins. As a result of strong spin–orbit interaction, photoexcited holes very quickly lose the initial orientation of their angular momenta (at the rate of momentum scattering, $1/\tau_p = 10^{13}$ s) and enter the recombination process disoriented. Besides, investigations on the optical orientation were conducted predominantly on p-type materials having large concentrations of non-oriented equilibrium holes.

D'yakonov and Perel'[22] have shown that the degree of circular polarization of luminescence $\mathcal{P}$ due to orientation of electron spins is given by a scalar product of the vector of the average spin **S** and the observation vector $\mathbf{S}_1$,

$$\mathcal{P} = 4\mathbf{S}\mathbf{S}_1 \qquad (5)$$

As usual, the degree of polarization $\mathcal{P}$ is defined by the ratio $(I_+ - I_-)/(I_+ + I_-)$, where I_+ and I_- are, respectively, the intensities of recombination radiation with polarizations σ^+ and σ^-. In nondeformed crystals for the transitions Γ_8-Γ_6, we have $\mathbf{S}_1 = \mathbf{n}_1/4$ where $\mathbf{n}_1$ is the unit vector in the direction of observation of luminescence. In uniaxially stressed crystals the $\mathbf{S}_1$ vector is dependent on the specific valence subband from which the hole comes that recombines with an electron. If the recombination involves a light hole, the $\mathbf{S}_1$ vector is defined by expression (2) upon substituting vectors **p** for ν and $\mathbf{n}_1$ for **n**. At recombination with heavy holes, the $\mathbf{S}_1$ vector is defined in a similar manner by expression (1). In the general case, the directions of vectors **S** and $\mathbf{S}_1$ do not coincide.

2.1.2. Depolarizing Action of a Magnetic Field—The Hanle Effect. The simple relationship (5) shows that in all cases the degree of circular polarization of luminescence is defined as a projection of the average spin **S** on the observation vector $\mathbf{S}_1$. This makes it clear that a magnetic field must affect the polarization of luminescence. Its influence consists in forcing the average spin off its initial orientation and decreasing its absolute value due to precession of the electron spins in a magnetic field. The effect of a magnetic field will evidently be most pronounced when the field is directed perpendicular to the average spin. The behavior of the average spin in a magnetic field **B** is described by the following equation that accounts for generation, recombination, and spin relaxation processes:

$$\frac{d\mathbf{S}}{dt} = -[\mathbf{S}\mathbf{\Omega}_L] - \frac{\mathbf{S}}{T} + \frac{\mathbf{S}_0}{\tau} \qquad (6)$$

Here $\mathbf{\Omega}_L = g_e\mu_B\mathbf{B}$ is the frequency of Larmor precession of spins, g_e is the g factor of electrons, μ_B is the Bohr magneton, while $1/T = 1/\tau + 1/\tau_s$ is the reciprocal lifetime of oriented spins in the conduction band and is determined both by spin relaxation processes and by recombination that moves electrons out of the conduction band. It is assumed that the average electron spin is oriented by the light along the z axis and the magnetic field is applied along the x axis. Then, in a quasi-equilibrium case, the solution of equation (6) has the form

$$S_z(\mathbf{B}) = S(0) \frac{1}{1 + \Omega_L^2 T^2}, \quad S_y(\mathbf{B}) = S(0) \frac{\Omega_L T}{1 + \Omega_L^2 T^2}, \quad S_x(\mathbf{B}) = 0 \quad (7)$$

In the experiments on optical orientation a beam of light is usually directed perpendicular to the crystal surface and observation of luminescence is carried out from the opposite direction. As a result, in nondeformed crystals the average spin $\mathbf{S}$ will also become oriented perpendicular to the surface. Therefore, a magnetic field $\mathbf{B}$ is applied parallel to the surface. In this geometry, according to equation (5), the variation of the luminescence polarization in a magnetic field will be determined by the changes in the average spin component normal to the surface and in this case, equations (4) and (7) yield $\mathcal{P}(\mathbf{B})$ defined by

$$\mathcal{P}(\mathbf{B}) = \mathcal{P}(0) \frac{1}{1 + \Omega_L^2 T^2} \quad (8)$$

The resulting curve $\mathcal{P}(\mathbf{B})$ has a Lorentzian shape with half-width $\Delta \mathbf{B}$ determined from the condition $\Omega_L T = 1$,

$$\left| \frac{g_e \mu_B}{\hbar} \right| \Delta \mathbf{B} = \frac{1}{T} = \frac{1}{\tau} + \frac{1}{\tau_s} \quad (9)$$

By analogy with atomic physics the curve $\mathcal{P}(\mathbf{B})$ is commonly called the Hanle curve. The Hanle effect is widely used in research on optical spin orientation. It is employed as an unequivocal check on whether or not the optical spin orientation really occurs in the experiment. Measurements of the degree of luminescence polarization $\mathcal{P}$ together with the half-width ΔB of the Hanle curve separately yield the value of the lifetime and spin relaxation time of electrons in the conduction band. In what follows, these and a number of other useful applications of the Hanle effect will be described. These applications are based on the fact that the shape of the depolarization curve $\mathcal{P}(\mathbf{B})$ changes dramatically if a more detailed account is taken in equation (6) of the real conditions encountered in the crystal. What we mean is, first of all, the occurrence of local deformations, different rates of recombination and relaxation processes in the bulk and on the crystal surface, and the existence of diffusion of the oriented electrons into the crystal depth. In the following sections we consider some important instances of changes in the shape of the depolarization curve $\mathcal{P}(\mathbf{B})$ brought about by the above-mentioned factors.

2.1.3. The Hanle Effect in Uniaxially Deformed Crystals. As already noted above, an arbitrarily oriented uniaxial stress causes deviation of the average spin S from the direction **n** of the light beam. The only exceptions are represented by the two extreme orientations of stress **p**, i.e., parallel or perpendicular to **n**, when the vectors S and **n** remain collinear. In all other geometries the collinearity vanishes. In deformed crystals the degree of circular polarization is also determined by the projection of the average spin S on the observation vector S_1, their orientation in this case being dependent on the type of excitation and recombination transitions involved. The vectors S and S_1 are defined by the same equations (1) and (2) but with vector **p** substituted for ν and with appropriate substitution of the antiparallel vectors **n** and $\mathbf{n}_1$. From this it follows, in particular, that vectors S and S_1, forming an angle α, always lie in a plane passing through vectors **n** and **p**. The magnetic field B applied perpendicular to this plane will cause the projection of the S vector on the observation vector S_1 to vary. The variation of this projection $S(B)$ with the magnetic field is easily found using equation (6). Then, for a magnetic-field dependence of the degree of circular polarization of luminescence we obtain

$$\mathscr{P}(\mathbf{B}) = 4S(\mathbf{B})S_1 = 4|S||S_1|\frac{1}{1+\tau/\tau_s}\frac{\cos\alpha - \Omega_L T \sin\alpha}{1+\Omega_L^2 T^2} \qquad (10)$$

An analysis of this equation shows that in deformed crystals the curve $\mathscr{P}(\mathbf{B})$ becomes essentially asymmetric. In contrast to nondeformed crystals, on application of a magnetic field possessing appropriate direction the degree of polarization $\mathscr{P}(\mathbf{B})$ may even increase, reaching a maximum value in a field $\mathbf{B}_0$ where

$$|\mathbf{B}_0| = \frac{\hbar}{\mu_B g_e}\frac{1}{T}\tan(\alpha/2) \qquad (11)$$

and then decreasing to zero. For an opposite magnetic-field orientation the degree of polarization, on the contrary, drops rapidly from the very beginning, then changes sign at the field value $\mathbf{B} = \hbar \cotan\alpha/\mu_B g_e T$, and only then decreasing to zero.

An asymmetric shape of the curve $\mathscr{P}(\mathbf{B})$ in a deformed crystal has been observed experimentally by Marushchak and Titkov.[23] Figure 2 shows the curves $\mathscr{P}(\mathbf{B})$, obtained for two luminescence lines due to recombination of electrons with the light and heavy holes in a uniaxially compressed [100] GaSb crystal. Excitation of electrons was carried out only from the upper split-off light-hole subband. An insert in Figure 2 shows orientations of all the important vectors **n**, $\mathbf{n}_1$, **p**, $\mathbf{S}^{1/2}$, $\mathbf{S}_1^{1/2}$, and $\mathbf{S}_1^{3/2}$ as determined by the authors. At recombination of electrons with the light holes the vectors $\mathbf{S}^{1/2}$

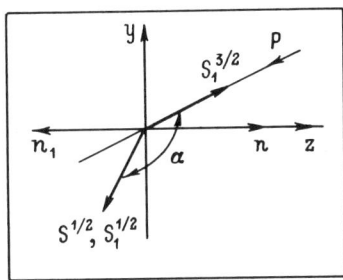

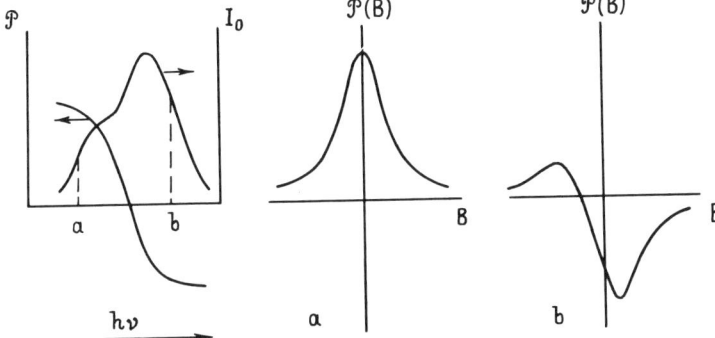

FIGURE 2. Luminescence and polarization spectra and depolarization curves $\mathscr{P}(\mathbf{B})$, obtained for the two luminescence lines due to recombination of electrons with light (a) and heavy (b) holes in uniaxially compressed GaSb. An insert shows the orientations of all the important vectors (from Maruschak and Titkov[23]).

and $S_1^{1/2}$ are collinear, and the curve $\mathscr{P}(\mathbf{B})$ retains a Lorentzian shape. The existence of the initial angle α between vectors $\mathbf{S}^{1/2}$ and $\mathbf{S}_1^{3/2}$, when recombination occurs with heavy holes, entails an appearance of asymmetry in the curve $\mathscr{P}(\mathbf{B})$ considered above.

The fact that stress affects the shape of the depolarization curve $\mathscr{P}(\mathbf{B})$ opens up an interesting possibility for detecting local strains in crystals, and determining their character (tensile or compressive) as well as their orientation.

2.1.4. The Hanle Effect in n-Type Crystals. In n-type crystals, optical orientation of nonequilibrium electrons causes an essential orientation of the entire electron system. This is explained by the fact that nonequilibrium holes recombine predominantly with equilibrium electrons, which are in the majority. Hence there occurs an accumulation of oriented electrons in the conduction band. And what is important is that this accumulation occurs at the crystal surface. Diffusion of spin-oriented electrons from the surface leads to changes in the shape of the depolarization curve $\mathscr{P}(\mathbf{B})$. Usually, the spin diffusion length L_s noticeably exceeds the depth at which the light is absorbed as well as the diffusion length L_h of the nonequilibrium holes. Under these conditions only the electrons originating from the rather thin layer nearest the surface of thickness L_h take part in recombination. Diffusion of spin-oriented electrons into the crystal depth in this case serves as an additional channel of spin relaxation. It is noteworthy that this channel is field-dependent because the magnetic field affects the gradient of the electron spin distribution.

An expression for the depolarization curve $\mathscr{P}(\mathbf{B})$ in the case under consideration may be obtained from an analysis of the diffusion equation for electron spins,

$$\frac{\partial \mathbf{S}}{\partial t} = D_e \frac{\partial^2 \mathbf{S}}{\partial t^2} - \frac{\mathbf{S}}{\tau_s} + [\boldsymbol{\Omega}_L \mathbf{S}] \tag{12}$$

Boundary conditions necessary for solving this equation are obtained by examining the balance between processes occurring in a thin layer L_h nearest the surface where generation and recombination of nonequilibrium carriers take place. Assuming $L_h \ll L_s$, we have

$$G\mathbf{S}_0 = -n_0 D_e \frac{\partial \mathbf{S}}{\partial z} + G\mathbf{S} \tag{13}$$

where G is the number of electron–hole pairs generated by light per unit area and in unit time, n_0 is the density of equilibrium electrons, and D_e is the coefficient of electron diffusion. The term on the left-hand side of equation (13) determines generation of the light-oriented spins; the two terms on the right-hand side account, respectively, for diffusion of spins into the crystal depth and their elimination through recombination.

D'yakonov and Perel'[24] solved equation (12), subject to boundary conditions (13), for the S_z component of the average spin at the crystal surface. They employed relation (5) to obtain the shape of the depolarization curve $\mathscr{P}(\mathbf{B})$ in the form

$$\mathscr{P}(\mathbf{B}) = \frac{\mathbf{S}_0}{1 + a} F(\mathbf{B}) \tag{14}$$

where

$$F(\mathbf{B}) = \frac{(1+a)(1+a\sqrt{(1+\eta)/2})}{1+a^2\eta + 2a\sqrt{(1+\eta)/2}}$$

$$\eta = (1+\Omega_L^2\tau_s^2)^{1/2} \quad \text{and} \quad a = \frac{n_0 L_s}{G\tau_s} = \frac{\tau_i}{\tau_s}$$

Here τ_i is the time required for all the equilibrium electrons in the layer of thickness L_h to be replaced by photoexcited electrons. It follows from this expression, in particular, that the shape of curve $\mathscr{P}(\mathbf{B})$ in n-type crystals depends also on the generation rate of nonequilibrium carriers, G. Let us consider the two limiting cases, $a \gg 1$ and $a \ll 1$. In the first case of low excitation level

$$F(\mathbf{B}) = \left[\frac{1+\sqrt{1+\Omega_L^2\tau_s^2}}{2(1+\Omega_L^2\tau_s^2)}\right]^{1/2} \tag{15}$$

As expected, the half-width of the curve $\mathscr{P}(\mathbf{B})$ depends almost exclusively on τ_s because the time τ_i, which plays the role of an effective lifetime of electrons, is much longer.

In the opposite case of high excitation level we have

$$F(\mathbf{B}) = \frac{1+(\Omega_L\tau^*)^{1/2}}{1+2(\Omega_L\tau^*)^{1/2}+\Omega_L\tau^*} \tag{16}$$

where

$$\tau^* = a^2\tau_s/2 = \Omega_L^2 D_e/2G^2$$

In this case the half-width of the curve $\mathscr{P}(\mathbf{B})$ is determined by the time τ^* and should be broadened proportionally to the square of the generation rate G.

Depolarization of luminescence by a magnetic field in n-type crystals has been observed experimentally by Vekua et al.[25] Their data, obtained for n-GaAs, are presented in Figure 3. The experimental dependence is seen to fall between theoretical curves calculated according to equations (15) and (16). Such a result evidently indicates that an intermediate value of the generation rate G has been attained in this experiment. Another indication is the experimentally observed broadening of the curve $\mathscr{P}(\mathbf{B})$ with increasing generation rate, though the dependence was slower than

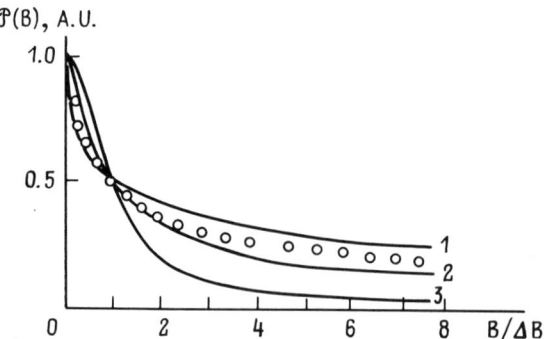

FIGURE 3. Depolarization of luminescence by a magnetic field in n-GaAlAs (circles). Curves 1 and 2 are calculated according to equations (15) and (16), respectively. Curve 3 is the Lorentz contour with half-width ΔB equal to the half-width of the experimental depolarization curve $\mathscr{P}(B)$ (from Vekua et al.[25]).

quadratic in G. In general, it should be stressed that the shape of depolarization curves in n-type crystals is essentially different from the Lorentzian of the same half-width, the most obvious difference being the existence of the slow decreasing tail in the high-field region.

Recently, Bakun et al.[26,27] have used the Hanle effect in n-type crystals to detect spin-dependent scattering of electrons by charged impurities in solids. The spin–orbit interaction is known to cause asymmetry of electron scattering by a charged center relative to the plane containing the vectors of momentum and spin. In semiconductors, the constant of spin–orbit interaction is by several orders of magnitude greater than for classical free electrons, hence this relativistic effect could be expected to show up even for the electrons at thermal velocities.

In the experiment an ordered flow of optically spin-oriented electrons was provided by their diffusion into the depth of an n-GaAs crystal. The magnetic field B applied parallel to the surface caused a deviation in the average spin of the diffusing electrons from its initial direction, normal to the surface, and thus gave rise to a component of the average spin perpendicular to the direction of diffusion. An insert in Figure 4 shows the geometry of the experiment. Asymmetry of the electron scattering by charged impurities revealed itself through the emergence of a photo-emf and, correspondingly, a photocurrent directed along the applied magnetic field. Figure 4 shows dependences of the photocurrent on magnetic field as well as depolarization curves $\mathscr{P}(B)$ for the edge luminescence upon excitation with σ^+ or σ^- polarized light. A change of the excitation light polarization from σ^+ to σ^- leads to the opposite orientation of electron spins and, as a result, causes the change of sign of the photocurrent. The application of a magnetic field initially causes an increase in the photocurrent due to an increase in

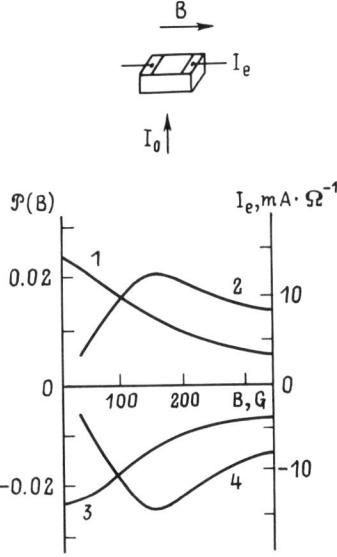

FIGURE 4. Depolarization curves $\mathscr{P}(\mathbf{B})$ (1, 3) and photocurrent $\mathbf{I}_e$ (2, 4) dependences on the magnetic field in n-GaAs for excitation by σ^+ (1, 2) and σ^- (3, 4) polarized light. An insert shows the geometry of the experiment (from Bakun et al.[26]).

the perpendicular projection of the average electron spin. Then, however, the photocurrent begins to decrease because of depolarization of electron spins by the magnetic field.

2.1.5. The Role of Surface Recombination and Electron Diffusion in p-Type Crystals. In crystals there is efficient reabsorption of recombination radiation. Consequently luminescence light can be emitted only from a rather thin surface layer of width $\alpha^{-1}(\lambda)$, where $\alpha(\lambda)$ is the absorption coefficient at the luminescence wavelength. In many cases the width of such a layer is comparable to or even less than the electron diffusion length. Then recombination and relaxation of electron spins in such a layer are determined not only by the bulk processes, but also by the processes at the surface and electron diffusion into the crystal depth. An account of these processes should change the Lorentzian shape of the depolarization curve. The degree to which the depolarization curve becomes distorted is usually dependent on a wavelength within the spectral range covered by the luminescence band, because absorption coefficients are different through the spectrum. For the most extensively studied bands of the edge luminescence distortions of the depolarization curves are larger at the short wavelength side, where the absorption coefficients are largest.

The first examination of the Hanle effect with allowance for reabsorption, the processes at the crystal surface, and electron diffusion was undertaken recently by Dzhioev et al.[28] These authors sought a solution to the diffusion equation in the form of equation (12) with boundary conditions accounting for recombination and spin relaxation of electrons on the surface, and obtained the following expression for the depolarization curve in the case of a semi-infinite crystal:

$$\mathcal{P}(\lambda, \mathbf{\Omega}_L) = |S_0| \frac{T[1 + \alpha(\lambda)L](1 + \eta(\tau/L))b}{\tau} \frac{}{b^2 + c^2} \quad (17)$$

where

$$b = \sqrt{(\sqrt{1 + \mathbf{\Omega}_L^2 T^2} + 1)/2} \, [\eta_s(T/L_s) + \alpha(\lambda)L_s] + \eta\alpha(\lambda)T + 1$$

and

$$c = \sqrt{(\sqrt{1 + \mathbf{\Omega}_L T^2} - 1)/2} \, [\eta_s(T/L_s) + \alpha(\lambda)L_s] + \mathbf{\Omega}_L T$$

Here η is the surface recombination rate, η_s the surface spin relaxation rate, $L = \sqrt{D_e \tau}$ the electron diffusion length, and $L_s = \sqrt{D_e T}$ the spin diffusion length. Equation (17) was derived neglecting the possibility of reemission of the absorbed light, and is therefore applicable in cases where recombination in the bulk is for the most part radiationless. Under the conditions $\alpha(\lambda) \ll L, L_s, \eta \ll L/\tau$, and $\eta_s \ll L_s/T$, equation (17) reduces to the usual Lorentzian.

Equation (17) contains many parameters relating to the transport and recombination of electrons in a crystal and its qualitative analysis is difficult. Nevertheless, its use for a quantitative description of the experimental curves $\mathcal{P}(\lambda, \mathbf{\Omega}_L)$ obtained at various wavelengths opens up a new possibility of determining all the parameters involved.[28]

2.1.6. Effect of Magnetic Field on the Polarized Luminescence of Graded-Band Crystals. Graded-band crystals are characterized by the dependence of the forbidden band gap E_g on the distance from the surface. If the band gap decreases with increasing distance from the surface, the photoelectrons excited at the surface penetrate into the crystal not only because of diffusion, but also because of a drift in the built-in quasi-electric field. The reduction in the band gap across the crystal thickness makes the edge luminescence band appreciably spread out compared with the spectrum of crystals whose band gap is uniform. Each spectral part of the luminescence band represents the recombination of electrons at a certain distance from the surface. The deeper the point where recombination occurs, the smaller the energy of emitted photons. Thus an analysis of the intensity and spectral distribution of the edge luminescence band allows one to determine the electron distribution through the crystal depth.

The possibilities of an analysis are substantially expanded when we study recombination of optically oriented electrons and the Hanle effect is employed. First, the degree of circular polarization of luminescence is not the same throughout the edge luminescence band, but decreases in the direction of lower energies owing to spin relaxation of electrons during their drift. Precession of the electron spins in a transverse magnetic field **B** leads to a dependence on the depth of the average spin orientation. As a result, the projection of the average spin **S** along the observation axis experiences sign-reversal oscillations as the electrons travel away from the surface. This causes the emergence of analogous oscillations of the degree of luminescence polarization throughout the edge luminescence band.

A theoretical description of the behavior of the electron spins and of the spectrum of polarized luminescence $(I^+ - I^-)$ in the graded-band crystals, given by Volkov et al.,[29,30] is based on solving a system of equations for the average spin projections S_z and S_y in the presence of a magnetic field **B** applied along the x axis,

$$L_s^2 \frac{d^2 S_{z,y}}{dz^2} - l_s \frac{dS_{z,y}}{dz} - S_{z,y} \mp \Omega_L T S_{y,z} = 0 \qquad (18)$$

with $\Omega_L = g_e \mu_B B$, $1/T = 1/\tau + 1/\tau_s$, $L_s = \sqrt{D_e T}$, $l_s = (\mu_0 T/e) dE_g/dz$ is the spin drift length, and μ_0 is the electron mobility. Boundary conditions, analogous to relation (13), account for the generation of the average spin component S_z at the surface ($z = 0$) and for the zero value of the average spin components S_z and S_y at infinity.

The angular velocity of precession Ω_L in graded-band crystals turns out to be a function of the distance from the surface because the g-factor of the electrons varies with depth. In many cases it suffices to assume a linear relationship between Ω_L and the depth z, namely $\Omega_{L,a} = \Omega_{L,0}(1 + z\, dg_e/g_e dz)$, where $dg_e/dz = \text{const}$.

The first experimental investigations of the polarization spectra in graded-band crystals were conducted by Volkov et al.[31,32] Figures 5a and 5b present a comparison[32] between the calculated variation in the average spin projection S_z with distance from the surface, and the experimentally obtained spectra of polarized luminescence $I^+ - I^-$ from the graded-band $Ga_{1-X}Al_X As$ crystals with X varying from 0.28 at the surface to 0.01 at a depth of 10 μm. This range of crystal composition is interesting in that the electron g-factor initially decreases with distance from the surface, then becomes zero at some depth z_0, and at still greater depths has a reversed sign. In this case the dependences given in Figures 5a and 5b along with

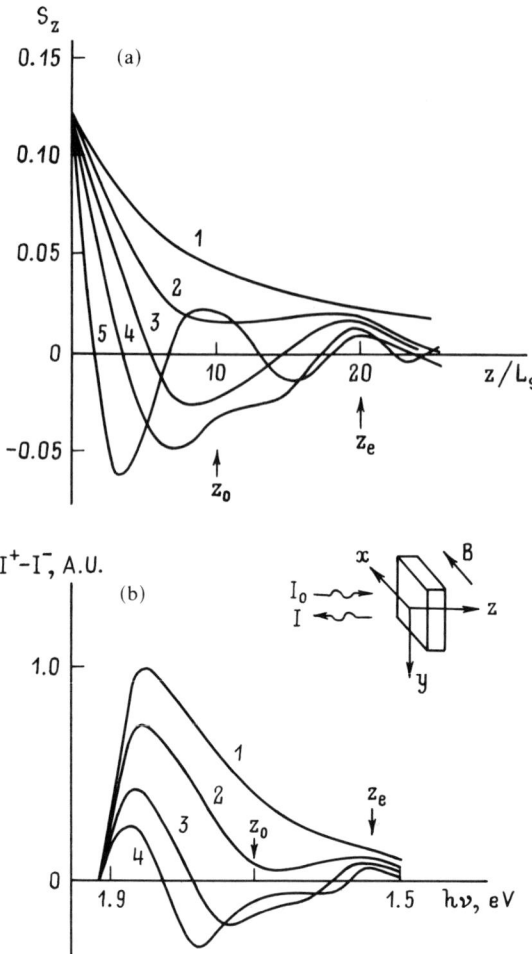

FIGURE 5. (a) Calculated variation of the average spin projection S_z with distance from the crystal surface in a transverse magnetic field **B**: 1, $\Omega T = 0$; 2, $\Omega T = 2$; 3, $\Omega T = 4$; 4, $\Omega T = 6$; 5, $\Omega T = 8$. It was assumed that $l_s/2L_s = 4$ and $L_s\, dg_e/g_e\, dz = 0.1$. (b) Spectra of polarized luminescence $I^+ - I^-$ of the graded-band GaAlAs in a transverse magnetic field **B**: 1, 0.2 T; 2, 0.4 T; 3, 0.6 T; 4, 0.8 T (from Volkov et al.[32]).

the characteristic attributes of the behavior of polarizational spectra in graded-band crystals allow one to observe another interesting effect of the "spin echo" type. Its essence is that, at depth z_0, the electron spin precession reverses the direction of rotation becasuse the g-factor changes sign and, as a consequence, at some greater depth z_e ($z_e = 2z_0$ if dE_g/dz and dg/dz are constant) the orientation of the average spin **S** happens to be the same as it was at the surface for any value of the applied magnetic field. In Figure 5b this effect is evident in that the polarization sign at $h\nu = 1.53$ eV is the same for all the curves presented. The complicated shape of the polarized luminescence spectra in Figure 5b displays the unusual behavior of the depolarization curves in graded-band crystals. Figure 6 shows the depolarization curves computed from the data in Figure 5b for spectral regions corresponding to electron recombination at the surface (curve 1) and at depths $z_0/2$ and z_0 (curves 2 and 3). A theoretical treatment of such depolarization curves has been given elsewhere.[30]

2.2. Occurrence of Electron-Nucleus Interaction in Polarized Luminescence

The effect of a magnetic field on luminescence polarization opens up interesting possibilities for investigating electron-nucleus interaction in crystals. In the case of optical orientation of electron spin, electron-nucleus interaction plays a dual role. On the one hand, it leads to the dynamic polarization of the nuclear spin. On the other hand, the polarization of nuclei and electrons creates effective magnetic fields of nuclei, $\mathbf{B}_N$, and of

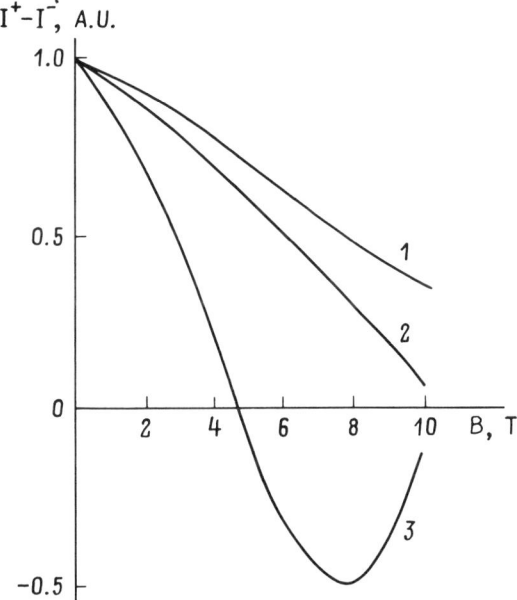

FIGURE 6. Depolarization of luminescence in graded-band GaAlAs crystals at wavelengths corresponding to electron recombination for different distances from the crystal surfaces: 1, $z = 0$; 2, $z = z_0/2$; 3, $z = z_0$.

electrons, $\mathbf{B}_e$, acting on electron and nuclear spins, respectively. These fields may contribute to all the effects associated with the behavior of spin in a magnetic field. By combining to form an external field $\mathbf{B}$, the fields $\mathbf{B}_N$ and $\mathbf{B}_e$ may also play a role in experiments on the depolarization of luminescence during recombination of oriented electrons.

A dominant interaction with nuclei for electrons in the conduction band of S-type is the contact hyperfine Fermi interaction described by the Hamiltonian[33]

$$H = \frac{16\pi}{3I_N} \mu_B \mu_I |\psi(\mathbf{R})|^2 \hat{\mathbf{I}} \hat{\mathbf{S}} \tag{19}$$

where $\psi(\mathbf{R})$ is the electron wave function at the nucleus, $\mathbf{I}_N$ the nuclear spin, μ_I the nuclear magnetic moment, while $\hat{\mathbf{I}}$ and $\hat{\mathbf{S}}$ are the operators for nuclear and electron spins. The expressions giving the magnitude of fields $\mathbf{B}_N$ and $\mathbf{B}_e$ for this case have been derived by D'yakonov and Perel',[34] Paget,[35] and their co-workers. For the uniformly polarized nuclei of only one species these expressions have the form

$$\mathbf{B}_e = b_e \mathbf{S} \quad \text{where} \quad b_e = \frac{16\pi}{3} n_e \mu_b \xi \tag{20}$$

n_e being the density of the oriented electrons, and

$$\mathbf{B}_N = b_N \mathbf{I}_{N,0}/I_N \quad \text{where} \quad b_N = \frac{16\pi}{3g_e} N \mu_I \xi \tag{21}$$

$\mathbf{I}_{N,0}$ being the induced mean nuclear spin, N the density of the nuclei, and ξ a parameter defining the localization of the electron wave function at the nucleus. The quantity b_N is the nuclear field, which would appear in the case of complete nuclear polarization. An idea of the range of b_N values may be obtained from estimates for GaAs carried out by Paget et al.[35]: $b_N(^{75}\text{As}) = -27.6$ T, $b_N(^{69}\text{Ga}) = -13.7$ T, and $b_N(^{71}\text{Ga}) = -11.7$ T. The sign of b_N (at $\mu_I > 0$) is defined by the sign of the electron g-factor. In GaAs, $g_e = -0.44$.

We note that nuclear spins experience strong relaxation due to the spin nonconserving dipole–dipole interaction between neighboring nuclei. This interaction is characterized by a local magnetic field $\mathbf{B}_L$, which is of the order of μ_I/a_0, a_0 being the lattice constant. the field $\mathbf{B}_L$ is estimated at a few gauss. This is why an efficient polarization of nuclei through the contact electron–nucleus interaction is possible only in the presence of an external magnetic field B exceeding the field $\mathbf{B}_\mathrm{L}$. The relation between the average nuclear $\mathbf{L}_{N,0}$ and electron $\mathbf{S}$ spins in an external magnetic field $\mathbf{B}$ has the form

$$\mathbf{I}_{N,0} = \tfrac{4}{3}\mathbf{I}_N(\mathbf{I}_N + 1)\frac{(\mathbf{S}\cdot\mathbf{B})\mathbf{B}}{\mathbf{B}^2}\frac{\mathbf{B}^2}{\mathbf{B}^2 + \mathbf{B}_\mathrm{L}^2} \qquad (22)$$

It follows from equation (22) that at arbitrary orientations of the average spin $\mathbf{S}$ and field $\mathbf{B}$ the induced average nuclear spin $\mathbf{I}_{N,0}$ conserves only its component along the field $\mathbf{B}$, i.e., nuclear spins are oriented along the field $\mathbf{B}$ but not the spin $\mathbf{S}$.

Now let us consider how fields $\mathbf{B}_N$ and $\mathbf{B}_e$ reveal themselves in optical orientation experiments with different relative orientations of the average spin $\mathbf{S}$ and external magnetic field $\mathbf{B}$. The magnetic field directed along the spin $\mathbf{S}$ may suppress the electron spin relaxation, thus increasing $\mathbf{S}$ and, consequently, the degree of circular polarization of luminescence. Berkovits et al.[36] investigated this effect and found that at fixed orientation of field $\mathbf{B}$ the increase in $\mathcal{P}$ with the field was essentially different for the electron spin oriented along or opposite to the field. These two dependences are shown in Figure 7 by curves 1 and 2. The difference in their courses is direct evidence of the effect of the nuclear field $\mathbf{B}_N$. A reversal of the electron spin orientation is accompanied by a change in the orientation of the field $\mathbf{B}_N$. As a result, the magnetic field $\mathbf{B} + \mathbf{B}_N$ acting on the electron spins is changed by a quantity $2\mathbf{B}_N$, which explains the different courses of curves 1 and 2. The difference between the external field values chosen so as to cause the same increase in $\mathcal{P}(\mathbf{B})$ for curves 1 and 2 is equal to double the nuclear field $\mathbf{B}_N$. Under the conditions of this experiment the field $\mathbf{B}_N$ reached the value 0.4 T.

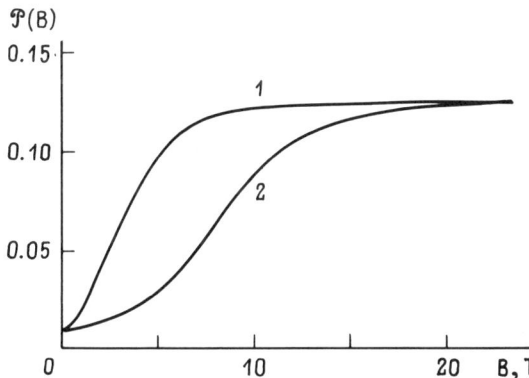

FIGURE 7. The longitudinal-magnetic-field dependences of the degree of circular polarization of luminescence in GaAlAs for excitation by σ^+ (1) and σ^- (2) polarized light (from Berkovits et al.[36]).

Variation in the degree of luminescence polarization in a longitudinal magnetic field is not observed under all experimental conditions. A more universal method for observing fields $\mathbf{B}_N$ and $\mathbf{B}_e$ is to study the luminescence depolarization curves in an oblique field $\mathbf{B}$ forming an angle θ with the average spin **S**. The role of fields $\mathbf{B}$ and $\mathbf{B}_N$ in the oblique geometry is seen from the theoretical dependences $\mathcal{P}(\mathbf{B})$ given in Figure 8a. In the absence of nuclear polarization the external field $\mathbf{B}$ causes depolarization of the electron spins. However, there remains the spin component $S \cdot \cos \theta$ along the field. Since, according to equation (5), the degree of luminescence polarization $\mathcal{P}$ is determined by the spin component along its initial direction, the value of $\mathcal{P}(\mathbf{B})$ decreases monotonically, regardless of the field orientation, down to a value $\mathcal{P}(0) \cos^2 \theta$. The presence of the nuclear field, which is always oriented along the external field, forces the bell-shaped depolarization curve $\mathcal{P}(\mathbf{B})$ to shift to the right or left by the value $\mathbf{B}_N$ relative

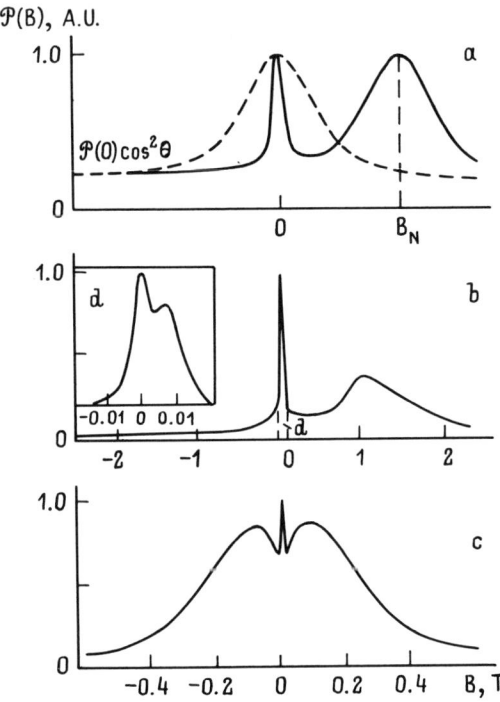

FIGURE 8. Depolarization of luminescence in an oblique magnetic field: (a) schematic presentation of the depolarization curves in the absence (dashed curve) and presence (solid curve) of nuclear polarization; (b) depolarization of luminescence of GaAs in an oblique field, $\theta = 70°$; (c) depolarization of luminescence of GaAs in a transverse field, $\theta = 90°$ (from Zakharchenya et al.[37]).

to the zero of the external field. Besides the shift of the depolarization curve $\mathcal{P}(\mathbf{B})$, a narrow line appears at $\mathbf{B} = 0$ since, according to equation (22), in a zero external field the nuclear field also vanishes.

Such behavior of the depolarization curve $\mathcal{P}(\mathbf{B})$ in the oblique external field was first observed by D'yakonov et al.[34] More detailed investigations of this effect have been undertaken by Zakharchenya et al.[37] on a number of III-V compounds. Figure 8b shows the dependence $\mathcal{P}(\mathbf{B})$ taken from their work for GaAs at $\theta = 70°$. The shift in the bell-shaped curve $\mathcal{P}(\mathbf{B})$ is a clear demonstration of the nuclear spin polarization and gives some estimate of the magnitude of the nuclear field $\mathbf{B}_N$. A narrow line is also observed at zero external field. This line has been found to have a fine structure (see insert d). Here we are faced with the appearance of the electron field $\mathbf{B}_e$.

The effect of the $\mathbf{B}_e$ field may be more conveniently studied in a different experimental geometry, i.e., with the external field $\mathbf{B}$ exactly normal to the average spin $\mathbf{S}$. One would think that in this case the nuclear field should not arise and the external field, if it is weak, should not appreciably affect the luminescence polarization. Nevertheless, it is seen in Figure 8c that in this geometry the fine structure also arises. The appearance of the structure is explained by the fact that under an electron field $\mathbf{B}_e$ greater than the local field $\mathbf{B}_L$ the polarization of the nuclei becomes possible even in the absence of the external field $\mathbf{B}$. When a weak external field $\mathbf{B} \approx \mathbf{B}_e$ is applied, the resultant field $\mathbf{B} + \mathbf{B}_e$ acting on the nuclear spin is deflected from the direction of the spin $\mathbf{S}$ causing the same deflection of the induced field $\mathbf{B}_N$. The experimental geometry becomes oblique and the strong field $\mathbf{B}_N$ lowers the polarization $\mathcal{P}(\mathbf{B})$. As the field $\mathbf{B}$ is further increased, the total field $\mathbf{B} + \mathbf{B}_e$ becomes practically perpendicular to the spin $\mathbf{S}$. Thus, according to equation (22), the nuclear field decreases while the degree of polarization again increases. The depolarizing effect of the external field $\mathbf{B}$ attains still greater values, comparable to the half-width of depolarization curve $\mathcal{P}(\mathbf{B})$.

The above examples of the interaction of fields $\mathbf{B}$, $\mathbf{B}_N$, and $\mathbf{B}_e$ certainly do not represent all the possible experimental situations, but they give some idea of how fields $\mathbf{B}_N$ and $\mathbf{B}_e$ might be evident in experiments on polarized luminescence. Interesting experiments are performed by employing radio-frequency fields. Depending on the conditions of monitoring the nuclear magnetic resonance, variations in both the magnitude and direction of the field $\mathbf{B}_N$ can take place thus making possible optical detection of NMR. Figures 9 and 10 provide examples of the optical observation of NMR, respectively, in GaAlAs at longitudinal and in GaAs at transverse orientations of the external field $\mathbf{B}$ relative to the average spin $\mathbf{S}$. Quadrupole nuclear magnetic resonance in GaAlAs has also been detected optically.[38,39]

The diverse ways in which electron-nucleus interaction arises in the polarized luminescence has been used for detailed investigations of the

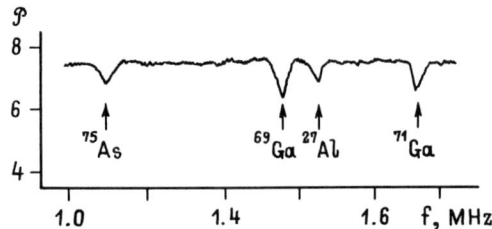

FIGURE 9. Optically detected NMR in GaAlAs in a longitudinal magnetic field $B = 0.15$ T (from D'yakonov and Perel'[16]).

coupled electron-nuclear system in semiconductors. Among other problems, serious attention was paid to ultralow cooling of nuclear spins through their exchange with the optically oriented electron spins, to the behavior of coupled electron-nuclear spins in crystals with low symmetry, and to instabilities in tightly coupled spin systems and their relaxation. Detailed analysis of research in this field can be found in a review by Fleisher and Merkulov.[19]

2.3. Optical Alignment of Electron Momenta in a Magnetic Field

Zemskii et al.[40] studied recombination luminescence in GaAs and found that when a crystal is excited by linearly polarized light, the "hot" luminescence due to the recombination of nonthermalyzed photoelectrons is also linearly polarized. This effect implies that photoelectrons somehow preserve a "memory" of the polarization of the exciting light. This was explained by Dymnikov et al.[41] as due to the alignment of electron

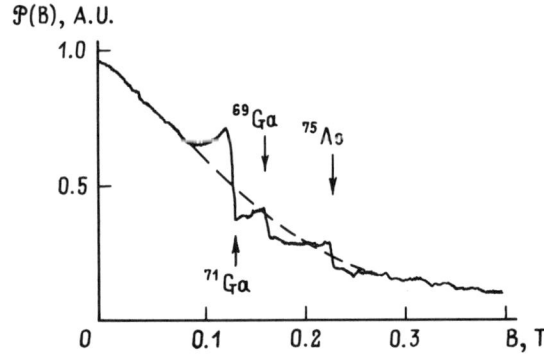

FIGURE 10. Optically detected NMR in GaAs in a transverse magnetic field at a radio frequency of 170 kHz (from Fleisher and Merkulov[19]).

momenta. In GaAs and other crystals with degenerate valence band, according to the selection rules for band-to-band transitions the distribution function of photoelectron momenta at the instant of excitation is highly anisotropic (see, for example, Bir and Pikus[42]). For transitions from the heavy- and light-hole subbands and neglecting the valence band anisotropy, the distribution function of photoelectron momenta has the form

$$F_i(\mathbf{p}) = F_{0i}(\mathscr{E})[1 + \alpha_{0i} P_2(\cos \varphi)] \qquad (23)$$

where $F_{0i}(\mathbf{p})$ is the symmetrical part of the distribution function, φ the angle between the momentum $\mathbf{p}$ and the polarization vector of the exciting light $\mathbf{e}$, while P_2 is the second Legendre polynominal. For excitation from the light-hole subband the parameter α_{01} equals $+1$, and in the case of excitation from the heavy-hole subband $\alpha_{02} = -1$.

The distribution functions for these two cases are shown in Figure 11. Anisotropy of the electron momenta distribution, according to the same selection rules for band-to-band transitions, causes the linear polarization of the recombination luminescence. Scattering processes bring about a change in the initial electron momenta distribution and, correspondingly, a decrease in the degree of anisotropy defined by the parameter α_{0i}. Simultaneously, there occurs relaxation of the initial photoelectron energy $\mathscr{E}_{01,2} = (\hbar\omega - \mathscr{E}_0) m_{1,2}/(m_{1,2} + m_e)$, where $\hbar\omega$ is the photon energy of the exciting light, m_e the electron effective mass, and $m_{1,2}$ is the effective mass, respectively, of the heavy (1) and light (2) hole. If the initial electron energy $\mathscr{E}_{01,2}$ is substantially greater than kT but less than the optical phonon energy, the energy relaxation down to energy values $\mathscr{E} \geq (2-3) kT$ originates from emission of acoustic phonons, each phonon taking away a small portion

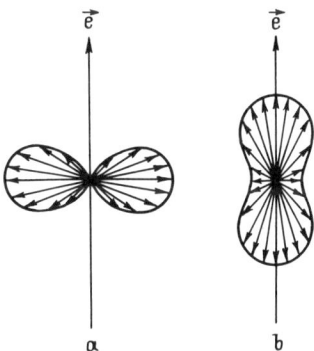

FIGURE 11. Momentum distribution function of photocreated electrons during excitation by linearly polarized light: (a) hh-c transitions; (b) lh-c transitions.

of energy. In this case the electrons attain energy $\mathscr{E}$ after time

$$\tau(\mathscr{E}, \mathscr{E}_0) = \int_{\mathscr{E}}^{\mathscr{E}_0} \tau_{\mathscr{E}'} \frac{d\mathscr{E}'}{\mathscr{E}'} \tag{24}$$

where $\tau_{\mathscr{E}}$ is the energy relaxation time defined by $d\mathscr{E}/dt = -\mathscr{E}/\tau_{\mathscr{E}}$. The decrease in parameter α during this time is given by

$$\alpha(\mathscr{E}) = \alpha_0 \exp\left(-\int_{\mathscr{E}}^{\mathscr{E}_0} \frac{\tau_{\mathscr{E}'}}{\tau_{p2}} \frac{d\mathscr{E}'}{\mathscr{E}'}\right) \tag{25}$$

where τ_{p2} is the relaxation time of the momentum anisotropy described by the polynomial $P_2(\cos \varphi)$.

Upon recombination of electrons excited from an ith valence subband with holes from a jth valence subband, the degree of luminescence polarization in x,y coordinates is described by[43]

$$\mathscr{P}_{\text{lin},ij} = \frac{I_{x'} - I_{y'}}{I_{x'} + I_{y'}}$$

$$= \tfrac{3}{4}\alpha_{0j} \int d\Omega \, (\cos^2 \varphi' - \cos^2 \varphi'')[1 + \alpha_i(\varepsilon)P_2(\cos \varphi)] \tag{26}$$

$$\times \int d\Omega \{1 + \alpha_{0j}[P_2(\cos \varphi') + P_2(\cos \varphi'')]\}[1 + \alpha_i(\varepsilon)P_2(\cos \varphi)]$$

where φ' and φ'' are the angles between the electron momentum $\mathbf{p}$ and axes x and y, respectively. Integration in equation (26) is carried out over all the momentum directions. For radiation propagating along or opposite to the exciting light beam with $\mathbf{e} \parallel \mathbf{x}$, equation (26) is reduced to

$$\mathscr{P}_{\text{lin},ij} = \frac{3\alpha_{0j}\alpha_i(\mathscr{E})}{20 + \alpha_{0j}\alpha_i(\mathscr{E})} \tag{27}$$

In experiments, one usually observes the recombination of electrons with holes trapped on acceptors for which the selection rules are practically the same as for the heavy holes ($i = 1$). In accordance with equation (27) for the electrons excited from the heavy-hole subband ($j = 1$) in the absence of scattering, the degree of luminescence polarization equals 0.14. If the anisotropy of the heavy-hole subband is taken into account the degree of luminescence polarization is found to depend also on the direction of radiation propagation and the orientation of the polarization plane relative to the crystal axes. In the case of radiation propagating along the [001] axis and $\mathbf{e} \parallel [110]$, the degree of luminescence polarization reaches the maximum value of 0.25.

An external magnetic field affects the degree of linear polarization of luminescence through the Lorentz force, which makes electrons change the direction of their motion. The accompanying changes in the electron momenta distribution are most easily analyzed in experiments using the Faraday geometry when the magnetic field **B** is oriented along the light beam. In this geometry the symmetry axis of the momentum distribution, initially parallel to **e**, rotates about the magnetic field **B** with the cyclotron frequency Ω_C, remaining all the time at right angles to the field. Experimental evidence that the symmetry axis rotates in this way may be found in Figure 12, which shows the variation with the field **B** of the degree of linear polarization for the radiation from GaAs propagating in the direction opposite to the exciting light beam. Here $\mathcal{P}_{\text{lin}}$ is the degree of linear polarization of luminescence in x, y coordinates with $\mathbf{x} \| \mathbf{e}$, and $\mathcal{P}'_{\text{lin}}$ is the same in the x', y' coordinate system rotated from **e** through $\pi/4$ in the plane perpendicular to the field B. It is seen that the decrease in $\mathcal{P}_{\text{lin}}$ is accompanied by an increase in $\mathcal{P}'_{\text{lin}}$. If the energy relaxation time $\tau(\mathcal{E}, \mathcal{E}_0)$ is practically the same for all the recombining electrons with energy $\mathcal{E}$ and is given by equation (24), then functions $\mathcal{P}_{\text{lin}}(\mathbf{B})$ and $\mathcal{P}'_{\text{lin}}(\mathbf{B})$ satisfy the expressions

$$\mathcal{P}_{\text{lin}}(\mathbf{B}) = \mathcal{P}_{\text{lin}}(0) \cos 2\Omega_C \tau(\mathcal{E}, \mathcal{E}_0)$$

and

$$\mathcal{P}'_{\text{lin}}(\mathbf{B}) = \mathcal{P}_{\text{lin}}(0) \sin 2\Omega_C \tau(\mathcal{E}, \mathcal{E}_0) \tag{28}$$

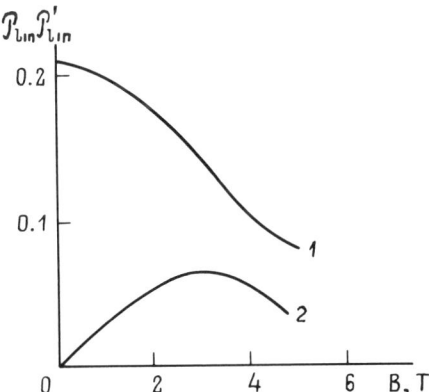

FIGURE 12. Magnetic field effect on the linear polarizations of hot luminescence $\mathcal{P}_{\text{lin}}(\mathbf{B})$ (1) and $\mathcal{P}'_{\text{lin}}(\mathbf{B})$ (2) in p-GaAs for excitation light propagating along the z axis (from Zakharchenya et al.[43]).

For electrons with energy $\mathscr{E}$ close to the initial energy $\mathscr{E}_0$, when $\mathscr{E}_0 - \mathscr{E}$ is less than or comparable to the characteristic energy lost at one step of relaxation, the fluctuations in the time $\tau(\mathscr{E}, \mathscr{E}_0)$ are large and there occurs a change in the nature of magnetic depolarization. Now $\tau(\mathscr{E}, \mathscr{E}_0)$ has the meaning of electron residence time in the range between $\mathscr{E}$ and $\mathscr{E}_0$, and for $\mathscr{E} \to \mathscr{E}_0$ it is simply the residence time in the initial state, τ_0. The probability for τ_0 to be in the interval $d\tau_0$ is given by

$$w(\tau_0)\, d\tau_0 = \frac{1}{\tau} \exp\left(-\tau_0/\tau\right) d\tau_0$$

where τ is the lifetime of electrons in the initial state with energy $\mathscr{E}_0$. To obtain functions $\mathscr{P}_{\text{lin}}(\mathbf{B})$ and $\mathscr{P}'_{\text{lin}}(\mathbf{B})$ in this case, one must average equation (28) over times $\tau(\mathscr{E}, \mathscr{E}_0) = \tau_0$, the average being weighted by the probability given above. Then one obtains

$$\mathscr{P}_{\text{lin}}(\mathbf{B}) = \mathscr{P}_{\text{lin}}(0) \frac{1}{1 + 4\Omega_C^2\tau^2} \quad \text{and} \quad \mathscr{P}'_{\text{lin}}(\mathbf{B}) = \mathscr{P}_{\text{lin}}(0) \frac{2\Omega_C\tau}{1 + 4\Omega_C^2\tau^2} \quad (29)$$

Figure 13 shows functions $\mathscr{P}_{\text{lin}}(\mathbf{B})$ measured for different parts of the hot luminescence spectrum in GaAs.[44] To the smaller photon energy of the luminescence light there corresponds a greater number of energy relaxation steps and a longer relaxation time $\tau(\mathscr{E}, \mathscr{E}_0)$ with a corresponding increase in the rotation angle of the anisotropy axis of the momentum distribution. This relation shows up as a more abrupt decrease in $\mathscr{P}_{\text{lin}}(\mathbf{B})$ with $\mathbf{B}$ for smaller photon energies in the luminescence spectrum. From the curves presented in Figure 13 it is possible to calculate the time $\tau(\mathscr{E}, \mathscr{E}_0)$ using expressions (28) or (29). These $\tau(\mathscr{E}, \mathscr{E}_0)$ values have, in one limiting

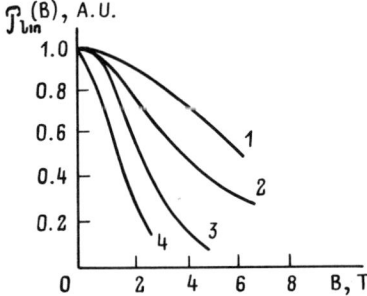

FIGURE 13. Magnetic field effect on the linear polarization of hot luminescence $\mathscr{P}_{\text{lin}}(\mathbf{B})$ in GaAs for various energies in the HL spectrum: 1, 1.89 eV; 2, 1.85 eV; 3, 1.81 eV; 4, 1.77 eV. The photon energy of the excitation light is 1.96 eV (from Mirlin et al.[44]).

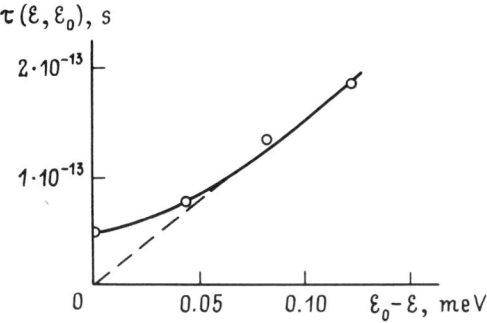

FIGURE 14. Electron energy relaxation time $\tau(\mathscr{E}, \mathscr{E}_0)$ as a function of energy $\mathscr{E}_0 - \mathscr{E}$ in p-GaAs. At $\mathscr{E} \to \mathscr{E}_0$, $\tau = 0.5 \times 10^{-13}$ s and represents the electron lifetime in the initial state (out-scattering time) (from Mirlin et al.[44]).

case, the meaning of lifetime in the initial state, and in the other case, the meaning of energy relaxation time. The resulting dependence of $\tau(\mathscr{E}, \mathscr{E}_0)$ on the dissipated energy $\mathscr{E}_0 - \mathscr{E}$ is presented in Figure 14. This example shows that the use of the Hanle effect in experiments on the optical alignment of electron momenta presents the possibility of measuring very short times on the femtosecond scale and of studying the dependences of τ_ε and τ_p on the electron energy. This method turned out to be very effective in studies of spectral features and scattering mechanisms of carriers at high energies.[18,43]

3. THE HANLE EFFECT ON EXCITONS

The Hanle effect on free excitons is substantially influenced by the crystal symmetry and the position of extrema of the conduction and valence bands in the Brillouin zone. For bound excitons the nature of the optical orientation and of the Hanle effect is also largely dependent on the type of impurity. The excitons bound to charged centers and isoelectronic traps behave like Hg atoms,[45-47] i.e., they may exhibit both polarization and alignment, the latter at resonant excitation. When the exciton becomes bound to a neutral donor (N_D) or neutral acceptor (N_A), a three-particle complex is formed: two electrons and a hole, or an electron and two holes, respectively. It is usual in this situation that because of the e-e or h-h exchange interaction, carriers of the same type are found in a state with antiparallel spins so that only the third carrier will be oriented. This kind of orientation occurs under excitation with circularly polarized light, as is the case for excitation of the D_1 line of Na[48] ($1S$-$2P_2$ transition).

In cubic semiconductors with a degenerate Γ_8 valence band, subject to resonant excitation by linearly polarized light, excitons bound to N_D would also experience an alignment similar to the orientation of Na atoms possessing a single S electron due to $1S-2P_1$ transitions[45-47] (D_2 line). Joos[49] and Hanle[48] have shown that the maximum degree of linear polarization in this case reaches 0.6. In cubic crystals a similar alignment under resonant excitation could be observed also for excitons bound to N_A, provided that the ground state of the two holes is a fivefold-degenerate symmetric state with momentum $\mathbf{J} = 2$ or a superposition of this state and the singlet state $\mathbf{J} = 0$. In the latter case the maximum possible degree of polarization of luminescence upon excitation with linearly polarized light equals 0.5. This kind of alignment of the excitons bound to N_D or N_A has not yet been observed.

When a free exciton (or electron–hole pair) generated by circularly polarized light becomes bound, the hole, which quickly lost its orientation while it was free, again becomes oriented due to an exchange interaction with an oriented electron in a binding process. If binding is preceded by the formation of the free exciton, then the sign and degree of the hole polarization in the exciton bound to N_D depend on the ratio of the free exciton lifetimes in optically active and inactive states, respectively.[21,50]

Below we examine features of the Hanle effect on free and bound excitons in III-V cubic crystals having zincblend structure (T_d—$\bar{4}3$ m class), in II-VI hexagonal crystals with wurtzite structure (C_{6v}—6 mm class), and II-VI crystals of various modifications (D_{6h}—6 mm, C_{3v}—3 m, and D_{3h}—$\bar{6}$ m2 classes).

3.1. The $\Gamma_8 \times \Gamma_6$ and $\Gamma_7 \times \Gamma_6$ Excitons in Cubic Crystals

As indicated above, in the direct band III-V crystals of T_d symmetry class the conduction band is simple, i.e., degenerate only in the spin, and the corresponding wave functions transform according to the Γ_6 representation. The Γ_{15} valence band is threefold degenerate if the spin is disregarded, and the spin-orbit interaction splits it into the fourfold degenerate Γ_8 band and the twofold degenerate Γ_7 band, the latter being the lower of the two in most crystals. As a consequence, excitons of the two types, $\Gamma_8 \times \Gamma_6$ and $\Gamma_7 \times \Gamma_6$, can occur in these crystals. The exchange interaction splits the $\Gamma_8 \times \Gamma_6$ state further into two threefold degenerate states, Γ_{15} and Γ_{25}, and a twofold degenerate Γ_{12} state, the Γ_{15} state being electric-dipole-active. The $\Gamma_7 \times \Gamma_6$ state splits into an electric-dipole active Γ_{15} state and an inactive singlet Γ_2 state. For free excitons, alongside the electron–hole exchange interaction, the annihilation interaction is also essential as it causes longitudinal-transverse splitting of the electric-dipole-active states, i.e., into states of a longitudinal exciton with dipole momentum **d** parallel to the exciton wave vector **k**, and of a transverse exciton with **d** perpendicular to **k**. At large longitudinal-transverse splitting, $\hbar\omega_{LT}$, in a spectral range where exciton and photon branches cross mixed exciton-photon modes, called polaritons, form and substantially affect the character of the Hanle effect at resonant excitation of excitons. For bound excitons annihilation interaction makes an additional contribution to electron-hole exchange constants.

3.1.1. The Density Matrix Method. This method may be employed for calculation of the Hanle effect on bound excitons; the components of the matrix are dependent on the spin indices of the electrons and holes forming the exciton. The same method is applicable to free excitons as well, if the lifetime τ and spin relaxation time τ_s far exceed the momentum relaxation time τ_p so that the complete density matrix ρ may be expressed as the product of the momentum distribution function, which is independent of the spin, and the spin density matrix, which is independent of the exciton momentum.

An equation for the density matrix has the form

$$\frac{\partial \rho}{\partial t} = \left(\frac{\partial \rho}{\partial t}\right)_{rec} + \left(\frac{\partial \rho}{\partial t}\right)_{sr} - \frac{i}{\hbar}[H_{ex}\rho] - \frac{i}{\hbar}[H_{ext}\rho] + G \qquad (30)$$

The first term on the right-hand side of equation (30) describes how ρ is affected by the recombination of electrons and holes (or by the capture of an exciton on a center), and the second term takes account of spin relaxation; H_{ex} and H_{ext} are the Hamiltonians which represent, respectively, the electron-hole exchange and the interaction with external fields, such as a magnetic field or deformation; G is the generation function.

Under resonant excitation of the exciton

$$G_{mm'} \sim \sum_{\alpha,\beta} J_m^\alpha J_{m'}^{\beta*} d_{\alpha\beta}^0 \qquad (31)$$

where J_m^α is the matrix element of the J_α component of the current operator corresponding to an exciton excitation into the mth state, $d_{\alpha\beta}^0 = \langle e_\alpha e_\beta^* \rangle$ is the polarization tensor for the exciting light, and **e** is the polarization vector.

If excitons are formed through binding of free electrons and holes, and the binding probability is independent of their spin indices, then

$$G = G_e \times G_h \qquad (32)$$

where G_e and G_h are the density matrices for electrons and holes at the moment of the binding into an exciton. Having calculated ρ, one may obtain the polarization tensor of the excitonic emission,

$$d_{\alpha\beta} \sim \sum_{m,m'} J_m^{\alpha*} J_{m'}^\beta \rho_{mm'} \qquad (33)$$

3.1.2. The $\Gamma_7 \times \Gamma_6$ Exciton. In order to illustrate the influence of the exchange interaction on the Hanle effect, let us consider first the simplest case of the $\Gamma_7 \times \Gamma_6$ exciton. Corresponding to the dipole-active Γ_{15} state are three functions with total momentum $J = 1$ and $J_z = 0, \pm 1$. In the following these states are denoted by $|1, 1\rangle, |1, 0\rangle$, and $|1, -1\rangle$. To the inactive Γ_2 state, denoted by $|0, 0\rangle$ and split off from the Γ_{15} state by a quantity Δ, there corresponds $J = J_z = 0$. Accordingly, the nonzero components of the current operator are

$$\mathbf{J}_{1,1}^+ = -\mathbf{J}_{1,-1}^- = \mathbf{J}_{1,0}^z = \mathbf{J}_0: \qquad \mathbf{J}^{\pm} = \frac{1}{\sqrt{2}}(\mathbf{J}_x \pm i\mathbf{J}_y) \tag{34}$$

The interaction with the magnetic field is described by a Hamiltonian similar to equation (6) and determined by the g-factors for electrons and holes, g_e and g_h, respectively. The magnetic field $\mathbf{B}$ mixes the Γ_{15} and Γ_2 states. In writing $(\partial \rho/\partial t)_{\text{rec}}$, it is necessary to take into account the disparity between the lifetimes τ_0 and τ_1 for the states Γ_{12} and Γ_{15}, respectively, as the former may possess a substantial contribution from radiative recombination. Generally, spin relaxation in the $\Gamma_7 \times \Gamma_6$ exciton is characterized by six relaxation times, τ_{sl}, which determine the relaxation rates for the ρ_{mm} components that transform according to the different irreducible representations. If the spin relaxation of the exciton comes about as an independent relaxation of the electron and hole spins, determined by the respective times τ_{se} and τ_{sh}, then for $\Delta \ll kT$ we have[51]

$$\left(\frac{\partial \rho_{mm',nn'}}{\partial t}\right)_{\text{sr}} = -\frac{1}{\tau_{sh}}\left(\rho_{mm',nn'} - \frac{1}{r_h}\delta_{mm'}\sum_{m''}\rho_{m''m'',nn'}\right)$$

$$-\frac{1}{\tau_{se}}\left(\rho_{mm',nn'} - \frac{1}{r_e}\delta_{nn'}\sum_{n''}\rho_{mm',n''n''}\right) \tag{35}$$

where n and m are the spin indices of the electrons and holes, while r_e and r_h are the multiplicity of degeneracy in spin. For the $\Gamma_7 \times \Gamma_6$ exciton, $r_e = r_h = 2$.

The degree of polarization of radiation propagating along the z axis is determined by the components $\rho_{1,1}, \rho_{-1,-1}$, and $\rho_{1,-1} = \rho_{-1,1}^*$ as follows:

$$\mathcal{P}_{\text{circ}} = (\rho_{1,1} - \rho_{-1,-1})/(\rho_{1,1} + \rho_{-1,-1}) \tag{36}$$

$$\mathcal{P}_{\text{lin}} = 2|\rho_{1,-1}|/(\rho_{1,1} + \rho_{-1,-1}) \tag{37}$$

The angle θ between the polarization plane and the direction x, chosen in order to have the desired basic functions $|1 \pm 1\rangle = \mp(X \pm iY)/\sqrt{2}$, is set through a phase of the component $\rho_{1,-1} = |\rho_{1,-1}| \exp i\varphi_{1,-1}$ and is given by

$$\theta = \tfrac{1}{2}\varphi_{1,-1} \tag{38}$$

Under this condition the degrees of polarization $\mathcal{P}'_{\text{lin}}$ and $\mathcal{P}''_{\text{lin}}$, respectively, in coordinate axes x, y and axes x^1, y^1 rotated from the former about the z axis by $\pi/4$, are

$$\mathcal{P}'_{\text{lin}} = 2 \operatorname{Re} \rho_{1,-1}/(\rho_{1,1} + \rho_{-1,-1}) \quad \text{and} \quad \mathcal{P}''_{\text{lin}} = 2 \operatorname{Im} \rho_{1,-1}/(\rho_{1,1} + \rho_{-1,-1}) \tag{39}$$

In order to illustrate how the exchange interaction modifies the Hanle effect, we shall present the expressions for $\mathcal{P}_{\text{circ}}$ and $\mathcal{P}_{\text{lin}}$ in the simplest case of no spin relaxation, i.e., $\tau_1 = \tau_0 = \tau$, and subject to resonant excitation of the exciton by, respectively, circularly, or linearly polarized light with $\mathbf{e} \parallel \mathbf{x}$.[21] Here and subsequently it is assumed that incident and emitted light propagate along the z axis (in uniaxial crystals, along the principal axis) in opposite directions.

In a longitudinal magnetic field ($\mathbf{B} \parallel \mathbf{z}$), $\mathcal{P}_{\text{circ}}$ in the case under consideration does not change while $\mathcal{P}_{\text{lin}}$ and the angle of rotation of the polarization plane θ vary as follows:

$$\mathcal{P}_{\text{lin}}(\mathbf{B}) = \frac{1}{(1 + \Omega_+^2 \tau^2)^{1/2}} \tag{40}$$

$$\theta(\mathbf{B}) = \tfrac{1}{2} \arctan \Omega_+ \tau \tag{41}$$

where $\hbar \Omega_\pm = \hbar(\Omega_e \pm \Omega_h) = g_\pm \mu_B \mathbf{B}$ and $g_\pm = g_e \pm g_h$. These equations hold irrespective of the numerical relation between Δ and $\hbar/\tau$.

In a transverse magnetic field $\mathbf{B} \parallel \mathbf{e}$ or $\mathbf{B} \perp \mathbf{e}, \mathbf{z}$, quantity $\mathcal{P}_{\text{lin}}$ remains unaffected. However, if $\mathbf{B} \perp \mathbf{z}$ and $\mathbf{B}$ makes an angle of $\pi/4$ with $\mathbf{e}$, the Hanle effect depends on the relation between Δ and $\hbar/\tau$. When $\Delta \ll \hbar/\tau$,

$$\mathcal{P}'_{\text{lin}} = 2(L_e + L_h)/(2 + L_+ + L_-) \tag{42}$$

$$\mathcal{P}''_{\text{lin}} = (L_+ - L_-)/(2 + L_+ + L_-) \tag{43}$$

where $L_\pm = (1 + \Omega_\pm^2 \tau^2)^{-1}$ and $L_{e,h} = (1 + \Omega_{e,h}^2 \tau^2)^{-1}$. When $kT \gg \Delta \gg \hbar/\tau$, $\hbar \Omega_{e,h}$,

$$\mathcal{P}'_{\text{lin}} = \frac{4}{4 + 3\Omega_+^2 \tau^2} \tag{44}$$

$$\mathcal{P}''_{\text{lin}} = -\frac{\Omega_+^2 \tau^2}{4 + 3\Omega_+^2 \tau^2} \tag{45}$$

At resonant excitation with circularly polarized light in the presence of a transverse magnetic field $\mathbf{B} \| \mathbf{x}$, the reduction in the degree of circular polarization $\mathscr{P}_{\text{circ}}$ is accompanied by the appearance of linear polarization, $\mathscr{P}_{\text{lin}}$. Under these conditions the dependence $\mathscr{P}_{\text{circ}}(B)$ is given by equations which coincide with expression (43) or (45), while the equations for $\mathscr{P}_{\text{lin}}(B)$ coincide with expression (42) or (44). An analogous linear polarization of emission in a plane parallel to $\mathbf{B}$ arises also during illumination by unpolarized light. When $\Delta \gg \hbar/\tau$ the $\Gamma_7 \times \Gamma_6$ exciton in a longitudinal magnetic field may be treated as a classical dipole being excited by linearly polarized light that rotates about the field direction $\mathbf{B} \| \mathbf{z}$ with angular velocity $\Omega_+/2$. A similar model was proposed for the Hg atom by Hanle.[52] With its aid Eldridge[53] derived expressions similar to equations (40) and (41). A projection on the x' axis of the dipole moment $\mathbf{d}_0$ induced by the light with $\mathbf{e} \| \mathbf{x}$ at $t = 0$ is

$$\mathbf{d}_{x'}(t) = \mathbf{d}_0 \cos\left(\frac{\Omega_+ t}{2} - \chi\right) \tag{46}$$

where χ is the angle between the x and x' axes.

The degree of linear polarization of luminescence in the coordinate system x', y' rotated about x by χ or $\chi + \pi/2$ is

$$\mathscr{P}'_{\text{lin}} = (\overline{|\mathbf{d}_{x'}|^2} - \overline{|\mathbf{d}_{y'}|^2})/(\overline{|\mathbf{d}_{x'}|^2} + \overline{|\mathbf{d}_{y'}|^2}) \tag{47}$$

where

$$\overline{|\mathbf{d}_i|^2} = \frac{\int_0^\infty \exp(-t/\tau)|\mathbf{d}_i(t)|^2 \, dt}{\int_0^\infty \exp(-t/\tau) \, dt} \tag{48}$$

An analogous expression may be easily obtained also for the case of a transverse magnetic field and excitation by linearly or circularly polarized light. In the latter case an angular momentum $\mathbf{J}$ precesses about $\mathbf{B}$ with angular velocity Ω_+. In the same manner one may obtain an expression for the case $\Delta \ll \hbar/\tau$, when the electron and hole spins precess about $\mathbf{B}$ independently with angular velocities Ω_e and Ω_h.

3.1.3. The $\Gamma_8 \times \Gamma_6$ Exciton. As noted earlier, in nearly all cubic crystals of the III-V group the top of the valence band corresponds to the fourfold degenerzte Γ_8 representation. The exchange interaction operator for the $\Gamma_8 \times \Gamma_6$ exciton has the form[54]

$$H_{\text{ex}} = \tfrac{1}{4}\Delta(\mathbf{J}\boldsymbol{\sigma}) + \Delta_1(\mathbf{J}_x^3\sigma_x + \mathbf{J}_y^3\sigma_y + \mathbf{J}_z^3\sigma_z) \tag{49}$$

where $\boldsymbol{\sigma}_i$ are Pauli matrices for the electrons and $\mathbf{J}_i$ are operator matrices for the holes in the basis of the function $\mathbf{J} = \frac{3}{2}$, $J_z = \pm\frac{3}{2}, \pm\frac{1}{2}$. The interaction of the hole with the magnetic field B is defined by the operator

$$H_\mathrm{B} = \mu_\mathrm{B}[g_1(\mathbf{J} \cdot \mathbf{B}) + g_2(\mathbf{J}_x^3 \mathbf{B}_x + \mathbf{J}_y^3 \mathbf{B}_y + \mathbf{J}_z^3 \mathbf{B}_z)] \tag{50}$$

The nonzero components of $\mathbf{J}_\alpha$ for the optically active representation Γ_{15}, as well as for the $\Gamma_7 \times \Gamma_6$ exciton, are given by equation (34) and, correspondingly, equations (36)-(39) remain valid.

When $\Delta, \Delta_1 \ll kT$, the spin relaxation operator is still defined by equation (35), the only difference being that for the $\Gamma_8 \times \Gamma_6$ exciton $r_\mathrm{h} = 4$. Usually, the strong spin-orbit interaction leads to $\tau_\mathrm{sh} \ll \tau$, while τ_se is determined by the exchange interaction with holes.

When $\Delta_1 \ll \Delta$,

$$\tau_\mathrm{se} = \tfrac{1}{5}T\left[3 + 8\left(\frac{\Delta T}{\hbar}\right)^{-2}\right] \tag{51}$$

where $T^{-1} = \tau_\mathrm{sh}^{-1} + \bar{\tau}^{-1}$, $\bar{\tau}$ is the mean lifetime such that $\bar{\tau}^{-1} = (3\tau_1^{-1} + 5\tau_2^{-1})/8$, while τ_1 and τ_2 are the lifetimes in the optically active Γ_{15} state and in optically inactive states Γ_{25} and Γ_{12}, respectively. For small exchange splitting, $\Delta \ll \hbar/T_1$, where $T_1^{-1} = \tau_1^{-1} + \tau_\mathrm{sh}^{-1}$ and $\tau_\mathrm{sh} \ll \tau_1$, only electrons are oriented, and the variation in circular polarization with transverse magnetic field will be given by the same equation (8) as in the instance of free electrons. In the opposite limiting case, $kT \gg \Delta \gg \hbar T_1$, the behavior of $\mathscr{P}_\mathrm{circ}(\mathbf{B})$ is more complicated,[51]

$$\mathscr{P}_\mathrm{circ}(\mathbf{B}_\perp) = -\tfrac{1}{2}P_\mathrm{e}\frac{\bar{T}}{\bar{\tau}}\frac{(1 - \chi_1 - \Omega_1\tau_1\chi_2)}{(1 + \Omega_1^2 T_1^2)[(1 - \chi_1^2) + \chi_2^2]} \tag{52}$$

where

$$\chi_1 = \frac{1}{16\tau_\mathrm{sh}}\left[\frac{T_1}{(1 + \Omega_1^2 T_1^2)} + \frac{5T_2}{(1 + \Omega_2^2 T_2^2)}\right]$$

$$\chi_2 = \frac{1}{16\tau_\mathrm{sh}}\left[\frac{\Omega_1 T_1^2}{(1 + \Omega_1^2 T_1^2)} + \frac{5\Omega_2 T_2^2}{(1 + \Omega_2^2 T_2^2)}\right]$$

$$\Omega_{1,2} = g_{1,2}\mu_\mathrm{B}B, \qquad g_1 = \tfrac{1}{4}(-g_\mathrm{e} + 5g_\mathrm{h}), \qquad g_2 = \tfrac{1}{4}(g_\mathrm{e} + 3g_\mathrm{h})$$

$$\bar{T} = \tfrac{1}{8}(3T_1 + 5T_2), \qquad \bar{\tau} = \tfrac{1}{8}(3\tau_1 + 5\tau_2), \qquad T_{1,2}^{-1} = \tau_{1,2}^{-1} + \tau_\mathrm{sh}^{-1}$$

P_e is the initial degree of orientation of the electron spins. The Hanle curves are seen to have Lorentzian shape only for $\tau_{sh} \gg \tau_{1,2}$ when transitions between the states Γ_{15} and Γ_{25} or Γ_{12} may be neglected. If the exchange interaction is strong, the dependence of the degree of linear polarization on the longitudinal magnetic field is described by the standard equations (40) and (41), also only for $\tau_{sh} \gg \tau_{1,2}$.

The Hanle effect on thermalized free and bound excitons in III-V crystals was observed in a number of investigations. Besides those mentioned in the Introduction we especially note here the papers by Fishman et al.[55,56] performed on InP, GaAs, and CdTe. They established that when an exciton is formed via binding of electrons and holes generated by polarized light, the half-width of the Hanle curve has the same value for all lines corresponding to the recombination of excitons bound at different centers. This result shows that the longest time among all involved times is that needed for binding an electron, and it is just this time that determines the polarization quenching by the magnetic field.

3.2. The $\Gamma_9 \times \Gamma_7$ and $\Gamma_7 \times \Gamma_7$ Excitons in Hexagonal II-VI Crystals with Wurtzite Structure

The band structure of II-VI crystals of the C_{6v} symmetry class differs from that of the T_d class cubic crystals in that the former has the valence band Γ_8 split into two bands, Γ_9 and Γ_7, degenerate only in the spin, while the Γ_7 band of the cubic crystal corresponds to the lower of the two, i.e., to the Γ_7 band. In both Γ_7 bands the momentum of the holes is $J_z = \pm\frac{1}{2}$, while for Γ_9 $J_z = \pm\frac{3}{2}$. An exchange interaction splits the $\Gamma_9 \times \Gamma_7$ exciton into two terms, Γ_5 and Γ_6, and the $\Gamma_7 \times \Gamma_7$ exciton into the terms $\Gamma_1 + \Gamma_2 + \Gamma_6$. With polarization $\mathbf{e} \perp \mathbf{z}$ the dipole active states is the Γ_5 state, with total angular momentum $J_z = \pm 1$, while if $\mathbf{e} \parallel \mathbf{z}$ the dipole active state is a Γ_1 state. In uniaxial crystals the electron and hole g-factors are anisotropic and have two components each, $g_{e\parallel}, g_{h\parallel}$ and $g_{e\perp}, g_{h\perp}$. For the $\Gamma_9 \times \Gamma_7$ exciton, $g_{h\perp} = 0$. The longitudinal magnetic field $\mathbf{B} \parallel \mathbf{z}$ splits the Γ_5 state as in cubic crystals. Therefore, the variation in the degree of linear polarization with the longitudinal magnetic field is defined by expressions which differ from equations (40) and (41) only in that g_e, g_h are substituted for $g_{e\parallel}, g_{h\perp}$. The transverse magnetic field $\mathbf{B} \perp \mathbf{z}$ mixes the Γ_5 state with Γ_1 and Γ_2 or with Γ_6. Therefore, if $g_{e,h}\mu_B B$ is small compared with separations Δ_1 and Δ_2 between the terms Γ_5 and Γ_1, Γ_2 of the $\Gamma_7 \times \Gamma_7$ exciton, then the splitting of the term Γ_5 is quadratic in $\mathbf{B}$ and equals

$$\Delta E_\perp = 2\gamma B^2 \tag{53}$$

where

$$\gamma = \tfrac{1}{8}\mu_B^2 \left[-\frac{(g_e + g_h)^2}{\Delta_1} + \frac{(g_e - g_h)^2}{\Delta_2} \right]$$

Correspondingly, the quenching of circular polarization in a transverse magnetic field is given by the expression[52,54]

$$\mathcal{P}_{\text{circ}}(\mathbf{B}) = \mathcal{P}_{\text{circ}}(0)/(1 + \kappa^2) \tag{54}$$

$$\kappa = (T_1 T_2)^{1/2} \Delta E_\perp / \hbar, \qquad T_i^{-1} = \tau^{-1} + \tau_{si}^{-1}$$

where τ is the lifetime and $\tau_{s1,2}$ the spin relaxation time of the density matrix components, $\rho_{1,1} - \rho_{-1,-1}$ and $\rho_{1,-1}$, respectively. In addition, in a transverse magnetic field there arises a linear polarization of luminescence in the plane making an angle of $\pi/4$ with the direction of $\mathbf{B}$:

$$\mathcal{P}_{\text{lin}}(\mathbf{B}) = -\mathcal{P}_{\text{circ}}(0) \frac{\kappa (T_2/T_1)^{1/2}}{1 + \kappa^2} \tag{55}$$

As distinct from cubic crystals, linear polarization is possible for excitation by circularly polarized light.

At resonant excitation by linearly polarized light, and in the presence of a transverse magnetic field applied at an angle of $\pi/4$ to the polarization plane, the degree of linear polarization decreases according to a relationship analogous to equation (54), while a circular polarization arises obeying an expression similar to equation (55). For the $\Gamma_9 \times \Gamma_7$ exciton we have $g_{h\perp} = 0$ and $\Delta_1 = \Delta_2$; therefore, according to equation (53), i.e., on the assumption of a quadratic dependence on $\mathbf{B}$, $\Delta E_\perp = 0$ and the Hanle effect in a transverse magnetic field does not take place. The Hanle effect on bound excitons, as well as on thermalized free excitons in II–VI crystals, has been observed by many authors. Besides the works mentioned in the Introduction we note the studies by Benoit et al.[57] on CdS, by Oka and Kushida[58] on ZnTe and ZnSe, and by Dzhioev et al.[59] on HgI_2. In particular, in the last study it was found that, under excitation by circularly polarized light, the polarization happened to be of different signs for centers of a different type. Consequently, for a certain type of center the sign was opposite to that for a free exciton. This happens when the center preferentially captures excitons in the optically inactive state with $\mathbf{J} = 2$.[21]

3.3. The $\Gamma_7 \times \Gamma_8$ Excitons in III–VI Crystals with Symmetry Class D_{3h}

Of all III–VI crystals we shall examine GaSe on which thorough experimental investigations have been performed. GaSe crystallizes in three modifications: β, γ, and ε. The respective symmetry classes are D_{6h}, C_{3v}, and D_{3h}. All the modifications have essentially the same band structure and excitonic spectra, so only the ε modification will be considered. In GaSe the valence and conduction bands correspond to the representations Γ_1 and Γ_4, which transform into Γ_7 and Γ_8, respectively, if spin is taken into account. As a result of spin–orbit mixing of the Γ_1 band with the lower valence band Γ_5, transitions $\Gamma_7 \to \Gamma_8$ become allowed for the transverse polarization $\mathbf{e} \perp \mathbf{z}$ with rather weak oscillator strength. As regards representations Γ_1 and Γ_4, the transition $\Gamma_1 \to \Gamma_4$ is allowed only for $\mathbf{e} \| \mathbf{z}$. An exchange interaction splits the $\Gamma_7 \times \Gamma_8$ exciton into three terms: Γ_3, Γ_4, and Γ_6. The optically active one among these, for the polarization $\mathbf{e} \| \mathbf{z}$, is the Γ_4 state with $\mathbf{J} = \mathbf{J}_z = 0$ which lies $\Delta_1 \simeq 2$ meV higher than the Γ_6 term. The Γ_6 term is optically active for a polarization $\mathbf{e} \perp \mathbf{z}$. The matrix elements of the current operator for the two functions of the representation Γ_6 with $\mathbf{J} = 1$ and $\mathbf{J}_z = \pm 1$ are identical to relations (34). The Γ_3 state with $\mathbf{J} = 1$ and $\mathbf{J}_z = 0$ is optically inactive and lies Δ above the Γ_6 state, Δ being much less than Δ_1.

As in hexagonal II–VI crystals, the splitting of the terms in a magnetic field is determined by the four g-factors: $g_{e,h\|}$ and $g_{e,h\perp}$. For GaSe, Δ is comparable to $\hbar/T$ and $\Delta_1 \gg \hbar/T$, T being the exciton state relaxation time. Here, in contrast to the $\Gamma_7 \times \Gamma_7$ exciton treated above, it is necessary to accurately take into account intermixing of the states Γ_6 and Γ_3 in the transverse magnetic field, while mixing of these states with the Γ_4 state may be neglected. As in the case of II–VI crystals, the dependence of the degree of linear polarization on the longitudinal magnetic field $\mathbf{B} \| \mathbf{z}$ is given by relationships like equations (40) and (41). In a transverse magnetic field under excitation with circularly polarized light we have[21]

$$\mathscr{P}_{\text{circ}}(\mathbf{B}) = \mathscr{P}_{\text{circ}}(0) \frac{2[1 + (\tfrac{1}{4}\Omega_\perp^2 + \Delta^2/\hbar^2)\tau^2]}{(1 + L_\perp)(1 + \tfrac{1}{4}\Omega_\perp^2 \tau^2)^2 + (\Delta\tau/\hbar)^2} \tag{56}$$

where

$$L_\perp = 1 - \frac{\Omega_\perp^2 \tau^2}{2[1 + (\Omega_\perp^2 + \Delta^2/\hbar^2)\tau^2]} \quad \text{and} \quad \hbar\Omega_\perp = \mu_B B(g_{e\perp} + g_{h\perp})$$

Under the same conditions a transverse magnetic field also gives rise to a linear polarization of luminescence in the plane which makes an angle of $\pi/4$ with the direction of $\mathbf{B}$:

$$\mathscr{P}_{\text{lin}}(\mathbf{B}) = 2\mathscr{P}_{\text{circ}}(0) \frac{\Delta\tau(\Omega_\perp \tau)^2}{\hbar(1 + L_\perp)[1 + \tfrac{1}{4}\Omega_\perp^2 \tau^2)^2 + (\Delta\tau/\hbar)^2]} \tag{57}$$

At resonant excitation, the polarization in the plane parallel to **B** occurs upon excitation by unpolarized or circularly polarized light and by linearly polarized light having **e** directed at an angle of $\pi/4$ to **B**:

$$\mathscr{P}'_{\text{lin}}(\mathbf{B}) = -\frac{\Omega_\perp^2 \tau^2}{4[1 + (\Delta^2/\hbar^2 + \tfrac{3}{4}\Omega_\perp^2)\tau^2]} \tag{58}$$

Equations (56)–(58) were derived under the assumption that the lifetimes τ_0 and τ_1 of the states Γ_4 and Γ_6, respectively, are the same and equal to τ, and that spin relaxation does not occur. If $\tau_0 \neq \tau_1$ the linear polarization also arises when the exciton is formed by the binding of a free electron and hole and may reach 0.3 when $\tau_1 \ll \tau_0$ and $\hbar\Omega_\perp \gg \Delta(8\tau_1/3\tau_2)^{1/2}$. For resonant excitation by linearly polarized light the polarization in a magnetic field **B** directed at an angle of $\pi/4$ relative to the polarization plane varies according to an equation analogous to relation (56) and the circular polarization produced follows an equation similar to expression (57).

3.3.1. Bound Excitons. Experiments on GaSe by Gamarts *et al.*[60] have shown that the alignment of bound excitons for resonant excitation by linearly polarized light may be observed only in a sufficiently strong transverse magnetic field in configuration $\mathbf{B} \parallel \mathbf{e}$ or $\mathbf{B} \perp \mathbf{e}$. This is explained by the fact that the spin relaxation time τ_s of the $\rho_{1,-1}$ component of density matrix is small and increases with the magnetic field. The behavior of $\mathscr{P}_{\text{lin}}(\mathbf{B})$ was described assuming

$$\tau_s(B) = \tau_{s0}/(1 + B_0^2/B^2) \tag{59}$$

and taking $B_0 = 3.1$ kG and $\tau_{s0} = 1.1 \times 10^{-8}$ s. An appearance of circular polarization was also detected, however, in an oblique magnetic field forming an angle $\varphi = 75°$ with the z axis, i.e., having components $\mathbf{B}_z$ and $\mathbf{B}_x \parallel \mathbf{e}_0$ (see Figure 15). The $\mathbf{B}_x$ component causes the increase in τ_s while the $\mathbf{H}_z$ component produces circular polarization because of the splitting Δ, and the disparity of lifetimes in the states Γ_3 and Γ_6 as well as of spin relaxation times for different components, $\rho_{i,j}$, of the density matrix. No linear polarization was found for the orientation of the transverse component of magnetic field at an angle of $\pi/4$ with **e**, since the magnetic field that brings about a sufficient increase in τ_s is at the same time strong enough to suppress $\mathscr{P}_{\text{lin}}$. For excitation with circularly polarized light the oblique magnetic field was found to cause a rise in the linear polarization with $\mathbf{e} \parallel \mathbf{B}_x$. In all these cases the sign of polarization changed following a reversal of the direction of **B** or a change in sign of the polarization of the excitation light. Excitation with unpolarized light produced an emission that was unpolarized. These experiments made it possible to obtain a value $\Delta = -0.032$ meV, which is

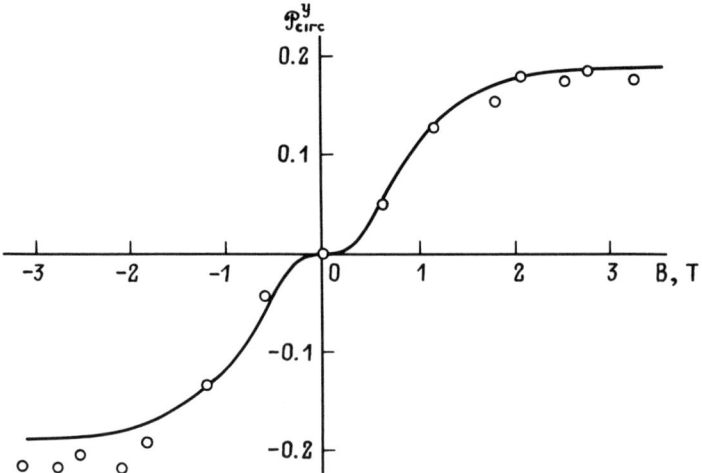

FIGURE 15. Appearance of the circular polarization of bound exciton luminescence in GaSe in an oblique magnetic field at resonant excitation by linearly polarized light (from Gamarts et al.[60]).

a value that cannot be measured by known spectroscopic methods. This value of Δ nearly coincides with the value -0.0357 meV, which has been determined by Cavenett et al.[61] using the method of optical detection of EPR for bound excitons.

3.3.2. Free Excitons. GaSe is peculiar in that it possesses an indirect minimum of the conduction band somewhat lower than the direct Γ_4 minimum. As a consequence, the lifetime of the $\Gamma_7 \times \Gamma_8$ free excitons happens to be very short, so that $\hbar/\tau \gg \Delta$, and splitting of the terms Γ_3 and Γ_6 may be ignored. The splitting Δ_1 still remains greater than $\hbar/\tau$, so the Γ_4 state need not be considered. According to equations (56) and (57), and for $\Delta \ll \hbar/\tau$, in a transverse field we obtain

$$\frac{\mathcal{P}_{\text{circ}}(\mathbf{B})}{\mathcal{P}_{\text{circ}}(0)} = \frac{\mathcal{P}_{\text{lin}}(\mathbf{B})}{\mathcal{P}_{\text{lin}}(0)} = \frac{1}{L(\mathbf{B}_\perp)} \frac{1}{1 + (\Omega_\perp T)^2} \quad (60)$$

where

$$L(\mathbf{B}_\perp) = \frac{1}{4}\left[3 + \frac{1}{1 + (\Omega_\perp \tau)^2}\right]$$

In the case of linearly polarized excitation it is again assumed, as above, that the angle between $\mathbf{e}_0$ and $\mathbf{B}$ equals $\pi/4$.

A deviation in equation (60) from the usual Lorentzian shape is connected with the peculiar band structure of GaSe and is due to the fact that precession of the dipole momentum in the presence of the transverse magnetic field **B** occurs not about **B**, but about a direction that is perpendicular to **B** and to the z axis. In these circumstances the d_z component associated with the inactive Γ_3 term is radiationless. Hence for $\mathbf{B} \parallel \mathbf{x}$, $d_x = d_{0x} \cos \Omega_\perp t/2$ and $d_y = d_{0y}$, and consequently according to equation (47)

$$I_x \sim \overline{d_x^2} = \tfrac{1}{2} d_{0x}^2 \left(1 + \frac{1}{1 + (\Omega_\perp \tau)^2}\right) \quad \text{and} \quad I_y \sim \overline{d_y^2} = d_{0y}^2$$

when exciting light is unpolarized and $|d_{0x}|^2 = |d_{0y}|^2$, a linearly polarized emission arises for which

$$\mathscr{P}'_{\text{lin}} = -\frac{(\Omega_\perp \tau)^2}{4 + 3(\Omega_\perp \tau)^2} \tag{61}$$

For excitation with linearly polarized light when the angle between $\mathbf{e}_0$ and **B** is equal to $\pi/4$, function $\mathscr{P}_{\text{lin}}(\mathbf{B})$, in accordance with equation (47), is defined by equation (60) and the quantity $L(\mathbf{B}_\perp)$ in the denominator represents a relative change of intensity $I \sim (\overline{d_x^2} + \overline{d_y^2})$ at $d_{0x} = d_{0y}$. However, an experiment by Razbirin et al.[62] has shown that the behavior of the depolarization curves for GaSe deviates from equation (60) in that the polarization does not drop monotonically but exhibits a change of sign, and only at a still higher field drops to zero. Such nonmonotonic curves were observed during cascade excitation of atoms. If, for example, an electron is initially excited to level 1 and, after a mean time τ_1, goes over to level 2, and next, after a mean time τ_2, radiatively returns into the ground state, then the number of electrons in level 2 at any time is given by the following equation, assuming that excitation ceases at $t = 0$;

$$\Psi(t) = \frac{1}{\tau_1 - \tau_2} [\exp(-t/\tau_1) - \exp(-t/\tau_2)] \tag{62}$$

Substitution into equation (48) of function (62) instead of $\exp(-t/\tau)$ makes $\mathscr{P}(\mathbf{B})$ a nonmonotonic function. In the cascade scattering of excitons the states with different values of the wave vector **k** will be intermediate. In the back-scattering geometry, if excitons with $\mathbf{k} = \mathbf{q}_0$ are generated, the emission is due to the annihilation of excitons with $\mathbf{k} = -\mathbf{q}_0$. Transition between these states occurs through multiple scattering. The secondary emission observed in this case is described by the theory of resonant scattering of light, which takes into account the contributions from all the excitons which, prior to annihilation, experienced a different number N of collisions from

$N = 1$ up to $N = \infty$. For elastic scattering of excitons on impurity atoms, all the excitons contribute to the emission at the same frequency, which is the frequency of the exciting light. The average number of scattering acts is $\bar{N} = 1 + \tau_0/\tau_p$, where τ_0 is the lifetime and τ_p the momentum relaxation time. When $\tau_0 \gg \tau_p$, namely $\bar{N} \gg 1$, the density matrix becomes factorized and the Hanle effect may be described by the same expressions as in the case of bound excitons. In GaSe, τ_0 is small enough to be compared to τ_p. A calculation by Ivchenko et al.[63] using a diagram technique gave the following expression for the Hanle curves by excitation with linearly polarized light propagating along the principal z axis with $\mathbf{e} \parallel \mathbf{x}$ and $\mathbf{B} \parallel \mathbf{z}$:

$$\mathcal{P}'_{\text{lin}}(B_\parallel) = \frac{1}{1 - \tau/2\tau_0} \Phi_1(\Omega_\parallel, \tau_0, \tau) \tag{63}$$

$$\mathcal{P}^\parallel_{\text{lin}}(B_\parallel) = \frac{1}{1 - \tau/2\tau_0} \Phi_2(\Omega_\parallel, \tau_0, \tau) \tag{64}$$

where $1/\tau = 1/\tau_0 + 1/\tau_p$,

$$\Phi_1(\Omega, \tau_0, \tau) = \frac{1}{1 + \Omega^2\tau^2} \left(\frac{1 - \Omega^2\tau\tau_0}{1 + \Omega^2\tau_0^2} - \frac{\tau}{2\tau_0} \frac{1 - \Omega^2\tau^2}{1 + \Omega^2\tau^2} \right)$$

$$\Phi_2(\Omega, \tau_0, \tau) = \frac{1}{1 + \Omega^2\tau^2} \left(\frac{\Omega(\tau + \tau_0)}{1 + \Omega^2\tau_0^2} - \frac{\Omega\tau^2}{\tau_0(1 + \Omega^2\tau^2)} \right) \tag{65}$$

When $\tau_0 \gg \tau_p$, equations (63)–(65) reduce to familiar expressions (40) and (41) in which $\Omega = \Omega_\parallel$. In a transverse field B_x making an angle of $\pi/4$ with the polarization plane,

$$\mathcal{P}_{\text{lin}}(B_\perp) = \mathcal{P}_{\text{lin}}(0) \frac{1}{1 - \tau/2\tau_0} \frac{\Phi_1(\Omega_\perp/2, \tau_0, \tau)}{1 + \Phi_3(\Omega_\perp, \tau_0, \tau)} \tag{66}$$

where

$$\Phi_3(\Omega, \tau_0, \tau) = 1 - \frac{1}{2(1 + \Omega^2\tau^2)^{1/2}} + \frac{\Phi_1(\Omega, \tau_0, \tau)}{1 - \tau/2\tau_0} \left[1 - \frac{(1 + \Omega^2\tau^2)^{1/2}}{2} \right] \tag{67}$$

An analogous relationship describes the variation in the circular polarization of radiation in a transverse magnetic field.

Equation (64) allows for the Faraday rotation of the polarization plane of radiation prior to the generation and after the annihilation of excitons, and equation (66) for birefringence and dichroism in a transverse field. For $\tau_0 \gg \tau_p$, when the linewidth of $\hbar/\tau$ is great compared to $\hbar/\tau_0$, these contributions become insignificant. This is also the case for strong inhomogeneous broadening, i.e., $\delta\omega \gg 1/\tau_0$. Excitation with unpolarized light also produces linearly polarized emission

$$\mathcal{P}_{\text{lin}}(\mathbf{B}_\perp) = \frac{\Phi_3(\Omega_\perp, \tau_0, \tau) - 1}{\Phi_3(\Omega_\perp, \tau_0, \tau) + 1} \tag{68}$$

The Hanle effect on free excitons has been studied in detail by Gamarts et al.[64] Figure 16 shows the function $\mathcal{P}'_{\text{lin}}(\mathbf{B}_\parallel)$ and Figures 17a and b the functions $\mathcal{P}'_{\text{lin}}(\mathbf{B}_\perp)$ and $\mathcal{P}_{\text{circ}}(\mathbf{B}_\perp)$. Theoretical curves in these figures were calculated using relations (63)-(66) for $g_\parallel = 2.7$, $g_\perp = 1.9$, $\tau_0 = 1.6 \times 10^{-11}$ s, and $\tau = 0.4 \times 10^{-11}$ s. Thus, these experiments made it possible to measure $\tau_p = 0.5 \times 10^{-11}$ s. The measurements of the degree of linear polarization under excitation with unpolarized light enabled the role of dichroism to be established directly: when $\delta\omega < 1/\tau_0$ the behavior of $\mathcal{P}_{\text{lin}}(\mathbf{B}_\perp)$ is described by equation (68). Curve 1 in Figure 18a was calculated using this equation. Curve 2 is a plot of $\mathcal{P}_{\text{lin}}(\mathbf{B}_\perp)$ when $\delta\omega \gg 1/\tau_0$. The experimental data in Figure 18b agree with equation (68).

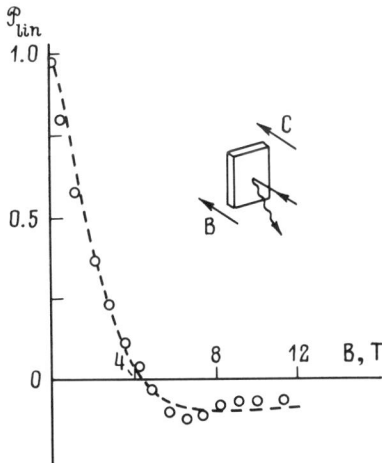

FIGURE 16. The degree of linear polarization $\mathcal{P}_{\text{lin}}$ as a function of the longitudinal magnetic field for GaSe (from Gamarts et al.[64]).

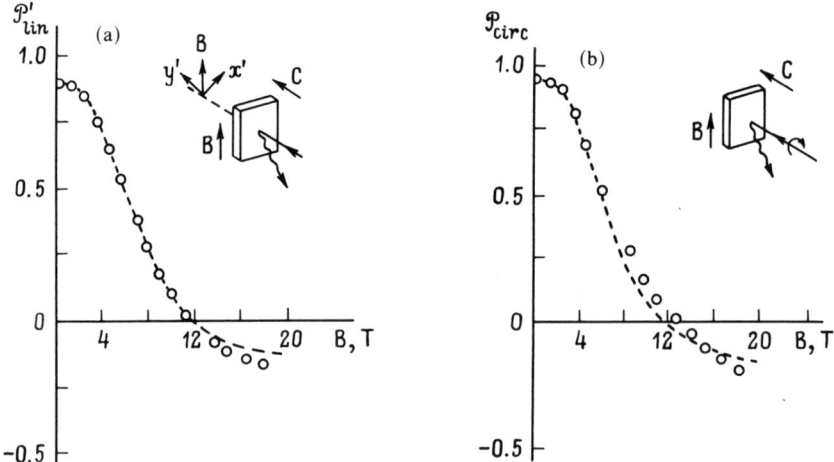

FIGURE 17. Linear polarization of free exciton luminescence in GaSe for resonant excitation by linearly polarized light (a) and circular polarization for excitation with circularly polarized light (b) as functions of the transverse magnetic field (from Gamarts et al.[64]).

3.4. The Influence of Reemission on the Hanle Effect

The theory presented in the previous section was based on the assumption that the exciton radiative lifetime τ_{rad} exceeds by far the nonradiative lifetime τ_0. This assumption sets the following restriction on the value of longitudinal-transverse splitting $\hbar\omega_{\text{LT}}$ which determines τ_{rad}:

$$\omega_{\text{LT}} \ll (\omega_q \tau \tau_0)^{-1} \tag{69}$$

where $\hbar\omega_q$ is the kinetic energy of excitons excited by photons of wave vector $\mathbf{q}_0$, $\tau^{-1} = \tau_p^{-1} + \tau_1^{-1}$, and $\tau_1^{-1} = \tau_0^{-1} + \tau_{\text{rad}}^{-1}$.

If criterion (69) is satisfied, the following factors may be ignored: (a) exciton diffusion during its lifetime and (b) creation of mixed exciton–photon modes, called polaritons. Criteria for the fulfillment of the last two conditions are

$$\omega_{\text{LT}} \ll (\omega_q \tau \tau_0)^{-1} \left(\frac{\tau_0}{\tau}\right)^{1/2} \tag{70a}$$

and

$$\omega_{\text{LT}} \ll (\omega_q \tau \tau_0)^{-1} \frac{\tau_0}{\tau} \tag{70b}$$

If criterion (69) is not satisfied, it becomes necessary to take into consideration the reabsorption of radiation that leads to depolarization. The high degree of polarization of the emission from GaSe indicates that criterion (69) is satisfied. The smallness of ω_{LT} in GaSe is due to the above-mentioned fact that optical transitions into the Γ_6 state are allowed only in cases for which spin-orbit interaction mixes the states Γ_1 and Γ_5. In the papers by Ivchenko et al.[65] and Sobirov and Yuldashev[66] calculations were conducted on radiation depolarization, and the effect of a longitudinal magnetic field on the linear polarization $\mathcal{P}_{lin}(\mathbf{B})$ for the $\Gamma_7 \times \Gamma_6$ free exciton in a cubic crystal, with allowance for reemission. In this calculation it was further assumed that $\tau = \tau_p \ll \tau_1$, $\Delta \gg \hbar/\tau_1$, and that the scattering of excitons was elastic and inequalities (70a) and (70b) were valid. The above assumptions make this problem equivalent to a problem of radiation transport in a diffusely scattering media encountered in astrophysics. The case of a solid

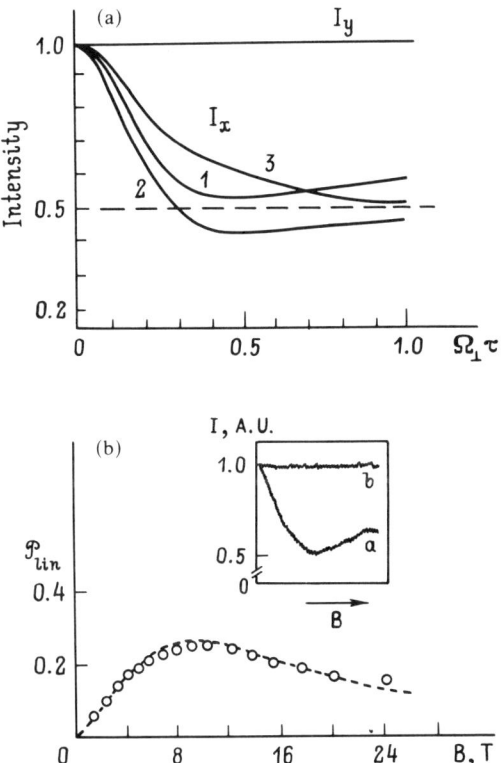

FIGURE 18. Degree of linear polarization of exciton luminescence as a function of the transverse magnetic field for unpolarized resonant excitation: (a) theoretical functions $I_x(\mathbf{B})$ and $I_y(\mathbf{B})$ at $\mathbf{B} \| \mathbf{x}$; (b) experimental functions $I_x(\mathbf{B})$, $I_y(\mathbf{B})$, and $\mathcal{P}_{lin}(\mathbf{B})$.

poses the additional problem of taking into account reflections of the emission from the crystal inner boundary, which substantially lessens the degree of polarization of the outgoing radiation for the albedo $\tilde{\omega}_0 = \tau_1/\tau_{\text{rad}}$ nearing unity, i.e., for $\tau_{\text{rad}} > \tau_0$. The problem of radiation transport has been solved by the method of Chandrasekhar. The reemission may be viewed as a mechanism of spin relaxation and characterized by the time $\bar{\tau}_s$, which is dependent on $\tilde{\omega}_0$ and defined by expressions similar to equations (4) and (9),

$$\mathcal{P}_{\text{lin}}(0) = \frac{\bar{\tau}_s}{\bar{\tau} + \bar{\tau}_1}, \qquad g\mu_B \Delta B = \hbar\left(\frac{1}{\bar{\tau}_s} + \frac{1}{\bar{\tau}_1}\right) \qquad (71)$$

Figure 19 shows plots of $\tau_1/\bar{\tau}_1$, $\tau_1/\bar{\tau}_s$, and $\tau_1/\bar{\tau}$, where $1/\bar{\tau} = 1/\bar{\tau}_s + 1/\bar{\tau}_1$, as functions of the albedo $\tilde{\omega}_0$. It is seen that as $\tilde{\omega}_0$ is increased, the effective lifetime $\bar{\tau}_1$ increases and $\bar{\tau}_s$ decreases. The deviation from the exact Hanle curves of the curves calculated using equations (40) and (41) with the τ_1 and τ_s values determined from (71) is illustrated in Figures 20 and 21, which show plots of $\mathcal{P}'_{\text{lin}}(z)/\mathcal{P}'_{\text{lin}}(0)$ and $\mathcal{P}^{\|}_{\text{lin}}(z)/\mathcal{P}'_{\text{lin}}(0)$ as functions of $z = B/\Delta B$. The deviations are seen to be not great. It must be stressed that when reemission is highly effective, the calculation of $\bar{\tau}_1$ and $\bar{\tau}_s$ values will require an accurate solution to the problem of radiation propagation in a semi-infinite medium under conditions of reabsorption and reflection from the boundary. Such a solution can only be obtained by numerical methods.

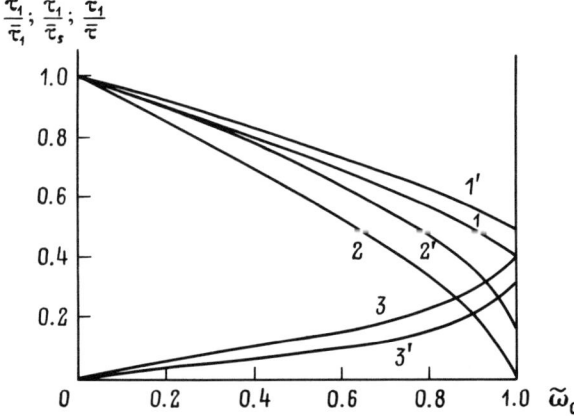

FIGURE 19. Plots of $\tau_1/\bar{\tau}_1$ (curves 1, 1'), $\tau_1/\bar{\tau}_s$ (curves 2, 2'), and $\tau_1/\bar{\tau}$ (curves 3, 3') as functions of albedo $\tilde{\omega}_0$ when $n^2 = 10$ (curves 1, 2, 3) and $n^2 = 1$ (curves 1', 2', 3'), n being the refractive index (from Sobirov and Yuldashev[66]).

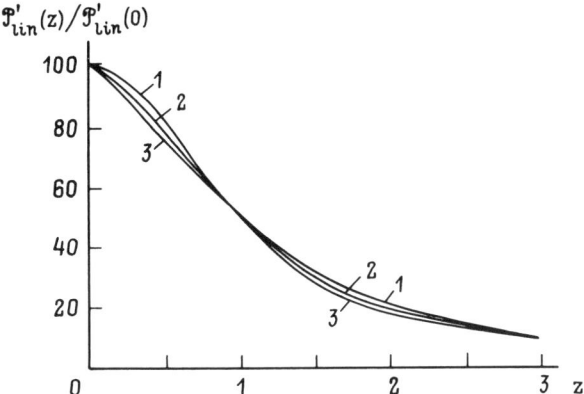

FIGURE 20. Plots of $\mathcal{P}'_{\text{lin}}(z)/\mathcal{P}'_{\text{lin}}(0)$ as functions of $z = B/\Delta B$ when $n^2 = 10$: 1, the exact Hanle curve; 2, $\tilde{\omega}_0 = 0.6$; 3, $\tilde{\omega}_0 = 1.0$ (from Sobirov and Yuldashev[66]).

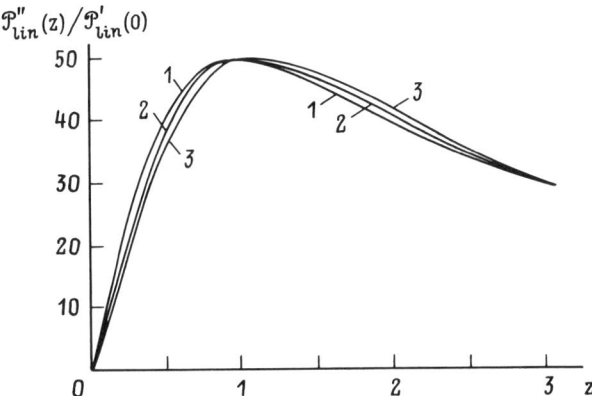

FIGURE 21. Plots of $\mathcal{P}''_{\text{lin}}(z)/\mathcal{P}'_{\text{lin}}(0)$ as functions of $z = B/\Delta B$ when $n^2 = 10$: 1, the exact Hanle curve; 2, $\tilde{\omega}_0 = 0.6$; 3, $\tilde{\omega}_0 = 1.0$ (from Sobirov and Yuldashev[66]).

3.5. Hot Excitons and Polaritons

In many polar crystals, excitation by light of photon energy considerably exceeding the forbidden gap produces hot excitons. Conservation of the total momentum is accomplished through simultaneous emission of longitudinal optical (LO) phonons. These hot excitons lose their energy through emission of several LO phonons until they reach the bottom of the excitonic band. Correspondingly, in the luminescence spectra one can observe a series of narrow lines at frequencies $\omega_N = \omega_0 - N\omega_{\text{LO}}$ ($N = 2, 3, \ldots$). Under polarized light excitation these lines are also partially polarized.[67] The degree of line polarization decreases with the number of emitted LO phonons. Bir et al.[68] and Ivchenko et al.[69] have shown that this depolarization is due to spin–orbit interaction, which in crystals with degenerate valence band or closely lying Γ_9 and Γ_7 bands changes the hole angular momentum when the direction of the hole wave vector is varied.

An investigation by Takunov[70] suggests that in a magnetic field, in principle, we should observe depolarization of these lines. However, because of strong interaction with LO phonons the lifetime of the hot exciton is short, so this kind of depolarization is possible only for excitons near the bottom of excitonic bands for which further thermalization is considerably retarded since their kinetic energy is less than $\hbar\omega_{LO}$. In this region, called the bottleneck region, an interaction between excitons and photons becomes essential, leading to the formation of mixed modes or polaritons. The polaritons that reach the surface can leave the crystal as photons at once or after emitting an additional number of LO phonons through virtual transitions, or slowly thermalize via interaction with acoustic phonons. These thermalized polaritons appear in luminescence spectra as wide bands adjoining the bottleneck polariton line or its phonon replicas. The degree of polarization in such bands decreases with increasing separation from the corresponding main line.

The Hanle effect on $\Gamma_9 \times \Gamma_7$ polaritons in CdSe has been observed by Nawrocki et al.[71] In these experiments the frequency of the excitation light ω_0 was chosen so as to exceed by $2\omega_{LO}$ the frequency corresponding to the bottleneck. Under excitation with linearly polarized light a line at $\hbar\omega = \hbar\omega_0 - 2\hbar\omega_{LO}$ happened to be strongly polarized. In a longitudinal magnetic field, the polarization decreased and the polarization plane was rotated according to the known relationships (40) and (41) with parameters $g = 0.69$ and $\tau = 6 \times 10^{-12}$ s. However, the mechanism governing the Hanle effect differs here and is due to Faraday rotation of the polarization plane of the polariton. If a polariton of frequency $\omega = \omega_0 - 2\omega_{LO}$ is excited at distance z from the crystal surface where the light intensity is $I(z) = I(0) \exp(-\alpha z)$, α being the light absorption coefficient, then its contribution to the emission is proportional to $\exp[-(\alpha + \alpha_p)z]$, where α_p is the damping coefficient of the polariton determined by its scattering. The angle through which the polarization plane is rotated along the path z is given by

$$\theta(z) = \tfrac{1}{2}(q_+ - q_-)z = \frac{\mathbf{\Omega}_\|}{2v_g} z \qquad (72)$$

where $q_\pm(\omega)$ are the wave vectors of the left- and right-hand circularly polarized polariton modes, and v_g is the group velocity. Hence

$$\mathscr{P}'_{lin}(\mathbf{B}) = \mathscr{P}'_{lin}(0) \overline{\cos 2\theta(z)} = \frac{\mathscr{P}'_{lin}(0)}{1 + (\mathbf{\Omega}_\| T)^2}$$

$$\mathscr{P}^\|_{lin}(B) = \mathscr{P}'_{lin}(0) \overline{\sin 2\theta(z)} = \frac{\mathscr{P}'_{lin}(0)\mathbf{\Omega}_\| T}{1 + (\mathbf{\Omega}_\| T)^2} \qquad (73)$$

where

$$T = \frac{\bar{z}}{v_g} = \frac{1}{(\alpha + \alpha_p)v_g}$$

and

$$\overline{F(z)} = (\alpha + \alpha_p) \int_0^{\sim \infty} F(z) \exp[-(\alpha + \alpha_p)z] \, dz$$

At high magnetic fields, $B > 5\,T$, a deviation from equation (73) occurred since the magnitude of the splitting $\hbar\Omega_\parallel$ approached the value of the exchange splitting of the terms Γ_5 and Γ_6 and caused their mixing.

In studies by Permogorov and Travnikov[72,73] the exciting light frequency was higher than the bottleneck frequency by ω_{LO}, so that the polariton was created right in the bottleneck region. Radiation was observed at a frequency of the third phonon replica, $\omega_0 - 3\omega_{LO}$. At this frequency the polariton wave is essentially a light wave and both its absorption and Faraday rotation can be ignored. Under excitation with linearly polarized light the degree of polarization was found to be 0.6, and in a longitudinal magnetic field it varied according to equation (73). The quantity T in this case is the lifetime of the polariton with an energy $\hbar(\omega - \omega_{LO})$, which is determined predominantly by scattering on acoustic phonons, i.e., by the term $1/\alpha_p v_g$. In different samples the T value varied from 1.5 to 3.0×10^{-12} s. This experiment demonstrates that, when two or three LO phonons are emitted in a Raman scattering of light with frequency ω_0, the main intermediate state is indeed the real polariton state with energy $\hbar(\omega_0 - N\omega_{LO})$ ($N = 1, 2, \ldots$). If the contribution of the virtual states with energy $\mathscr{E}$, markedly different from the resonance energy, was considerable, the depolarization would be determined not by the value of $\Omega_\parallel T$, but, instead, by the value of $\hbar\Omega_\parallel/[\mathscr{E} - \hbar(\omega_0 - N\omega_{LO})]$, i.e., it would be considerably less.

4. THE HANLE EFFECT ON IMPURITY CENTERS

It was noted in the Introduction that the optical orientation of electrons bound on impurity centers was observed as early as the beginning of the 1960s. Nevertheless, until recently studies on the effects of level crossing and anticrossing for bound electrons have been performed mainly with the use of the methods of radiospectroscopy. In the last decade these methods have been enriched by the optical detection of resonances. In this case a nonequilibrium population of the electron levels is created by excitation with polarized light and the onset of resonance is observed as a change in the intensity or polarization of polarized luminescence. If there is some initial splitting of the electron levels due, for instance, to a crystalline field, optical detection permits observation of resonances even in the absence of a magnetic field as demonstrated by Salib and Cavenett.[74] Optically created spin polarization of bound electrons, in its turn, can be observed as a change in the absorption or generation of a microwave radiation.[75,13]

Optical investigations of the bound electron spin polarization were mainly conducted with the aid of methods employing the measurement of dichroism and of Faraday rotation under excitation with circularly polarized light.[12,13,76-78] Thus, using measurements of the inertialess component of the Faraday rotation in a region of level anticrossing, Zapasskii and Kozlov[78] succeeded in obtaining unique data on the magnitudes of the minor components of the g-tensor for the ground state of different impurity centers in a crystal and Zapasskii et al.[79] obtained data on the degree of magnetic anisotropy of impurity centers in glasses. Circular dichroism measurements by Mollenauer et al.[80] revealed optical orientation of atomic nuclei in the vicinity of impurity centers.

The magnetic moment arising during excitation of impurity centers by circularly polarized light and its precession in a transverse magnetic field can be directly observed in the measurements of sample magnetization.[81-83]

4.1. The Hanle Effect on Sm^{2+} in CaF_2

Polarized luminescence of impurities is usually observed during linearly polarized excitation. As a rule, polarization of the luminescence in these conditions is caused not by the optical orientation, but is due to the nonuniform occupancy of the impurity centers which, in a cubic lattice, are found in a number of equivalent positions with noncubic local symmetry and different orientation of the local axes with regard to the principal axes of the crystal, the state termed "orientational degeneracy" by Feofilov.[84] Under orientational degeneracy a magnetic field, provided it is not too strong, has little if any, effect on polarization.

The optical orientation of impurity centers under excitation by linearly polarized light was first observed by Akimov et al.[85] on Sm^{2+} in CaF_2. Although the Sm^{2+} levels in CaF_2 experience considerable crystalline splitting, the occupancy of the centers with different orientation with regard to the principal axes of the crystal is, however, the same. As a result, the linear polarization of the emission due to transitions between the excited states Γ_{15}^- and Γ_{15}^+ is a consequence of nonequilibrium occupancy of the states X, Y, Z of the term Γ_{15}^-, i.e., in fact, of an electron alignment. A high degree of luminescence polarization, up to -0.57, was observed for nonresonant excitation with the light polarized along one of the principal axes [100]. The energy of the photons exceeded by 0.22 eV the energy of transition from the ground state Γ_1^+ to the excited state Γ_{15}^-. A further increase 0.43 eV in the photon energy caused a decrease in the degree of luminescence polarization down to -0.20. In the case that the excitation light polarization **e** is along one of the [111] axes, the polarization of luminescence did not arise. These features are related to the fact that fd electron states of the Sm^{2+} center practically do not interact with deformations of the type Γ_{15}^+ (ε_{xy}, ε_{xz}, ε_{yz}), while they interact strongly with deformations of the type Γ_{12}^+ (ε_{xx}, ε_{yy}, ε_{zz}), which do not mix the states X, Y, Z but cause rapid relaxation of the state $(X + Y + Z)/3$ that becomes excited at **e**∥[111]. A negative sign of the luminescence polarization is explained by the properties of the selection rules for transitions between the states X, Y, Z and X', Y', Z' of the terms Γ_{15}^- and Γ_{15}^+, respectively. If, under illumination by linearly polarized light propagating along the z axis with **e**∥[100], the X state is excited, then the only possible recombination transitions are those into the states Y' and Z' of the Γ_{15}^+ term, the emitted light being polarized with **e**∥**z** and **e**∥**y**, respectively. For this reason, in the absence of depolarization the luminescence light propagating along the z axis is completely polarized. A longitudinal magnetic field gave rise to a decrease in the luminescence polarization, but the half-width of the depolarization curve at B = 1.5 T was far in excess of that to be expected at $T = 2 \times 10^{-10}$ s. In addition, the luminescence line featured a large half-width, much greater than $1/T$. This proves that the large half-width of the depolarization curve originates in splitting of the Γ_{15}^- state caused by local deformations ε_{xx}, ε_{yy}, and ε_{zz}. Under these conditions polarization changes in a magnetic field $\mathbf{B}_z$ are due to splitting of the states $J_z = 1$ and $J_z = -1$ and are fitted by equation (8) in which $(\Omega T)^2$ is substituted for $(\hbar\Omega/\Delta)^2$, where Δ is the splitting of the terms X and Y by local deformations $\varepsilon_{xx} - \varepsilon_{yy}$. If $\Delta \gg \hbar/T$,

$$\mathscr{P}_{\text{lin}}(\mathbf{B}) = \mathscr{P}_{\text{lin}}(0)/[1 + (\hbar\Omega/\Delta)^2] \tag{74}$$

where $\hbar\Omega = g\mu_B \mathbf{B}_z$ is the splitting of the terms $J_z = \pm 1$. Akimov et al.[85] assumed that the quantity obeys a random Gaussian distribution and, by comparing calculated and experimentally obtained Hanle curves, found that $\sqrt{\overline{\Delta^2}} = 0.3$ meV.

4.2. The Hanle Effect on Eu^{2+} in CaF_2 and SrF_2

The Hanle effect under circularly polarized excitation was observed by Kaplyanskii et al.[15] and Ivchenko et al.[86] Electrons were excited into the

Γ_8^+ state belonging to the Γ_{12}^+ state by nonresonant light having photon energy greater than that of the Γ_8^+ state. Unlike the Γ_8 state originating from the Γ_{15} or Γ_{25} terms, in a magnetic field the state $\Gamma_8 = \Gamma_6 \times \Gamma_{12}$ splits into two terms instead of four and its Hamiltonian H_B is given by expression (50), where $g_1 = -13g_2/4 = 13g/12$. Mixing of this state with close-lying states belonging to the Γ_{15}^+ and Γ_{25}^+ terms may push the value of $g_1 + 13g_2/4$ slightly off zero. The features of the Eu^{2+} center are a low degree of luminescence polarization, 0.0003 in SrF_2 and 0.0006 in CaF_2, and a large value of the half-width of the depolarization curve $\Delta B \approx 100$ G. At the same time independent measurements of the spin relaxation time τ_s and lifetime τ_0 give values close to 6×10^{-7} s, which implies $\Delta B \approx 1$ G. Ivchenko et al.[86] explain these features of the E_u^{2+} center in terms of two relaxation times: one is the spin relaxation time τ_s and the other is an orbital relaxation time τ_L which determines the transition time between the states $\psi_1 = \sqrt{2/3}\,[Z^2 - (X^2 + Y^2)/2]$ and $\psi_2 = (X^2 - Y^2)/\sqrt{2}$ of the term Γ_{12}^+. Functions $\psi_i \varphi_k$ correspond to the representation Γ_8^+ where φ_k are spin functions with spin $S_z = k$ ($k = \pm\frac{1}{2}$). Correspondingly, one may introduce two components of the density matrix that characterize the degree of orientation at the moment of excitation,

$$\rho_0 = \frac{S_1 + S_2}{n} \quad \text{and} \quad \rho_0' = \frac{S_1 - S_2}{n}$$

where $S_1 = n_{1,1/2} + n_{1,-1/2}$, $S_2 = n_{2,1/2} - n_{2,-1/2}$, and $n = \sum_{i,k} n_{ik}$. Here n_{ik} is the occupancy of the states $\psi_i \varphi_k$. According to Ivchenko et al.[86] the polarization of luminescence due to the transition into the ground state is

$$\mathcal{P}_{\text{circ}} = -\frac{1}{2}\frac{\rho_0}{1 + \tau_0/\tau_s} + \frac{2}{7}\frac{\rho_0'}{1 + \tau_0/\tau_s + \tau_0/\tau_L} \tag{75}$$

In a transverse magnetic field, both terms of equation (75) decrease as $1/(1 + \Omega^2 \tau^2)$ where $\tau_1^{-1} = \tau_0^{-1} + \tau_s^{-1}$, $\tau_2^{-1} = \tau_0^{-1} + \tau_s^{-1} + \tau_L^{-1}$, and $\hbar\Omega = g\mu_B B$. The observed behavior of the luminescence polarization for the Eu^{2+} center in CaF_2 can be explained subject to the assumption that $g = 3.6$, $\rho_0 < 10^{-3}$, $\rho_0' \approx 1$, $\tau_0 \simeq \tau_s \approx 6 \times 10^{-7}$ s, and $\tau_L \simeq 10^{-9}$ s. The mechanism by which the occupancy of the states $(1, 1/2)$ and $(2, -\frac{1}{2})$ or $(1, -\frac{1}{2})$ and $(2, \frac{1}{2})$ is leveled out remains unclear.

5. SUMMARY

In this chapter it has been shown that the Hanle effect is responsible for a whole series of interesting phenomena in solid state optics. The luminescence depolarization by a magnetic field depends essentially on the nature of the processes which determine the excitation of carriers and excitons, and their behavior while they are in the excited state and subsequently experience recombination. Analysis of the shape of the depolarization curves offers ample opportunities for the quantitative characterization of the processes involved. The present review highlights the Hanle effect as a basis for one of the skillful techniques of magneto-optic studies of solids. To conclude, we shall mention a number of comprehensive investigations in which the application of the Hanle effect plays a key role.

The possibility of measuring short lifetimes and spin relaxation times of carriers in a steady-state regime was realized without resorting to an intricate pulse technique. In GaAs, GaSb, and InP crystals over wide ranges of doping and temperature, the spin relaxation time has been studied in an interval from 10^{-9} to 10^{-11} s, which is not accessible with the commonly used technique of EPR. As a result, two new mechanisms of spin relaxation of free electrons have been discovered.[87,88] The first is related to the spin splittings of the conduction band, linear and cubic in the wave vector $\mathbf{k}$. The second involves exchange interaction of electrons and holes. Appropriate constants of the spin splittings and exchange interaction have been determined.[89]

Very interesting results were obtained in studies both of electron lifetimes at the bottom of the conduction band and of energy and momentum relaxation times of carriers in the upper states of the conduction and valence bands. Measurements on heavily doped GaAs and GaSb demonstrated that the minimum values of the radiative lifetimes could be as low as 2-5 × 10^{-10} s[90] which proved, for these crystals, the dominant role of the nonradiative Auger recombination that cuts the electron lifetime down to values of the order of 10 ps.[91,92] Still smaller values, in the femtosecond range, have been found for the times of energy and momentum relaxation of hot electrons and holes in GaAs and InP,[93-95] as well as for the time of carrier scattering between equivalent band minima in GaAs.[96] The Hanle effect was used in research on transport and recombination of carriers in heterostructures of semiconductors[97,98] and, recently, in the modern structures with quantum wells.[99,100] In these structures the deviation of the luminescence depolarization curves from Lorentzian shape gives a measure of the strains developing at their interfaces. We also considered the application of the Hanle effect in investigations of the diffusion and transport of carriers in various experimental situations.

Studies of the Hanle effect on excitons made possible the first determinations of the short lifetimes and momentum relaxation times for excitons in CdSe[63,64] and for polaritons in II-VI crystals.[71-73] the Hanle effect on bound excitons provided a new approach to studies of the binding mechanisms and, in particular, to ascertaining whether a center binds the carriers one by one or in the form of the free exciton.[21,50]

These and other examples presented in this review demonstrate the great potentialities initiated by employing the Hanle effect in studies of solids. Further research will undoubtedly lead to new examples.

REFERENCES

1. G. LAMPEL, Nuclear Dynamic Polarization by Optical Electronic Saturation and Optical Pumping in Semiconductors, *Phys. Rev. Lett.* **20**(10), 491–493 (1968).
2. E. F. GROSS, A. I. EKIMOV, B. S. RAZBIRIN, AND V. I. SAFAROV, Optical Orientation of Free and Bound Excitons in Crystals with Hexagonal Structure, *Zh. Eksp. Teor. Fiz. Pis'ma* **14**(2), 108–112 (1971); Engl. transl. *JETP Lett.* **14** (2), 70–72 (1971).
3. R. R. PARSONS, Band-to-Band Optical Pumping in Solids and Polarized Photoluminescence, *Phys. Rev. Lett.* **23**(20), 1152–1154 (1969).
4. A. I. EKIMOV AND V. I. SAFAROV, Optical Orientation of Carriers at Band-to-Band Transitions in Semiconductors, *Zh. Eksp. Teor. Fiz. Pis'ma* **12**(6), 293–297 (1970); Engl. transl. *JETP Lett.* **12**(6), 198–202 (1970).
5. B. P. ZAKHARCHENYA, V. G. FLEISHER, R. I. DZHIOEV, YU. P. VESHCHUNOV, AND I. B. RUSANOV, Optical Orientation of Electron Spins in GaAs Crystal, *Zh. Eksp. Teor. Fiz. Pis'ma* **13**(4), 195–199 (1971); *Engl. transl. JETP Lett.* **13**(4), 137–139 (1971).
6. YU. P. VESHCHUNOV, B. P. ZAKHARCHENYA, AND E. M. LEONOV, Optical Orientation of excitons and Hanle Effect in GaSe, *Fiz. Tverd. Tela* **14**(9), 2678–2681 (1972); Engl. transl. *Sov. Phys. Solid State* **14**(9), 2312–2314 (1972).
7. A. BONNOT, R. PLANEL, C. BENOIT A LA GUILLAUME, AND G. LAMPEL, Spin Orientation by Optical Pumping in CdS, *Proc. 11th ICPS*, edited by M. Miasek (PWN, Polish Scientific Publ., Warsaw, 1972), pp. 1334–1340.
8. C. WEISBUCH AND G. LAMPEL, Spin Orientation by Optical Pumping in InP, *Proc. 11th ICSP*, edited by M. Miasek (PWN, Polish Scientific Publ., Warsaw, 1972), pp. 1327–1332.
9. G. L. BIR AND G. E. PIKUS, Effect of Magnetic Field and Deformation on the Optical Orientation of Excitons in Crystals with Wurtzite Structure, *Zh. Eksp. Teor. Fiz. Pis'ma* **15**(12), 730–733 (1972); Engl. transl. *JETP Lett.* **15**(12), 516–518 (1972).
10. A. BONNOT, R. PLANEL, AND C. BENOIT A LA GUILLAUME, Optical Orientation of Excitons in CdS, *Phys. Rev. B* **9**(2), 690–702 (1974).
11. S. A. PERMOGOROV, YA. V. MOROZENKO, AND B. A. KAZENNOV, Optical Orientation of Hot Excitons in A_2B_6 Crystals, *Fiz. Tverd. Tela* **17**(10), 2970–2979 (1975); *Engl. transl. Sov. Phys. Solid State* **17**(10), 1974–1978 (1975).
12. N. V. KARLOV, J. MARGERIE, AND Y. MERLE-D'AUBIGNE, Pompage optique des centers F dans KBr, *J. Phys. (Paris)* **24**(3), 717–723 (1963).
13. C. N. ANDERSON, H. A. WEAKLIEM, AND E. S. SABISKY, Selective Absorption of Circularly Polarized Light in Broad Bands by the Zeeman Components Tm^{2+} in CaF_2, *Phys. Rev.* **143**(1), 223–228 (1965).
14. G. F. HULL, J. T. SMITH, AND A. F. QUESADA, Alignment of Cr^{3+} in Ruby, *Appl. Opt.* **4**(9), 117–1120 (1965).
15. A. A. KAPLYANSKII, E. V. MAKSIMOV, AND V. I. MEDVEDEV, Optical Orientation of Excited Eu^{2+} Ions in CaF_2, *Fiz. Tverd. Tela* **17**(8), 1838–1843 (1975); Engl. transl. *Sov. Phys. Solid State* **17**(8), 1205–1207 (1975).
16. M. I. D'YAKONOV AND V. I. PEREL', Theory of Optical Orientation of Electron and Nuclei in Semiconductors, in *Optical Orientation*, edited by F. Meier and B. P. Zakharchenya (North-Holland, Amsterdam, 1984), pp. 11–72.

17. G. E. Pikus and A. N. Titkov, Spin Relaxation under Optical Orientation in Semiconductors, in *Optical Orientation*, edited by F. Meier and B. P. Zakharchénya (North-Holland, Amsterdam, 1984), pp. 73-132.
18. D. N. Mirlin, Optical Alignment of Electron Momenta in GaAs-Type Semiconductor, in *Optical Orientation*, edited by F. Meier and B. P. Zakharchenya (North-Holland, Amsterdam, 1984), pp. 133-172.
19. V. G. Fleisher and I. A. Merkulov, Optical Orientation of the Coupled Electron-Nuclear Spin-System of a Semiconductor, in *Optical Orientation*, edited by F. Meier and B. P. Zakharchenya (North-Holland, Amsterdam, 1984), pp. 173-258.
20. R. Planel· and C. Benoit a la Guillaume, Optical Orientation of Excitons, in *Optical Orientation*, edited by F. Meier and B. P. Zakharchenya (North-Holland, Amsterdam, 1984), pp. 353-380.
21. G. E. Pikus and E. L. Ivchenko, Optical Orientation and Polarized Luminescence of Excitons in Semiconducors, in *Excitons*, edited by E. I. Rashba and M. D. Sturge (North-Holland, Amsterdam, 1982), pp. 205-266.
22. M. I. D'yakonov and V. I. Perel', Spin Orientation of Electrons at Inter-Band Absorption of the Light in Semiconductors, *Zh. Eksp. Teor. Fiz.* **60**(5), 1954-1963 (1971); Engl. transl. *Sov. Phys. JETP* **33**(5), 1053-1057 (1971).
23. V. A. Marushchak and A. N. Titkov, Optical Orientation in Deformed Crystals with Any Direction of Deformation, *Fiz. Tverd. Tela* **27**(5), 1423-1428 (1985); Engl. transl. *Sov. Phys. Solid Stdate* **27**(5), 858-860 (1985).
24. M. I. D'yakonov and V. I. Perel', Influence of an Electric Field and Deformation on the Optical Orientation in Semiconductors, *Fiz. Tekhn. Poluprovodn.* **7**(12), 2335-2339 (1973); Engl. transl. *Sov. Phys. Semicond.* **7**(12), 1551-1553 (1974).
25. V. L. Vekua, R. I. Dzioev, E. L. Ivchenko, V. A. Fleisher, and B. P. Zakharchenya, Polarization of Luminescence and Splitting of Acceptor Levels Due to the Deformation in Cubic Crystals, *Fiz. Tverd. Tela* **17**(6), 1096-1103 (1975); Engl. transl. *Sov. Phys. Solid State* **17**(6), 696-700 (1975).
26. A. A. Bakun, B. P. Zakharchenya, A. A. Rogachev, M. N. Tkachuk, and V. G. Fleisher, Observation of the Surface Photocurrent Due to the Optical Orientation of Electrons, *Zh. Eksp. Teor. Fiz. Pis'ma* **40**(11), 464-466 (1984); Engl. transl. *JETP Lett.* **40**(11), 326-328 (1984).
27. A. A. Bakun, B. P. Zakharchenya, M. N. Tkachuk, and V. G. Fleisher, Surface Photocurrent Due to the Optical Orientation of Electrons in Semiconductors, *Izv. Akad. Nauk SSSR, Ser. Fiz.* **50**(2), 235-238 (1986).
28. R. I. Dzhioev and V. G. Fleisher, A Role of Surface Recombination and Electron Diffusion in Optical Orientation Experiments, *Fiz. Tekhn. Poluprovodn.* **23**(3), 365-369 (1989); Engl. transl. *Sov. Phys. Semicond.* **23**(3), to appear.
29. A. S. Volkov and G. V. Tsarenkov, Photoluminescence of Graded-Band Semiconductors, *Fiz. Tekhn. Poluprovodn.* **11**(9), 1709-1717 (1977); Engl. transl. *Sov. Phys. Semicond.* **11**(9), 1004-1009 (1977).
30. A. S. Volkov, A. L. Lipko, S. E. Minakov, and B. V. Tsarenkov, Polarized Photoluminescence of a Graded-Band Semiconductor with a Gradient of the Electron g-Factor-I. Theory, *Fiz. Tekhn. Poluprovodn.* **19**(7), 1277-1282 (1985); Engl. transl *Sov. Phys. Semicond.* **19**(7), 780-783 (1985).
31. A. S. Volkov, A. I. Ekimov, V. I. Safarov, B. V. Tsarenkov, and G. V. Tsarenkov, Oscillations in Magnetic Field of the Polarization of Recombinative Radiation in Graded-Band Semiconductors, *Zh. Eksp. Teor. Fiz. Pis'ma* **25**(12), 560-563 (1977); Engl. Transl. *JETP Lett.* **25**(12), 526-527 (1977).
32. A. S. Volkov, A. A. Lipko, Sh. M. Meretliev, and B. P. Tsarenkov, Effect of the "Spin Echo" in Graded-Band Semiconductor, *Zh. Eksp. Teor. Fiz. Pis'ma* **41**(11), 458-460 (1985); Engl. transl. *JETP Lett.* **41**(11), 328-329 (1985).

33. A. ABRAGAM, *The Principles of Nuclear Magnetism* (Clarendon Press, Oxford, 1961).
34. M. I. D'YAKONOV, V. I. PEREL', V. L. BERKOVITS, AND V. I. SAFAROV, Optical Effects Stimulated by the Polarization of Nuclei in Semiconductors, *Zh. Eksp. Teor. Fiz.* **67**(4), 1912-1923 (1974); Engl. transl. *Sov. Phys. JETP* **40**(11), 950-968 (1975).
35. D. PAGET, G. LAMPEL, B. SAPOVAL AND V. I. SAFAROV, Low Field Electron-Nuclear Spin Coupling in GaAs under optical Pumping Conditions, *Phys. Rev. B* **15**(12), 5780-5796 (1977).
36. V. L. BERKOVITS, A. I. EKIMOV, AND V. I. SAFAROV, Optical Orientation of the Electron-Nuclear Spin System in Semiconductors, *Zh. Eksp. Teor. Fiz.* **65**(1), 346-361 (1973); *Sov. Phys. JETP* **38**(1), 169-181 (1974).
37. B. P. ZAKHARCHENYA, V. K. KALEVICH, V. D. KUL'KOV, AND V. G. FLEISHER, Optical Orientation of the Electron-Nuclear Spin System in a Semiconductor in an Inclined Magnetic Field, *Fiz. Tverd. Tela* **23**(5), 1387-1394 (1981); Engl. transl. *Sov. Phys. Solid State* **23**(5), 810-813 (1981).
38. V. A. NOVIKOV AND V. G. FLEISHER, Effect of the Local Anisotropy on Optical Orientation of Electron and Nuclear Spins in Semiconductors, *Zh. Eksp. Teor. Fiz.* **74**(3), 1026-1042 (1978); Engl. transl. *Sov. Phys. JETP* **47**(3), 539-553 (1979).
39. V. L. BERKOVITS AND V. I. SAFAROV, Optical Observation of Nuclear Quadrupole Resonance in Doped Semiconductors, *Fiz. Tverd. Tela* **20**(8), 2536-2537 (1978); Engl. transl. *Sov. Phys. Solid State* **20**(8), 1468-1469 (1978).
40. V. I. ZEMSKII, B. P. ZAKHARCHENYA, AND D. N. MIRLIN, Polarization of Hot Luminescence in GaAs, *Zh. Eksp. Teor. Fiz. Pis'ma* **24**(2), 96-99 (1976); Engl. transl. *JETP Lett.* **24**(2), 82-83 (1977).
41. V. D. DYMNIKOV, M. I. D'YAKONOV, AND V. I. PEREL', Anisotropy of Momentum Distribution of Photoelectrons and Polarization of Hot Luminescence in Semiconductors, *Zh. Eksp. Teor. Fiz.* **71**(12), 2373-2380 (1976); Engl. Transl. *Sov. Phys. JETP* **44**(12), 1252-1257 (1977).
42. G. L. BIR AND G. E. PIKUS, *Symmetry and Strain-induced Effects in Semiconductors* (Wiley, New York, 1974).
43. B. P. ZAKHARCHENYA, D. N. MIRLIN, V. I. PEREL', AND I. I. RESHINA, Optical Alignment of Hot Carriers in Semiconductors, *Usp. Fiz. Nauk* **136**(3), 459-475 (1982); Engl. transl. *Sov. Phys. Usp.* **25**(3), 143-157 (1983).
44. D. N. MIRLIN, L. P. NIKITIN, I. I. RESHINA, AND V. F. SAPEGA, Depolarization of Hot Luminescence in GaAs by a Magnetic Field, *Zh. Eksp. Teor. Fiz. Pis'ma* **30**(7), 419-422 (1979); Engl. transl. *JETP Lett.* **30**(7), 392-394 (1980).
45. R. W. WOOD AND A. ELLETT, On the Influence of Magnetic Fields on the Polarization of Resonance Radiation, *Proc. R. Soc. London, Ser. A* **103**, 396-403 (1923).
46. R. W. WOOD AND A. ELLETT, Polarized Resonance Radiation in Weak Magnetic Fields, *Phys. Rev.* **24**, 243-254 (1924).
47. W. HANLE, Über magnetische Beeinflussung der Polarisation der Resonanzfluoreszenz, *Z. Phys.* **30**, 93 105 (1924).
48. W. HANLE, Die Polarisation der Resonanzfluoreszenz von Natriumdampf bei Anregung mit zirkular polarisiertem Licht, *Z. Phys.* **41**, 164-183 (1927).
49. G. JOOS, Der Einfluss eines Magnetfelds auf die Polarisation des Resonanzlicht, *Phys. Z.* **25**, 130-134 (1924).
50. C. WEISBUCH, Thesis, Paris, 1977, unpublished.
51. G. E. PIKUS AND G. L. BIR, Optical Orientation of Excitons in Cubic Crystals, *Zh. Eksp. Teor. Fiz.* **67**(8), 788-800 (1974); Engl. transl. *Sov. Phys. JETF* **40**(8), 390-399 (1975).
52. W. HANLE, Über dem Zeemaneffekt bei Resonanzfluoreszenz. *Die Naturwissensch.* **11**, 690 (1923).
53. J. A. ELDRIGE, Theoretical Interpretation of the Polarization Experiment of Wood and Ellett, *Phys. Rev.* **24**, 234-242 (1924).

54. G. E. PIKUS AND G. L. BIR, Exchange Interaction in Excitons in Semiconductor, *Zh. Eksp. Teor. Fiz.* **60**(1), 195-208 (1971); Engl. transl. *Sov. Phys. JETP* **33**(1), 143-151 (1972).
55. G. FISHMAN, C. HERMANN, C. WEISBUCH, AND G. LAMPEL, Pompage optique d'excitons dans les semiconducteurs cubiques, *J. Phys.* (*Paris*) **35**, C3-C7 (1974).
56. C. WEISBUCH, C. HERMANN, AND G. FISHMAN, Dynamics of Excitonic Complexes and Detection of Electron Spin Resonance by Optical Spin Orientation Techniques, *Proc. 12th ICPS*, Stuttgart, 1974, (ed. M. H. Pilkuhn) (B. G. Tlubner, Stuttgart, 1974), pp. 761-765.
57. C. BENOIT A LA GUILLAUME, Orientation d'excitons par pompage optique, *J. Phys.* (*Paris*) *C*3-*C*1 (1974).
58. Y. OKA AND T. KUSHIDA, Relaxation processes and Raman scattering and exciton luminescence of ZnTe and ZnSe, *Proc. 14th ICPS, Edinburgh*, 1978, (ed. B. L. H. Wilson) Inst. of Physics. Bristol and London pp. 1287-1290.
59. R. I. DZHIOEV, B. P. ZAKHARCHENYA, I. G. KUSRAEV, AND V. G. FLEISHER, Optical Orientation and Alignment of Excitons in HgI_2, *Izv. Akad. Nauk SSSR, Ser. Fiz.* **46**(3), 514-517 (1982).
60. E. M. GAMARTS, E. L. IVCHENKO, G. E. PIKUS, R. S. RAZBIRIN, A. N. STARUKHIN, AND V. I. SAFAROV, Magnetic Field Induced Transition Orientation—Alignment on Bound Excitons in GaSe, *Fiz. Tverd. Tela* **24**(8), 2325-2344 (1982); Engl. transl. *Sov. Phys. Solid State* **24**(8), 1320-1325 (1982).
61. B. C. CAVENETT, P. DAWSON, AND K. MORIGAKI, Triplet Exciton Resonances in Type II GaSe, *J. Phys. C* **12**(5), *L* 197-*L*202 (1979).
62. B. S. RAZBIRIN, V. P. MUSHINSKII, K. I. KARAMAN, A. N. STARUKHIN, AND E. M. GAMARTS, Optical Alignment of Excitons, *Zh. Eksp. Teor. Fiz. Pis'ma* **22**(4), 203-206 (1975); Engl. transl. *JETP Lett.* **22**(4), 94-96 (1975).
63. E. L. IVCHENKO, G. E. PIKUS, B. S. RAZBIRIN, AND A. I. STARUKHIN, Optical Orientation and Alignment of Free Excitons in GaSe on Resonant Excitation. Theory, *Zh. Eksp. Teor. Fiz.* **72**(6), 2230-2245 (1977); Engl. transl. *Sov. Phys. JETP* **45**(6), 1172-1184 (1978).
64. E. M. GAMARTS, E. L. IVCHENKO, G. E. PIKUS, B. S. RAZBIRIN, AND A. N. STARUKHIN, Optical Orientation and Alignment of Free Excitons in GaSe on Resonant Excitation. Experiment, *Zh. Eksp. Teor. Fiz.* **73**(9), 1113-1128 (1977); Engl. transl. *Sov. Phys. JETP* **46**(9), 590-602 (1978).
65. E. L. IVCHENKO, G. E. PIKUS, AND N. KH. YULDASHEV, Transfer of Polarized Radiation in Crystals in the Exciton Region of the Spectrum, *Zh. Eksp. Teor. Fiz.* **79**(10), 1573-1590 (1980); Engl. transl. *Sov. Phys. JETP* **52**(10), 1241-1256 (1981).
66. M. K. SOBIROV AND N. KH. YULDASHEV, Theory of transfer of polarized radiation in a longitudinal magnetic field, *Zh. Eksp. Teor. Fiz.* **87**(8), 677-690 (1984); Engl. transl. *Sov. Phys. JETP* **60**(8), 521-537 (1985).
67. S. A. PERMOGOROV, Optical Emission Due to Exciton Scattering by LO Phonon in Semiconductor, in *Excitons*, edited by E. I. Rashba and M. D. Sturge (North-Holland, Amsterdam, 1982), pp. 177-204.
68. G. L. BIR, G. E. PIKUS, AND E. L. IVCHENKO, Alignment and Orientation of Hot Electrons and Polarized Luminescence, *Izv. Akad. Nauk SSSR, Ser. Fiz.* **40**(9), 1866-1871 (1976).
69. E. L. IVCHENKO, G. E. PIKUS, AND L. V. TAKUNOV, Alignment and Orientation of Hot Excitons in Semiconductors, *Fiz. Tverd. Tela* **20**(9), 2598-2609 (1978); Engl. transl. *Sov. Phys. Solid State* **20**(9), 1502-1508 (1978).
70. L. V. TAKUNOV, Influence of a Magnetic Field on the Optical Orientation of Hot Excitons, *Fiz. Tekhn. Poluprovodn.* **17**(6), 1102-1107 (1983); Engl. transl. *Sov. Phys. Semicond.* **17**(6), 692-694 (1983).

71. M. NAWROCKI, R. PLANEL, AND C. BENOIT A LA GUILLAUME, Rotation of Linearly Oriented Polaritons in a Magnetic Field, *Phys. Rev. Lett.* **36**, 1343-1346 (1976).
72. S. A. PERMOGOROV AND V. V. TRAVNIKOV, Optical Alignment of Hot Excitons in Crystalline CdS, *Fiz. Tverd. Tela* **22**(9), 2651-2657 (1980); Engl. transl. *Sov. Phys. Solid State* **22**(9), 1547-1551 (1980).
73. S. A. PERMOGOROV AND V. V. TRAVNIKOV, Temporal Evolution of Exciton Resonant Secondary Emission Spectrum, *Solid State Commun.* **29**(8), 615-620 (1979).
74. E. H. SALIB AND B. C. CAVENETT, Zero-Field Optically Detected Magnetic Resonance (ZF-ODNR) in Semiconductors, *J. Phys. C* **17**, CL 251 (1984).
75. G. M. ZVEREV, A. M. PROHOROV, AND A. K. SHEVCHENKO, Generation of Microwave Radiation in Ruby Under Optical Pumping, *Zh. Eksp. Teor. Fiz.* **44**(4), 1415-1418 (1963); Engl. transl. *Sov. Phys. JETP* **17**(4), 947-949 (1964).
76. V. S. ZAPASSKII, G. G. KOZLOV, AND V. A. MALYSHEV, Nonresonance Methods of the Optical Detection of the Energy Structure of Spin Systems, *Izv. Akad. Nauk SSSR, Ser. Fiz.* **50**(2), 216-219 (1986).
77. V. S. ZAPASSKII, G. G. KOZLOV, AND V. A. MALYSHEV, Pseudocrossing of Levels and the Van-Vleck Susceptibility of the Anisotropic Paramagnetic Center, *Fiz. Tverd. Tela* **28**(1), 119-129 (1986); Engl. transl. *Sov. Phys. Solid State* **28**(1), 64-69 (1986).
78. V. S. ZAPASSKII AND G. G. KOZLOV, Determination of Small Components of the g-tensor of Paramagnetic Centers in Crystals, *Fiz. Tverd. Tela* **29**(3), 899-903 (1987); Engl. transl. *Sov. Phys. Solid State* **29**(3), 514-515 (1987).
79. V. S. ZAPASSKII, G. G. KOZLOV, AND V. A. MALYSHEV, Artificial Van-Vleck Susceptibility of Amorphous Paramagnets, *Fiz. Tverd. Tela* **28**(1), 138-147 (1986); Engl. transl. *Sov. Phys. Solid State* **28**(1), 74-79 (1986).
80. L. F. MOLLENAUER, W. B. GRANT, AND C. D. JEFFRIES, Achievement of Significant Nuclear Polarizations in Solids by Optical Pumping, *Phys. Rev. Lett.* **20**(10), 488-490 (1968).
81. J. P. VAN DER ZIEL AND N. BLOEMBERGEN, Optically Induced Magnetization in Ruby, *Phys. Rev.* **138**(4A), 1287-1292 (1965).
82. Y. FUKUODA, Y. TAKAGI, K. YAMADA, AND T. HASHI, Optically Induced Precessing Magnetization in Ruby Near the Level Anticrossing Points, *J. Phys. Soc. Jpn.* **42**(3), 1061-1062 (1977).
83. Y. TAKAGI, K. YAMADA, Y. FUKUODA, AND T. HASHI, Optically Induced Transverse Magnetization in Ruby Near Zero Magnetic Field; Magnetically Detected Transient Hanle Effect, *Phys. Lett.* **98A**(5), 306-308 (1983).
84. P. P. FEOFILOV, *The Physical Basic of Polarized Emission* (Consultant's Bureau, New York, 1961).
85. A. V. AKIMOV, A. A. KAPLYANSKII, AND S. P. FEOFILOV, Polarized Luminescence of CaF_2-S_m^{2+} Crystals in Magnetic Field, *Opt. i Spectroscop.* **54**(2), 272-278 (1983).
86. E. L. IVCHENKO, E. V. MAKSIMOV, AND V. N. MEDVEDEV, Optical Orientation and Circular Polarization of Luminescence of the E_u^{2+} Center in CaF_2 and SrF_2, *Izv. Akad. Nauk SSSR, Ser. Fiz.* **40**(8), 1966-1975 (1976).
87. A. G. ARONOV, A. N. TITKOV, AND G. E. PIKUS, Spin Relaxation of Conductivity Electrons in III-V Compounds of the p-Type, *Zh. Eksp. Teor. Fiz.* **84**(3), 1170-1184 (1983); Engl. transl. *Sov. Phys. JETP* **57**(3), 680-689 (1983).
88. M. I. D'YAKONOV, V. A. MARUSHCHAK, V. I. PEREL', AND A. N. TITKOV, Effect of Deformation on Spin Relaxation of Conductivity Electrons in III-V Semiconductors, *Zh. Eksp. Teor. Fiz.* **90**(3), 1123-1133 (1986); Engl. transl. *Sov. Phys. JETP* **63**(3), 645-653 (1986).
89. G. E. PIKUS, V. A. MARUSHCHAK, AND A. N. TITKOV, Electron Spin Relaxation and Spin Splittings of the Conduction Band in III-V Semiconductors, *Fiz. Tekhn. Poluprovodn.* **22**(2), 453-464 (1988); Engl. transl. *Sov. Phys. Semicond.* **22**(2), (1988).

90. A. N. TITKOV, E. I. CHAIKINA, E. I. KOMOVA, AND N. G. ERMAKOVA, Low Temperature Luminescence in Degenerate p-Type Crystals, *Fiz. Tekhn. Pluprovodn.* **15**(2), 345-352 (1981); Engl. transl. *Sov. Phys. Semicond.* **15**(2), 198-201 (1981).
91. A. N. TITKOV, G. V. BENEMANSKAYA, B. L. GELMONT, AND G. N. ILURIDZE, Anger Recombination in p-Type GaSb, *J. Lumin.* **24/25**, 697-700 (1981).
92. A. N. TITKOV, V. N. CHEBAN, AND I. F. MIRONOV, Interband Anger Recombination in Doped III-V Compounds of p-type, *Izv. Akad. Nauk SSSR, Ser. Fiz.* **52**(4), 738-742 (1988).
93. D. N. MIRLIN, I. YA. KARLIK, AND I. I. RESHINA, Energy Relaxation Time of Hot Electrons in GaAs, *Solid State Commun.* **37**(9), 757-760 (1981).
94. I. YA. KARLIK, D. N. MIRLIN, AND V. F. SAPEGA, Recombination of Hot Electrons in InP, *Fiz. Tverd. Tela* **27**(7), 2210-2211 (1985); Engl. transl. *Sov. Phys. Solid State* **27**(7), 1326-1327 (1985).
95. W. KAUSCHKE, N. MESTRES, AND M. CARDONA, Spin Relaxation of Holes in the Split-Hole Band of InP and GaSb, *Phys. Rev. B* **35**(8), 3843-3853 (1987).
96. I. YA. KARLIK, D. N. MIRLIN, AND V. F. SAPEGA, The Probability of the Inter-band Γ-L Transitions in GaAs, *Fiz. Tekhn. Poluprovodn.* **21**(6), 1030-1032 (1987); Engl. transl. *Sov. Phys. Semicond.* **21**(6), (1987).
97. D. Z. GARBUZOV, I. A. MERKULOV, V. A. NOVIKOV, AND V. G. FLEISHER, Diffusion of Electrons with Oriented Spin in the Double Heterostructure, *Fiz. Tekhn. Poluprovodn.* **10**(5), 934-939 (1976); Engl. transl. *Sov. Phys. Semicond.* **10**(2), 552-557 (1976).
98. K. KALTENEGGER AND H. KRENN, Depolarization des Photomagnetismus in $Hg_{1-x}Mn_xTe$ in transversalen Magnetfeld, *Verh. Dtsch. Phys. Ges.* **22**(1), HL-218 (1987).
99. R. C. MILLER, D. A. KLEINMAN, W. A. NORDLAND, AND A. C. GOSSARD, Luminescence Studies of Optically Pumped Quantum Wells in GaAs-GaAlAs Multilayer Structures, *Phys. Rev. B* **22**(2), 863-878 (1980).
100. P. S. KOP'EV, V. P. KOCHERESCHKO, I. N. URAL'TSEV, AND D. N. YAKOVLEV, Influence of the Charge Impurities Potential on the Formation of Excitons in Quantum Wells, *Zh. Eksp. Teor. Fiz. Pis'ma* **46**(2), 74-77 (1987).

CHAPTER 7

QUANTUM THEORY OF THE HANLE LASER AND ITS USE AS A METRIC GRAVITY PROBE

JÁNOS A. BERGOU AND MARLAN O. SCULLY

1. INTRODUCTION

The experiment of Hanle[1] provides one of the clearest and most striking demonstrations of a situation where atomic coherence is important. In the present work we shall exploit an interesting aspect of the effect, namely, that it can be regarded as a possible tool for transferring atomic coherence to the coherence of an optical field.[2]

In the optical detection of ultrasmall changes of a certain physical quantity the change is converted into a phase shift (in passive systems) or a frequency shift (in active systems) of a laser by sending its light through or generating it in a cavity whose optical path length depends on the particular effect to be measured. The shift is then detected by beating the output light with that from a reference laser (often called the "local oscillator" in heterodyne or homodyne detection schemes). Typical examples of the measurement of ultrasmall optical displacements are gravity-wave detectors[3] and the laser gyroscope.[4]

The quantum noise limit of the active detection scheme is determined by the fluctuation, caused by the independent spontaneous emission events, in the relative phase between the two lasers.[5] In a recent paper by one of us,[6] it has been shown that the linewidth and the associated uncertainty in the relative phase may be eliminated by preparing the laser medium in

---

JÁNOS A. BERGOU AND MARLAN O. SCULLY • Center for Advanced Studies and Department of Physics and Astronomy, The University of New Mexico, Albuquerque, New Mexico 87131, USA, and Max-Planck-Institut für Quantenoptik, D-8046 Garching, West Germany.   János A. Bergou is presently on leave from the Central Research Institute for Physics, H-1525, Budapest, Hungary.

a coherent superposition of the two upper states. Two suggestions have been made: the coherent superposition can be achieved either via microwave coupling as in quantum beat experiments[7] or by coherent pumping as in the Hanle effect. The arguments of Ref. 6 were of a very general nature. In subsequent publications a complete theory of the quantum beat laser has been developed. First, in a linear theory an explicit expression for the diffusion constant of the relative phase of the two modes has been obtained and physical conditions under which this diffusion constant vanishes have been established.[8] Then, in the next step, a nonlinear theory has been given which is valid under the detuning and coupling conditions leading to correlated spontaneous emission in the linear theory.[9] The emerging physical picture can be described as follows. The strong classical coupling of the upper two levels establishes coherence between them on a time scale very short compared to the atomic lifetime. Thus, the active medium is, effectively, always in coherent superpositions of the upper levels, irrespective of the initial conditions. The detuning and coupling conditions of Ref. 8 will then select one of the two orthogonal coherent superpositions as upper level for the two laser transitions. Furthermore, as shown in Ref. 9, only a particular coherent superposition of the two modes (a dressed mode) will lase. In terms of this "dressed atom–dressed mode" picture, correlated spontaneous emission laser operation is essentially equivalent to one-mode laser operation with a two-level active medium.

The purpose of the present chapter is to bring the theory of the Hanle laser to the same level of sophistication as has been achieved in the quantum beat laser. This theory allows us to find the physical conditions under which correlated spontaneous emission laser operation (CEL) occurs in the Hanle laser. It turns out that, in contrast to the quantum beat laser, the Hanle laser is an initial value problem. As a consequence, these conditions are less restrictive for the detuning but quite sharp for the pumping and coupling. The conditions can, again, most easily be found in a linear theory. Working under these conditions the nonlinear theory delivers a similar picture to that of the quantum beat CEL in terms of appropriately defined dressed states and dressed modes.

The paper is organized as follows. In Section 2 we present the Hamiltonian model of our system. In Section 3 we develop the linear theory of the Hanle laser. From the requirement of the vanishing diffusion constant we find the conditions for pumping, coupling, and detuning under which CEL operation is possible. In Section 4 we work out the nonlinear theory for these conditions. We derive a master equation for the reduced density operator of the field and deduce the photon statistics. In Section 5 we show that the vanishing of the diffusion constant for the relative phase still persists in the nonlinear theory. In Section 6, we discuss the noise performance of the Hanle CEL when used as a detector of ultrasmall optical displacements.

For possible applications we review how the correlated spontaneous emission operation achieved in a Hanle laser can be used for the detection of gravity waves and in laser gyros. Finally, in Section 7 we briefly summarize the results of the paper together with a discussion of their interpretation and implications for future measurements.

2. THE MODEL

We consider the model of Ref. 6, namely, a system of three-level atoms, as shown in Figure 1, which are being pumped into a coherent superposition of the two upper states $|a\rangle$ and $|b\rangle$. The frequencies ν_1 and ν_2 of the $|a\rangle$-$|c\rangle$ and $|b\rangle$-$|c\rangle$ transitions are nearly degenerate (in fact, without an external magnetic field they are the same). The levels $|a\rangle$ and $|b\rangle$ are taken to be the "linear polarization" states (e.g., they correspond to the $|a\rangle = |m = +1\rangle + |m = -1\rangle$ and $|b\rangle = |m = +1\rangle - |m = -1\rangle$ linear combination of the Zeeman sublevels of a given degenerate upper state and $|c\rangle$ has the magnetic quantum number $m = 0$). The fields 1 and 2 emitted by the atoms of Figure 1 will thus differ in polarization. The doubly resonant cavity contains a polarization-sensitive mirror to separate the optical paths of the two orthogonal polarization modes. Except for the coherent initial condition there is no other coupling between $|a\rangle$ and $|b\rangle$. The transitions at ν_1 and ν_2 are treated quantum mechanically.

The Hamiltonian for the system is

$$H = H_0 + V \tag{1}$$

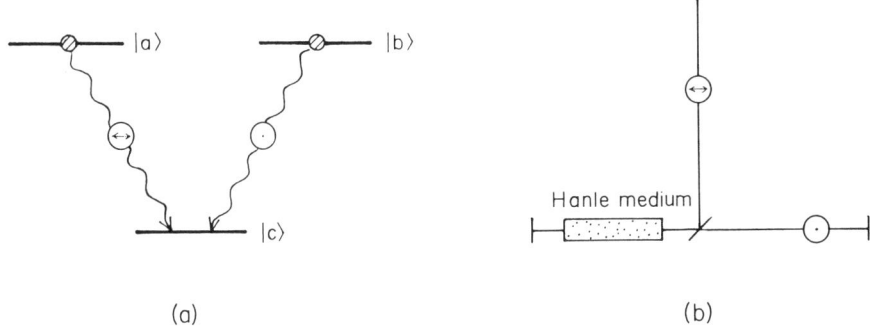

(a) (b)

FIGURE 1. (a) Three-level atom is prepared in a coherent superposition of states $|a\rangle$ and $|b\rangle$ via the Hanle effect. Transitions from these states to the common ground level $|c\rangle$ differ in their polarization. (b) Scheme of the Hanle laser using the coherently pumped atoms as active medium. A polarization-sensitive mirror separates the polarization modes in the doubly resonant cavity.

where

$$H_0 = \sum_{i=a,b,c} \hbar\omega_i |i\rangle\langle i| + \hbar\nu_1 a_1^+ a_1 + \hbar\nu_2 a_2^+ a_2 \qquad (2)$$

and

$$V = \hbar g_1(a_1|a\rangle\langle c| + a_1^+|c\rangle\langle a|) + \hbar g_2(a_2|b\rangle\langle c| + a_2^+|c\rangle\langle b|) \qquad (3)$$

Here a_1, a_2 (a_1^+, a_2^+) are annihilation (creation) operators of photons in the two orthogonally polarized modes 1 and 2, and g_1 and g_2 are the coupling constants for the transitions $|a\rangle$-$|c\rangle$ and $|b\rangle$-$|c\rangle$.

It is convenient to work in the interaction picture, defined as

$$V_I = \exp\left(\frac{i}{\hbar} H_0 t\right) V \exp\left(-\frac{i}{\hbar} H_0 t\right) \qquad (4)$$

With the help of a little algebra it is easy to show that the interaction Hamiltonian can be written, in obvious matrix notation, as

$$V_I = \hbar \begin{bmatrix} 0 & 0 & g_1 a_1 \exp(i\Delta_1 t) \\ 0 & 0 & g_2 a_2 \exp(i\Delta_2 t) \\ g_1 a_1^+ \exp(-i\Delta_1 t) & g_2 a_2^+ \exp(-i\Delta_2 t) & 0 \end{bmatrix} \qquad (5)$$

Here we introduced the detunings as $\Delta_1 = \omega_a - \omega_c - \nu_1$ and $\Delta_2 = \omega_b - \omega_c - \nu_2$. The starting point of our theory is this interaction Hamiltonian. In the next section we shall deal with the perturbative solution of the associated Schrödinger equation and, based on the lowest-order solution, we give a linear theory of the Hanle laser.

3. LINEAR THEORY

The theory of the Hanle laser is developed from the interaction Hamiltonian as given by equation (5). First we investigate the deterministic time evolution resulting from this Hamiltonian. As for the quantum beat laser, this consideration will help us, in the following derivation, to carry out the averaging with respect to the atomic variables in the master equation for the reduced density operator for the field.

The Schrödinger equation in the interaction picture can be written as

$$i\hbar\dot{\psi} = V_I \psi \qquad (6)$$

Here ψ is a column vector with components ψ_a, ψ_b, and ψ_c. Besides the atomic variables, each component is still a function of a_1, a_1^+, a_2, and a_2^+, i.e., any operator acting on the mode variables. When expressed in components, equation (6) represents three coupled equations:

$$i\dot\psi_a = g_1 a_1 \exp(i\Delta_1 t)\psi_c - i\frac{\gamma}{2}\psi_a$$

$$i\dot\psi_b = g_2 a_2 \exp(i\Delta_2 t)\psi_c - i\frac{\gamma}{2}\psi_b$$

$$i\dot\psi_c = g_1 a_1^+ \exp(-i\Delta_1 t)\psi_a + g_2 a_2^+ \exp(-i\Delta_2 t)\psi_b - i\frac{\gamma}{2}\psi_c \quad (7)$$

Here we have introduced the phenomenological decay constant γ for the levels a, b, and c (for simplicity we have taken them to be equal). If we make the substitution $\psi = \exp[-\gamma/2(t - t_0)]\tilde\psi$, then the components of $\tilde\psi$ satisfy an equation similar to system (7), only the decay terms on the right-hand side are missing. The perturbation approach becomes clearer if we rewrite the resulting equation in the form of an integral equation

$$\tilde\psi_a(t) = \tilde\psi_a(t_0) - ig_1 a_1 \int_{t_0}^{t} dt' \exp(i\Delta_1 t')\tilde\psi_c(t')$$

$$\tilde\psi_b(t) = \tilde\psi_b(t_0) - ig_2 a_2 \int_{t_0}^{t} dt' \exp(i\Delta_2 t')\tilde\psi_c(t')$$

$$\tilde\psi_c(t) = \tilde\psi_c(t_0) - ig_1 a_1^+ \int_{t_0}^{t} dt' \exp(-i\Delta_1 t')\tilde\psi_a(t')$$

$$- ig_2 a_2^+ \int_{t_0}^{t} dt' \exp(-i\Delta_2 t')\tilde\psi_b(t') \quad (8)$$

We assume the following initial conditions:

$$\psi_a(t_0) = \tilde\psi_a(t_0) = \cos\alpha(t_0)\exp[i\varphi_a(t_0)]\hat\psi_F(t_0)$$

$$\psi_b(t_0) = \tilde\psi_b(t_0) = \sin\alpha(t_0)\exp[i\varphi_b(t_0)]\hat\psi_F(t_0)$$

$$\psi_c(t_0) = \tilde\psi_c(t_0) = 0 \quad (9)$$

i.e., the two upper states are simultaneously pumped into a linear superposition of these states. We describe the state of the two-mode field at the moment of injection by $\hat{\psi}_F(t_0)$. The amplitudes in system (9) satisfy

$$|\cos \alpha(t_0) \exp[i\varphi_a(t_0)]|^2 + |\sin \alpha(t_0) \exp[i\varphi_b(t_0)]|^2 = 1 \tag{10}$$

If $\alpha(t_0) = \alpha$ and $\varphi_a(t_0) - \varphi_b(t_0) = \varphi_{ab}$, independent of t_0, then the initial condition corresponds to a coherent superposition of the upper states. If $\alpha(t_0)$, $\varphi_b(t_0)$, and $\varphi_b(t_0)$ are random variables of t_0, then system (9) describes an incoherent superposition of the upper states. As we shall see, these coherence properties will be transferred to the coherence properties of the two-mode field.

As usual, we develop successive approximations to equation (8) by substituting the initial values $\psi_a(t_0)$, $\psi_b(t_0)$, $\psi_c(t_0)$ into the integrals on the right-hand side. This gives the lowest-order perturbation result. If one is interested in the next order, then these lowest-order expressions should be substituted into the integrals on the right-hand side, etc. Here we confine ourselves to the lowest-order approximation. By using equations (9) in system (8), we find that the solution of equations (7), obtained by lowest-order perturbation theory, is given by

$$\psi_a = \exp\left[-\frac{\gamma}{2}(t - t_0)\right] \cos \alpha \, \exp(i\varphi_a)\hat{\psi}_F(t_0)$$

$$\psi_b = \exp\left[-\frac{\gamma}{2}(t - t_0)\right] \sin \alpha \, \exp(i\varphi_b)\hat{\psi}_F(t_0)$$

$$\psi_c = -i \exp\left[-\frac{\gamma}{2}(t - t_0)\right] g_1 a_1^+ \cos \alpha \, \exp(i\varphi_a) \int_{t_0}^{t} dt' \exp(-i\Delta_1 t')$$

$$+ g_2 a_2^+ \sin \alpha \, \exp(i\varphi_b) \int_{t_0}^{t} dt' \exp(-i\Delta_2 t') \bigg] \hat{\psi}_F(t_0) \tag{11}$$

We shall make explicit use of these amplitudes in the following derivation of the equation of motion for the reduced density operator ρ_F of the field.

The density operator of the coupled system (three-level atoms plus two-mode field) satisfies the following equation of motion in the interaction picture:

$$\dot{\rho} = -\frac{i}{\hbar}[V_1, \rho] \tag{12}$$

where the bracket stands for the commutator and V_1 is given by equation (5). The reduced density operator ρ_F for the field only is defined as the trace of ρ over the atoms,

$$\rho_F \equiv \text{Tr}_{\text{atom}}\rho \tag{13}$$

By employing expression (5) for V_1 in equation (12) and carrying out the trace operation, we find that ρ_F satisfies the following equation of motion:

$$\dot{\rho}_F = -i\{g_1 \exp(i\Delta_1 t)[a_1, \rho_{ca}] + g_1 \exp(-i\Delta_1 t)[a_1^+, \rho_{ac}]$$
$$+ g_2 \exp(i\Delta_2 t)[a_2, \rho_{cb}] + g_2 \exp(-i\Delta_2 t)[a_2^+, \rho_{bc}]\} + \text{loss} \tag{14}$$

For the purposes of the present section we need not specify the loss term. Its explicit expression will be given in the next section. To proceed further we need an expression for ρ_{ac}, ρ_{bc}, and their hermitian conjugates. The following procedure[5] is adopted to obtain this expression. We first calculate the contribution of one atom injected at time t_0 in the superposition of the upper states, as given by system (9), and then sum the contribution of all atoms which are injected at random times $t - \gamma^{-1} < t_0 < t$ at a rate r_a. In this way one obtains

$$\rho_{ac} = r_a \int_{t-\gamma^{-1}}^{t} dt_0\, \psi_a(t, t_0)\psi_c^+(t, t_0) \tag{15}$$

and a similar expression for ρ_{bc} with ψ_a replaced by ψ_b. We now substitute the expressions for ψ_a and ψ_c from equations (11) to yield

$$\rho_{ac} = ir_a \int_{t-\gamma^{-1}}^{t} dt_0 \int_{t_0}^{t} dt' \exp[-\gamma(t-t_0)] \cos\alpha \, \exp(i\varphi_a)\rho_F(t_0)$$
$$\times [g_1 a_1 \exp(i\Delta_1 t') \cos\alpha \, \exp(-i\varphi_a)$$
$$+ g_2 a_2 \exp(i\Delta_2 t') \sin\alpha \, \exp(-i\varphi_b)] \tag{16}$$

In this expression we can approximate $\rho_F(t_0)$ by $\rho_F(t)$ since the dynamics of the photon field is governed by the cavity lifetime γ_c^{-1}, which is much longer than γ^{-1}, and thus during the integration time ρ_F does not change appreciably. Then we can extend the lower limit of the t_0 integration to $-\infty$ since, due to the exponential damping factor in the integral, the contribution from $t_0 < t - \gamma^{-1}$ is negligible. After performing these steps it is easy to carry out time integrations in equation (16) and one can see that ρ_{ac} and ρ_{bc} have the following structure:

$$\rho_{ac} = T_{11}\rho_F a_1 + T_{12}\rho_F a_2 \quad \text{and} \quad \rho_{bc} = T_{22}\rho_F a_2 + T_{21}\rho_F a_1 \tag{17}$$

The coefficients can be found from equation (16) to be

$$T_{11} = i\frac{r_a g_1 \cos^2 \alpha}{\gamma(\gamma + i\Delta_1)} \exp(i\Delta_1 t) \tag{18a}$$

$$T_{12} = i\frac{r_a g_2 \sin \alpha \cos \alpha}{\gamma(\gamma + i\Delta_2)} \exp[i(\varphi_a - \varphi_b + \Delta_2 t)] \tag{18b}$$

$$T_{22} = i\frac{r_a g_2 \sin^2 \alpha}{\gamma(\gamma + i\Delta_2)} \exp(i\Delta_2 t) \tag{18c}$$

$$T_{21} = i\frac{r_a g_1 \sin \alpha \cos \alpha}{\gamma(\gamma + i\Delta_1)} \exp[i(\varphi_b - \varphi_a + \Delta_1 t)] \tag{18d}$$

Upon substituting for ρ_{ac}, ρ_{bc}, ρ_{ca}, and ρ_{cb} from equations (17) and its conjugate and taking into account system (18) in equation (14), one obtains the equation of motion for ρ_F in the form of a Liouville equation:

$$\dot{\rho}_F = \sum_{i,j=1}^{2} \mathscr{L}_{ij}(a_i, a_j^+) \rho_F \tag{19}$$

where

$$\mathscr{L}_{ii}\rho_F = -\tfrac{1}{2}[\alpha_{ii}^* a_i a_i^+ \rho_F + \alpha_{ii} \rho_F a_i a_i^+ - (\alpha_{ii} + \alpha_{ii}^*) a_i^+ \rho_F a_i] \tag{20a}$$

$$\mathscr{L}_{12}\rho_F = -\tfrac{1}{2}[\alpha_{21}^* a_1^+ a_2 \rho_F + \alpha_{12} \rho_F a_1^+ a_2 - (\alpha_{12} + \alpha_{21}^*) a_1^+ \rho_F a_2] e^{i\varphi} \tag{20b}$$

$$\mathscr{L}_{21}\rho_F = -\tfrac{1}{2}[\alpha_{12}^* a_1 a_2^+ \rho_F + \alpha_{21} \rho_F a_1 a_2^+ - (\alpha_{21} + \alpha_{12}^*) a_2^+ \rho_F a_1] e^{-i\varphi} \tag{20c}$$

In the above expressions

$$\alpha_{11} = \frac{2 r_a g_1^2 \cos^2 \alpha}{\gamma(\gamma + i\Delta_1)} \tag{21a}$$

$$\alpha_{12} = \frac{2 r_a g_1 g_2 \sin \alpha \cos \alpha}{\gamma(\gamma + i\Delta_2)} \tag{21b}$$

$$\varphi = (\nu_1 - \nu_2)t + \varphi_a - \varphi_b \tag{21c}$$

$$\alpha_{22} = \frac{2 r_a g_2^2 \sin^2 \alpha}{\gamma(\gamma + i\Delta_2)} \tag{21d}$$

$$\alpha_{21} = \frac{2r_a g_1 g_2 \sin\alpha \cos\alpha}{\gamma(\gamma + i\Delta_1)} \tag{21e}$$

We can now convert this Liouville equation into a Fokker–Planck equation by introducing the coherent state representation for a_1 and a_2 and the diagonal or P-representation for ρ_F.[5] A similar Liouville equation has been converted into a Fokker–Planck equation in Ref. 8. Here, instead of repeating the derivation, we simply borrow the results. If we define the coherent state as

$$a_i |v_1, v_2\rangle = v_i |v_1, v_2\rangle, \qquad i = 1, 2 \tag{22}$$

where v_i is an arbitrary complex number and we represent v_i as

$$v_i = \rho_i \exp(i\theta_i), \qquad i = 1, 2 \tag{23}$$

then the Fokker–Planck equation in terms of ρ_i, θ_i will contain a term which describes diffusion of the relative phase $\theta = \theta_1 - \theta_2$ as

$$\dot{P} = \frac{\partial^2}{\partial \theta^2}[D(\theta)P] \tag{24}$$

with

$$D(\theta) = \frac{1}{8}\left[\frac{\alpha_{11}}{\rho_1^2} + \frac{\alpha_{22}}{\rho_2^2} - \frac{\alpha_{12}}{\rho_1 \rho_2} e^{i\psi} - \frac{\alpha_{21}}{\rho_1 \rho_2} e^{-i\psi}\right] + \text{c.c.} \tag{25a}$$

and

$$\psi = \varphi - \theta \tag{25b}$$

We are interested in finding conditions under which the diffusion constant $D(\theta)$ for the relative phase angle $\theta = \theta_1 - \theta_2$ of the two modes vanishes. A possible set of conditions is the following (with $\rho_1 = \rho_2 = \rho$):

$$g_1 \cos\alpha = g_2 \sin\alpha \equiv g \tag{26a}$$

$$\Delta_1 = \Delta_2 \equiv \Delta \tag{26b}$$

The first of these conditions specifies the pumping, the second requires the equality of the detuning on the two transitions. The actual value of Δ is not fixed, e.g., the choice $\Delta = 0$ satisfies equation (26b). Under these conditions

$$D(\theta) = \frac{r_a g^2}{2(\gamma^2 + \Delta^2)\rho^2}(1 - \cos\psi) \qquad (27)$$

When $\psi = 0$ the vanishing of the diffusion constant takes place. A few remarks are in order here. When we calculated the contribution to ρ_{ca} of all atoms injected at random times t_0, it was crucial to assume that $\varphi_a - \varphi_b =$ const in equation (16). If φ_a and φ_b are independent and random, the second term in the integral averages to zero. Then the terms, proportional to α_{12} and α_{21} in equation (25a), would not appear and the diffusion constant would never vanish. In particular, this would lead to the vanishing of the $\cos\psi$ term from the bracket on the right-hand side of equation (27). Furthermore, if the initial amplitudes $\cos\alpha$ and $\sin\alpha$ of the upper states ψ_a and ψ_b are random, then $\cos^2\alpha$ and $\sin^2\alpha$ both average to $1/2$ in α_{11} and α_{22}. For this case equation (27) reduces, with zero detuning, to the Schawlow–Townes linewidth. If the pumping is partially coherent, then the spontaneous emission events into the two modes are partially correlated and the diffusion constant takes values between the Schawlow–Townes limit and zero. Thus, the Hanle laser is not simply a coherence transferring tool but a coherence increasing one. If, e.g., the pump has the Schawlow–Townes linewidth, then the Hanle laser can still have a much narrower relative linewidth for measurement times less than the inverse of the pump linewidth. It should be mentioned at this point that our analysis so far holds only for a homogeneously broadened active medium, i.e., for the case when atomic motion is neglected. The present treatment can, however, easily be generalized to take into account the effect of atomic motion and to properly deal with a Doppler-broadened active medium. Such a theory has been presented elsewhere,[10] but here we merely quote the results. All the conclusions of the present section are, essentially, maintained in an inhomogeneously broadened medium. The only difference is that, besides the equality of the detunings and coupling constants, one must assume coherence between the velocity groups that contribute to lasing. It is easiest to achieve CEL operation in an inhomogeneously broadened Hanle laser if only one velocity group contributes to both modes. Thus, CEL operation is possible, for example, if the modes have the same lasing frequency and differ in polarization only. This is precisely the case investigated in the present section.

In the next section we derive the nonlinear theory of the Hanle laser under the conditions of equations (26a) and (26b), and $\varphi_a = \varphi_b =$ const.

4. NONLINEAR THEORY

The starting point of the nonlinear theory is, again, the set of equations (7). This time, however, we solve them in a nonperturbative manner. First we make the substitution

$$\psi_{a,b} = e^{i\Delta t} \exp\left[-\frac{\gamma}{2}(t-t_0)\right]\tilde{\psi}_{a,b}(t) \quad \text{and}$$

$$\psi_c = \exp\left[-\frac{\gamma}{2}(t-t_0)\right]\tilde{\psi}_c(t) \tag{28}$$

where we have assumed $\Delta_1 = \Delta_2 = \Delta$, according to relation (26a). Then we introduce two orthogonal coherent superpositions from the components of $\tilde{\psi}$, as

$$\tilde{\psi}_+(t) = \cos\alpha \exp(-i\varphi_a)\tilde{\psi}_a + \sin\alpha \exp(-i\varphi_b)\tilde{\psi}_b$$

$$\tilde{\psi}_-(t) = \sin\alpha \exp(-i\varphi_a)\tilde{\psi}_a - \cos\alpha \exp(-i\varphi_b)\tilde{\psi}_b \tag{29}$$

In terms of these components the initial conditions specified by system (9) can be rewritten in the form

$$\tilde{\psi}_+(t_0) = \exp(-i\Delta t_0)\psi_F(t_0) \quad \text{and} \quad \tilde{\psi}_-(t_0) = \tilde{\psi}_c(t_0) = 0 \ . \tag{30}$$

The advantage of introducing these components is clear from here. Pumping occurs directly into the $\tilde{\psi}_+$ state. As we shall see later, under the noise quenching conditions (26) of the Hanle laser, $\tilde{\psi}_-$ is decoupled from the system. It is convenient to introduce, along with $\tilde{\psi}_+$ and $\tilde{\psi}_-$, two similar independent linear combinations of a_1 and a_2:

$$A = (g_1^2 \cos^2\alpha + g_2^2 \sin^2\alpha)^{-1/2}[g_1 \cos\alpha a_1 \exp(-i\varphi_a)$$
$$+ g_2 \sin\alpha a_2 \exp(-i\varphi_b)] \tag{31a}$$

$$B = (g_1^2 \cos^2\alpha + g_2^2 \sin^2\alpha)^{-1/2}[g_2 \sin\alpha a_1 \exp(-i\varphi_a)$$
$$- g_1 \cos\alpha a_2 \exp(-i\varphi_b)] \tag{31b}$$

with

$$[A, A^+] = [B, B^+] = 1 \quad \text{and} \quad [A, B] = [A, B^+] = 0 \tag{31c}$$

We can write system of equations (7) in terms of these components in the following way:

$$i\dot{\tilde{\psi}}_+ = \Delta\tilde{\psi}_+ + g'A\tilde{\psi}_c \tag{32a}$$

$$i\dot{\tilde{\psi}}_- = \Delta\tilde{\psi}_- + g'^{-1}[A \sin \alpha \cos \alpha (g_1^2 - g_2^2) + Bg_1g_2]\tilde{\psi}_c \tag{32b}$$

$$i\dot{\tilde{\psi}}_c = g'A^+\tilde{\psi}_+ + g'^{-1}[A^+ \sin \alpha \cos \alpha (g_1^2 - g_2^2) + B^+g_1g_2]\tilde{\psi}_- \tag{32c}$$

with $g' = (g_1^2 \cos^2 \alpha + g_2^2 \sin^2 \alpha)^{1/2}$.

From this set of equations we see the following behavior. Even if we start with $\tilde{\psi}_-(t_0) = 0$, a population will eventually build up in this state due to the interaction term in equation (32b). The phase of this state will carry the phase of $\tilde{\psi}_c$ which is completely random, since only $\varphi_a - \varphi_b$ is fixed in system (29). The overall phase of $\tilde{\psi}_+$ which, according to equation (32c), determines the phase of $\tilde{\psi}_c$ is a random quantity. Finally, the population in $\tilde{\psi}_-$ will lead to emission into the mode B due to the interaction term in equation (32c). As a result, the relative phases of the modes A and B are fluctuating quantities (just as those of a_1 and a_2) and no noise quenching occurs. The only exception is

$$g_1 = g_2 \equiv g \tag{33}$$

in which case there is no interaction between mode A and state $\tilde{\psi}_-$. In this case the initial condition

$$\tilde{\psi}_-(t_0) = 0$$

$$\tilde{\psi}_F(t_0) = |\psi(t_0)\rangle_A |0\rangle_B \quad \text{or} \quad \rho_F(t_0) = \rho^{(A)}(t_0)\delta_{n_B,0} \tag{34}$$

remains stable in the following sense:

$$\tilde{\psi}_-(t) = \tilde{\psi}_-(t_0) = 0, \qquad \tilde{\psi}_F(t) = |\psi(t)\rangle_A |0\rangle_B \tag{35}$$

By taking into account relation (33) in equation (26a) we find that

$$\cos \alpha = \sin \alpha = 1/\sqrt{2} \tag{36}$$

together with relations (33) and (26b) are the conditions for noise quenching. We now look for the solution of equations (32a-c) under these conditions. It is assumed that the solution is of the form

$$\tilde{\psi}_-(t) = 0, \qquad \tilde{\psi}_+(t) = \tilde{\psi}_+^{(A)}(t)|0\rangle_B, \qquad \tilde{\psi}_c = \tilde{\psi}_c^{(A)}(t)|0\rangle_B \tag{37}$$

Substitution of solution (37) into equations (32a–c) yields the following simple system of equations for $\tilde{\psi}_+^{(A)}$ and $\tilde{\psi}_c^{(A)}$:

$$i\dot{\tilde{\psi}}_+^{(A)} = \Delta \tilde{\psi}_+^{(A)} + gA\tilde{\psi}_c^{(A)} \tag{38a}$$

$$i\dot{\tilde{\psi}}_c^{(A)} = gA^+ \tilde{\psi}_+^{(A)} \tag{38b}$$

The solution of this system of coupled equations is straightforward. One can, for example, take the second derivative of equation (38a) and substitute $\tilde{\psi}_c^{(A)}$ from relation (38b). In this way, one finds a homogeneous linear second-order equation for $\tilde{\psi}_+^{(A)}$ (and, similarly, for $\tilde{\psi}_c^{(A)}$) which can always be solved by the substitution $\tilde{\psi}_+^{(A)} = C \exp[-i\lambda(t - t_0)]$. The two eigenfrequencies are

$$\lambda_\pm = \frac{\Delta}{2} \pm \sqrt{\left(\frac{\Delta}{2}\right)^2 + g^2 AA^+} \quad \left(\equiv \frac{\Delta}{2} \pm \omega_A\right) \tag{39}$$

and the solution can be written as

$$\tilde{\psi}_+^{(A)} = C_+ \exp[i\lambda_+(t - t_0)] + C_- \exp[-i\lambda_-(t - t_0)] \tag{40}$$

Equations (29) and (28) can be employed to transform this backward and obtain the solution to system of equations (7). The solution, satisfying the initial conditions (9) [or, equivalently, equations (30)] can then be written as

$$\psi_-(t) = 0 \tag{41a}$$

$$\psi_+(t) = \left[\cos \omega_A(t - t_0) - i\frac{\Delta}{2\omega_A} \sin \omega_A(t - t_0)\right]$$
$$\times \exp\left[\left(i\frac{\Delta}{2} - \frac{\gamma}{2}\right)(t - t_0)\right] \psi_A(t_0)|0\rangle_B \tag{41b}$$

$$\psi_c(t) = -iA^+ \frac{g}{\omega_A} \sin \omega_A(t - t_0) \exp\left[-\left(i\frac{\Delta}{2} + \frac{\gamma}{2}\right)(t - t_0) - i\Delta t_0\right] \psi_A(t_0)|0\rangle_B \tag{41c}$$

We shall not need the components ψ_c and ψ_b explicitly since, as we shall see, in the master equation for the reduced density operator for the field only the $+$ and $-$ components enter. From equations (38) and (41) we may conclude that the dynamics of the system is very similar to the dynamics of a simple two-level system coupled to one quantized mode of the radiation field, in terms of ψ_+, ψ_c, and A.

In the next step we shall derive the equation of motion for the density operator under the conditions of noise quenching, namely, equations (26), (33), and (36). Our starting point is equation (14) where we shall specify the loss term later. If we introduce A and B from equations (31), then equation (14) takes the form

$$i\dot{\rho}_F = -ig\{e^{i\Delta t}([A, \rho_{c+}] + [B, \rho_{c-}]) + e^{-i\Delta t}([A^+, \rho_{+c}] + [B^+, \rho_{-c}])\} + \text{loss} \quad (42)$$

where we have introduced the following components of the density operator:

$$\rho_{\pm c} = \frac{1}{\sqrt{2}}(\exp(-i\varphi_a)\rho_{ac} \pm \exp(-i\varphi_b)\rho_{bc}); \qquad \rho_{c\pm} = \rho_{\pm c}^\dagger \quad (43)$$

To proceed further we need an expression for ρ_{+c}, ρ_{-c}, and their hermitian conjugate. As in the previous section the following procedure is adopted to obtain this expression. We first calculate the contribution of one atom injected at time t_0 into the coherent superposition $\psi_+(t_0)$ and then sum the contribution of all atoms which are injected at random times $t - \gamma^{-1} < t_0 < t$ at a rate r_a. In this way we find

$$\rho_{\pm c} = r_a \int_{t-\gamma^{-1}}^{t} dt_0 \psi_\pm(t, t_0)\psi_c^+(t, t_0) \quad (44)$$

We now substitute the solutions for ψ_+, ψ_-, and ψ_c from system (41) which yields

$$\rho_{-c} = \rho_{c-} = 0 \quad (45)$$

and

$$\rho_{+c} = \rho_{c+}^+ = ir_a \int_{t-\gamma^{-1}}^{t} dt_0 \exp[-\gamma(t - t_0)]$$

$$\left[\cos \omega_A(t - t_0) - i\frac{\Delta}{2\omega_A}\sin \omega_A(t - t_0)\right]$$

$$\times \rho^{(A)}(t_0)\rho^{(B)}(t_0) \sin \omega_A(t - t_0) \frac{g}{\omega_A} A \quad (46)$$

where we have introduced $\rho_A(t_0) = \psi_a(t_0)\psi_a^+(t_0)$ and $\rho_B(t_0) = |0\rangle_{BB}\langle 0| = \delta_{n,0}$. For the sake of calculational simplicity we assume at this point that ρ_F separates as

$$\rho_F(t) = \rho^{(A)}\rho^{(B)} \tag{47}$$

and, furthermore, that

$$\rho^{(B)}(t) = \rho^{(B)}(t_0) = |0\rangle_{BB}\langle 0| \quad \text{or} \quad \rho^{(B)}_{n,n'} = \delta_{n,0} \tag{48}$$

This assumption can be justified if one takes equation (45) into account in equation (42), since then there is no gain in mode B. As we shall see, equations (47) and (48) remain valid if we include the loss terms in the master equation [see equation (51) below]. With the aid of equation (46) we can simplify both sides of equation (42) by $\rho^{(B)}(t_0)$ and obtain a master equation for $\rho^{(A)}$ only. The coefficients in this master equation can be found in the following way. In equation (46) $\rho^{(A)}(t_0)$ can be approximated by $\rho^{(A)}(t)$, since the dynamics of the photon field is governed by the cavity lifetime γ_c^{-1} which is much longer than γ^{-1} and thus, during the integration time, $\rho^{(A)}$ does not change appreciably. Then we can extend the lower limit of integration to $-\infty$ since, due to the exponential damping factor in the integrand, the contribution from $t_0 < t - \gamma^{-1}$ is negligible. After performing these steps and taking the n,n' matrix element of equation (46) it is easy to carry out time integration, yielding

$$(\rho_{+c}^{(A)})_{n,n'} = \tfrac{1}{2} i r_a g \frac{\sqrt{n'}}{\omega_{n'-1}} \left(R^+_{n,n'-1} - R^-_{n,n'-1} + i \frac{\Delta \gamma}{2\omega_n} S_{n,n'-1} \right) \tag{49}$$

where

$$R^{\pm}_{n,n'-1} = \frac{\omega_n \pm \omega_{n'-1}}{\gamma^2 + (\omega_n \pm \omega_{n'-1})^2}$$

$$S_{n,n'-1} = \frac{1}{\gamma^2 + (\omega_n + \omega_{n'-1})^2} - \frac{1}{\gamma^2 + (\omega_n - \omega_{n'-1})^2} \tag{50}$$

and ω_n can be obtained from ω_A [defined in equation (39)] by replacing AA^+ with $n+1$.

We now specify the loss term in equation (42) in the usual way (see Ref. 5):

$$\text{Loss} = -\frac{\nu_1}{2Q_1}(a_1^+ a_1 \rho_F + \rho_F a_1^+ a_1 - 2a_1 \rho a_1^+) - \frac{\nu_2}{2Q_2}(1 \to 2) \tag{51}$$

where Q_i is the Q-factor of the double cavity at frequency ν_i ($i = 1, 2$). If we assume, for simplicity, that

$$\frac{\nu_1}{Q_1} = \frac{\nu_2}{Q_2} = \gamma_c \tag{52}$$

i.e., the cavity lifetime is the same for both modes, then the loss term can be expressed in terms of the operators A and B by simply replacing a_1 with A and a_2 with B in expression (51). Since there is no interaction between modes A and B, the separation ansatz, equation (47), remains valid. Also, solution (48) for $\rho^{(B)}(t)$ will hold since mode B is coupled to a loss reservoir only. Taking into account equations (47) through (52) in equation (42) finally yields the following master equation for the matrix elements of $\rho^{(A)}$:

$$\dot{\rho}_{n,n'}^{(A)} = -\frac{\mathcal{N}'_{n,n'}\mathcal{A}}{1+\mathcal{N}_{n,n'}\mathcal{B}/\mathcal{A}}\rho_{n,n'}^{(A)} + \frac{\sqrt{nn'}\,\mathcal{A}}{1+\mathcal{N}_{n-1,n'-1}\mathcal{B}/\mathcal{A}}\rho_{n-1,n'-1}^{(A)}$$

$$-\frac{\gamma_c}{2}[(n+n')\rho_{n,n'}^{(A)} - 2\sqrt{(n+1)(n'+1)}\,\rho_{n+1,n'+1}^{(A)}] \tag{53}$$

Here we have introduced the following notation:

$$\mathcal{A} = 2r_a\left(\frac{g}{\gamma}\right)^2, \qquad \mathcal{B} = 4\left(\frac{g}{\gamma}\right)^2\mathcal{A} \tag{54a}$$

and

$$\mathcal{N}_{n,n'} = \frac{1}{2}\left[n+1+\left(\frac{\Delta}{2g}\right)^2 + n'+1+\left(\frac{\Delta}{2g}\right)^2\right] + \frac{1}{16}\frac{\mathcal{B}}{\mathcal{A}}(n-n') \tag{54b}$$

$$\mathcal{N}'_{n,n'} = \frac{1}{2}(n+1+n'+1) + \frac{1}{8}\frac{\mathcal{B}}{\mathcal{A}}\left(n-n'+i\frac{\Delta\gamma}{g^2}\right)(n-n') \tag{54c}$$

As usual $\mathcal{A}$ has the meaning of the linear gain and $\mathcal{B}$ is the self-saturation coefficient.

The central result of the paper is represented by equation (53). When written in the above form it is easy to see that the part corresponding to mode A is identical with the master equation of a one-mode laser with two-level active medium, with detuning Δ.[5] The steady-state photon statistics, in particular, can be obtained from equation (53) by setting $\partial/\partial t = 0$ and taking the diagonal ($n = n'$) elements of the equation. The result is formally identical with that of a one-mode laser, with detuning Δ, and with a two-level active medium. Therefore, in terms of the composite mode A, under the conditions when noise quenching occurs the Hanle laser exhibits the same type of behavior (photon statistics, threshold, saturation) as the one-mode laser. The detailed analysis of the steady-state version of the diagonal part of equation (53) can be found in the literature (see, e.g., Ref. 5) and we do not pursue it here any further. In the next section we shall elaborate on another consequence of equation (53), namely, the vanishing of the diffusion constant of and noise quenching from the relative phase between modes a_1 and a_2.

5. VANISHING OF DIFFUSION CONSTANT FOR THE RELATIVE PHASE

The beat signal between two modes of an ordinary laser or between two independent lasers decays to zero in time because, due to the independent spontaneous emission events, the relative phase fluctuates freely. The decay rate for the beat signal is, in principle, given by the Schawlow–Townes linewidth [see below, equation (61)].

In the following, we shall show that in the Hanle laser the beat signal has a nonvanishing part in steady state if the conditions of equations (26), (33), and (36) are satisfied. For this case, by using the inverse of expressions (31a) and (31b), the beat signal can be expressed as the real part of

$$\text{Tr}(a_1^+ a_2 \rho) = \tfrac{1}{2}\text{Tr}\{[A^+A - B^+B + AB^+ - A^+B]\rho_F\} e^{-i\varphi} \tag{55}$$

Here ρ_F is the reduced density operator for the two-mode field which, according to relation (47), can be written as a product of the density operators $\rho^{(A)}$ and $\rho^{(B)}$ of the two noninteracting modes A and B. Furthermore, $\rho^{(A)}$ satisfies equation (53) and $\rho^{(B)}$ satisfies a similar equation with only the loss terms present and all the gain terms (terms proportional to $\mathscr{A}$) missing.

By looking for time-dependent solutions of equation (53) one can show[5] that the general solution is of the form

$$\rho_{n,n}(t) = \sum_{j=0} \varphi_j(n) \exp(-\mu_j t) \tag{56}$$

for the diagonal elements and

$$\rho_{n,n+p}(t) = \sum_{j=0} \varphi_j(n, p) \exp(-\mu_j^{(p)} t) \tag{57}$$

for the off-diagonal elements. Furthermore,

$$\mu_j \geq 0 \quad \text{with } \mu_0 = 0 \tag{58a}$$

i.e., the lowest eigenvalue vanishes, allowing for a nonvanishing solution at steady state for the diagonal elements. Also,

$$\mu_j^{(p)} > 0 \quad \text{for } p \neq 0 \tag{58b}$$

i.e., the off-diagonal elements decay to zero for large times. In the case of an ordinary two-mode laser or two separate lasers, the density matrix factorizes in terms of the original a_1 and a_2 modes as

$$\rho(1,2) = \rho^{(1)}\rho^{(2)} \tag{59}$$

and the beat signal, defined in equation (55), will be proportional to the product of the off-diagonal elements of the type

$$\langle a_1^+ a_2 \rangle_{\text{ordinary laser}} \sim \sum_{n,n'} \rho^{(1)}_{n,n+1}\rho^{(2)}_{n',n'+1}\sqrt{(n+1)(n'+1)} + \text{c.c.} \tag{60}$$

According to relations (57) and (58b) these matrix elements decay at a rate $\mu_0^{(1)}$ (neglecting the more rapidly decaying terms). This defines the phase diffusion coefficient D as $\mu_0^{(1)} = \tfrac{1}{2}D$ and from equation (53) it can be shown (see Ref. 5) that

$$D = \gamma_c/2\langle n \rangle \tag{61}$$

which is the Schawlow–Townes linewidth.

The crucial difference between the Hanle laser and the ordinary two-mode laser is that the beat signal contains a part that is diagonal in the "true" eigenmodes A and B of the system,

$$\langle a_1^+ a_2 \rangle_{\text{Hanle laser}} \sim \sum_n n\rho^{(A)}_{n,n} + \text{other terms} \tag{62}$$

According to relations (56) and (58a) these matrix elements are nonzero in steady state (the lowest decay rate is $\mu_0 = 0$), i.e., there is always a nonvanishing part of the beat signal. The beat signal can be regarded as a measure of the "true" eigenmodes of the system. If we maintain the definition of the diffusion coefficient as twice the lowest decay rate in equation (55), then

$$D = 0 \tag{63}$$

for the Hanle laser and the beat signal will be given in the stationary case by

$$2\text{Re}\langle a_1^+ a_2 \rangle_{ss} = e^{-i\varphi}\bar{n}_A + \text{c.c.} \tag{64}$$

We can interpret this result in the following way. Spontaneous emission is, of course, always present in the system. But it is in mode A only, since the upper level for emission into mode B is empty. According to equations (31a) and (31b) both a_1 and a_2 can be expressed in terms of A and the spontaneous emission contribution will be common to a_1 and a_2. When beating these two modes this fluctuating, but common, contribution cancels from the beat signal.

6. APPLICATIONS AS A METRIC GRAVITY PROBE

In the first part of this section we show how the Hanle laser can be used to detect small phase differences between the two modes a_1 and a_2. Then we determine the noise associated with such a measurement, both inside and outside the cavity. From the requirement of the equality of the signal to the noise (ratio $S/N = 1$) we determine the magnitude of the minimum detectable phase difference in an extracavity beat measurement. We show that the "in principle" performance might exceed the standard quantum limit. Then we discuss practical limitations that arise mainly from phase-locking. We apply this general scheme for gravity wave detectors and laser gyroscopes following largely the lines of presentation of Refs. 11 and 12 for these cases.

Let us assume that in the second arm of the resonator in Figure 1b an additional phase shift ψ is accumulated as a result of some external perturbation to be specified later. We can account for this phase shift by simply replacing a_2 with $a_2 e^{i\psi}$ everywhere in the previous sections. In particular, we define two new independent linear combinations of a_1 and a_2 [the analogs of equations (31a) and (31b) for $g_1 = g_2$ and $\alpha = \pi/4$]:

$$\tilde{A} = \frac{1}{\sqrt{2}} \{a_1 \exp(-i\varphi_a) + a_2 \exp[i(\psi - \varphi_b)]\} \qquad (65a)$$

$$\tilde{B} = \frac{1}{\sqrt{2}} \{a_1 \exp(-i\varphi_a) - a_2 \exp[i(\psi - \varphi_b)]\} \qquad (65b)$$

Then, in the $\psi \neq 0$ case, the Hanle laser exhibits the same behavior in terms of $\tilde{A}$ and $\tilde{B}$ as the one found in Sections 4 and 5 for $\psi = 0$ in terms of A and B. One just replaces A and B by $\tilde{A}$ and $\tilde{B}$ everywhere, i.e., the dressed mode $\tilde{A}$ will now be described by the master equation (53) of a one-mode laser with two-level active medium, while mode $\tilde{B}$ will be in vacuum state as given by relation (48).

In a beat experiment one prepares a superposition of the two-mode field

$$Q(t) = c_1 a_1 e^{-i\nu t} - c_2 a_2 e^{-i\nu t} \qquad (66)$$

and measures the beat signal S given by

$$S = \langle Q^+(t) Q(t) \rangle \tag{67}$$

Here c_1 and c_2 are two arbitrary complex numbers describing the measurement apparatus. For simplicity we assumed $\nu_1 = \nu_2 = \nu$ and examine the simplest case of equal times for $Q^+(t)$ and $Q(t)$. The inverse of equations (65a) and (65b) enables equation (66) to be written in the form

$$Q(t) = (d_1 \tilde{A} + d_2 \tilde{B}) e^{-i\nu t} \tag{68a}$$

with

$$d_{1,2} = \frac{1}{\sqrt{2}} \{c_1 \exp(i\varphi_a) \mp c_2 \exp[i(\varphi_b - \psi)]\} \tag{68b}$$

Substitution of this and the distribution functions (48) and (53) in equation (67) yield the beat signal in the form

$$S = |d_1|^2 \langle \tilde{A}^+ \tilde{A} \rangle = |d_1|^2 \tilde{n} \tag{69}$$

where the notation $\tilde{n}$ has been introduced for the steady-state number of photons in mode $\tilde{A}$.

In a similar manner we can calculate the noise in this measurement as

$$(\delta S)^2 = \langle (Q^+ Q)^2 \rangle - \langle Q^+ Q \rangle^2$$
$$= |d_1|^4 [\langle (\tilde{A}^+ \tilde{A})^2 \rangle - \langle \tilde{A}^+ \tilde{A} \rangle^2] + |d_1|^2 |d_2|^2 \tilde{n} \tag{70}$$

For a laser above threshold and in steady state the first term on the right-hand side is $\tilde{n} |d_1|^4$ since in coherent state $\langle n^2 \rangle - \langle n \rangle^2 = \tilde{n}$. If we assume $c_1 = \tilde{c}_1 \exp(i\varphi_1)$ and $c_2 = \tilde{c}_2 \exp(i\varphi_2)$, where $\tilde{c}_1$ and $\tilde{c}_2$ are real and positive amplitudes and φ_1 and φ_2 are the phases, then equation (69) can be solved for $\tilde{\psi} = \psi + (\varphi_a + \varphi_1) - (\varphi_b + \varphi_2)$, yielding

$$\cos \tilde{\psi} = \frac{\tilde{c}_1^2 + \tilde{c}_2^2}{2 \tilde{c}_1 \tilde{c}_2} - \frac{S}{\tilde{n} \tilde{c}_1 \tilde{c}_2} \tag{71}$$

From here the uncertainty in the determination of ψ will then be related to the noise δS in the measurement of S:

$$(\delta \psi)^2 = (\delta \tilde{\psi})^2 = \frac{(\delta S)^2}{\tilde{n}^2 \tilde{c}_1^2 \tilde{c}_2^2 \sin^2 \tilde{\psi}} \tag{72}$$

QUANTUM THEORY OF THE HANLE LASER

If equation (70) is substituted in equation (72), we obtain

$$(\delta\psi)^2 = \frac{1}{2\tilde{n}\sin^2\tilde{\psi}} \frac{(1+y^2)^2 - 2y(1+y^2)\cos\tilde{\psi}}{y^2} \tag{73a}$$

with

$$y = \tilde{c}_2/\tilde{c}_1 \tag{73b}$$

In other words the noise depends only upon the ratio $y = \tilde{c}_2/\tilde{c}_1$, and equation (73a) has a minimum for $y = \pm 1$. This gives the minimum detectable phase shift. With the choice $y = +1$

$$\delta\psi = \tilde{\psi}_{\min} = \frac{1}{\sqrt{\tilde{n}}\,|\cos(\tilde{\psi}/2)|} \tag{74}$$

This choice is advantageous if $\tilde{\psi} = 0$. Similarly, with $y = -1$

$$\delta\psi = \tilde{\psi}_{\min} = \frac{1}{\sqrt{\tilde{n}}\,|\sin(\tilde{\psi}/2)|} \tag{75}$$

This choice too is advantageous if $\tilde{\psi} = \pi$. In both cases the minimum detectable phase shift is $\psi_0 = \tilde{n}^{-1/2}$, where $\tilde{n}$ is the number of photons in steady state in mode $\tilde{A}$. This quantity can be related to the detected power P if we use the relationships

$$\dot{n} = \gamma_c \tilde{n} \tag{76}$$

where $\dot{n}$ is the rate of photons leaking out from the cavity at steady state, and

$$P = \hbar\nu\dot{n} \tag{77}$$

In this last relation it is assumed that all photons escaping from the cavity are detected. Relations (76) and (77) yield

$$\tilde{n} = \frac{P}{\gamma_c \hbar\nu} \tag{78}$$

and, finally,

$$\psi_0 = \sqrt{\frac{\hbar\nu\gamma_c}{P}} \tag{79}$$

This result is now applied to determine the sensitivity of the Hanle laser as a probe for metric gravity effects.

First, we treat the problem of gravity wave detection. A gravity wave train is assumed to travel through the Hanle laser of Figure 1b with propagation direction parallel to one of its arms. It perturbs the length of the other arm, due to its transverse nature, in such a way that the relative change in length l is

$$\Delta l/l = h_0 \tag{80a}$$

This, in turn, leads to a frequency shift of the mode oscillating in this arm. The shift $\Delta \nu$ is given by

$$\Delta \nu = \nu h_0 \tag{80b}$$

The laser is now inside the cavity, so, in principle, we must use the expression for the associated phase shift characteristic of an active system,[11]

$$\psi_0 = \nu h_0 t_m \tag{80c}$$

In the above expression h_0 is the strength of the gravity wave and t_m is the measurement time. When equation (80c) is substituted into expression (79) and the resulting equation solved for h_0, we obtain the minimum detectable gravity wave strength in the form

$$h_0^{(\min)} = \left(\frac{\gamma_c}{\nu}\sqrt{\frac{\hbar\nu}{Pt_m}}\right)\frac{1}{\sqrt{\gamma_c t_m}} \tag{81}$$

The term in parentheses on the right-hand side represents the standard quantum limit while the factor

$$\varepsilon = \frac{1}{\sqrt{\gamma_c t_m}} \tag{82}$$

corresponds to the enhancement of sensitivity in this scheme. In the derivation we have assumed $\psi = \text{const}$ which, in view of equation (80c), means that t_m is limited by $t_m < \omega_g^{-1}$ where ω_g is the frequency of the gravity wave. Thus, the maximum enhancement factor is

$$\varepsilon_{\max} = \sqrt{\frac{\omega_g}{\gamma_c}} \tag{83}$$

If we assume $\omega_g = 10^2$ Hz and $\gamma_c = 10^6$ Hz, then $\varepsilon \sim 10^{-2}$.

Next, we apply expression (79) to the detection of an (effective) rotation rate Ω with the help of the (generalized) Sagnac effect as discussed in Ref. 12. For this case the expression for the phase shift in an active system is given[4,12,13] as

$$\psi_0 = S\Omega t_m \tag{84}$$

where S is a dimensionless scale factor, $S = 4A/\lambda p$, A being the area of the enclosed ring, p its perimeter, and λ the reduced wavelength. By inserting equation (84) into relation (79), we obtain the minimum detectable rotation rate:

$$\Omega_{min} = \left(S^{-1}\gamma_c\sqrt{\frac{\hbar\nu}{Pt_m}}\right)\frac{1}{\sqrt{\gamma_c t_m}} \tag{85}$$

The term in parentheses again represents the standard quantum limit and we find the same improvement factor as was given by equation (82). The measurement time is again limited by $t_m < \Omega^{-1}$, since ψ in the above derivation must remain constant. Thus, the maximum enhancement factor is

$$\varepsilon_{max} = \sqrt{\frac{\Omega}{\gamma_c}} \tag{86}$$

For $\Omega = 10^{-4}$ Hz (earth's rate of rotation) and $\gamma_c = 10^6$ Hz, $\varepsilon \sim 10^{-5}$. In relation (84) the effective rotation rate may be a result of several different effects like, e.g., mechanical rotation, preferred frame, the Lense-Thirring effect, etc.,[13] i.e., the generalized Sagnac effect is indeed a probe to test metric gravity effects.

It is clear from the considerations so far that the best one can hope for is an improvement of the sensitivity by a factor ε as given by equation (83) or (86). It should be noted at this point that such an improvement would render some of the above-mentioned effects observable. There are other sensitivity limiting factors, however. The origin of this improvement can be traced back to the fact [see equation (79) together with relations (80c) and (84)] that we equated a signal that is characteristic of an active system with a noise that is characteristic of a passive system. Due to the strong, coherent coupling of the two modes in a Hanle laser, the relative phase exhibits the phenomenon of phase locking. This means that the phase shift will not be proportional to the time as in relations (80c) and (84) but locks to a value that can be obtained from (80c) or (84) by replacing t_m with γ_c^{-1}. Then ε as given by expression (82) is unity and our results reproduce the standard quantum limit for this case.

Another, practical issue has also been addressed in Refs. 11 and 12. The total loss rate of the cavity can be thought of as the result of three different loss mechanisms:

$$\gamma_c = \gamma_a + \gamma_d + \gamma_t \tag{87}$$

Here γ_a is the rate of (unavoidable) mirror loss due to absorption and γ_d is the detection rate. These are irreversible losses. Quantity γ_t stands for all other, transmission-type losses. This latter one is, in principle, reversible. The specially designed feedback systems of Refs. 11 and 12, the so-called controlled backscattering, eliminates this extra noise source by the in-phase reinjection of the undetected and unabsorbed photons into the cavity. If one compares the sensitivity of a nonideal measurement (some transmitted photons are undetected and lost) to an ideal measurement (all photons transmitted through the output mirror are detected), then the effective improvement factor $\tilde{\varepsilon}$ is found to be

$$\tilde{\varepsilon} = \sqrt{\frac{\gamma_d}{\gamma_a + \gamma_d + \gamma_t}} \tag{88}$$

The feedback system indicates one way to realize an ideal measurement and is consequently of great practical importance. The ultimate sensitivity is seen from expression (88) to be limited by the (irreversible) absorption loss.

Thus for our model of detection, based on the use of the correlated emission Hanle laser, as discussed so far, sensitivity is given by the standard quantum limit. In a practical measurement this can be realized to the extent that γ_a is negligibly small as compared to γ_d and $\gamma_t = 0$. This is the best one can achieve in a single measurement. However, as noted above, the signal locks to its steady-state value on a time scale $t \sim \gamma_c^{-1}$. The measurement time t_m is, on the other hand, limited by the duration of the gravity wave train ω_g^{-1} or characteristic revolution time Ω^{-1}. During this total allocated time t_m the single measurement can be repeated N times where

$$N = t_m \gamma_c \tag{89}$$

This reduces the noise by $\sqrt{N}$. When applied to equation (79) it yields

$$\psi_0 = \sqrt{\frac{\hbar \nu}{P t_m}} \tag{90}$$

If ψ_0 is the phase locked signal [equation (80c) or (84) with $t_m = \gamma_c^{-1}$] this brings us back to the standard quantum limit over the longer measurement time $t_m = \omega_g^{-1}$ or Ω^{-1}.

The crucial difference between the detection schemes involving conventional lasers and CEL lasers can be seen in the following experimental setup. Imagine that with N' identical conventional lasers we carry out measurements of ψ_0 with each laser. This would not improve the overall sensitivity because the individual $\psi_0 - s$ still diffuse relative to each other at the usual laser phase diffusion rate, leaving the signal-to-noise ratio unchanged. On the other hand, we can carry out measurements of ψ_0 using N' CEL lasers. Since now all the relative phases are free from phase diffusion, the noise is reduced by $\sqrt{N'}$. It is easy to see that the optimum number N'_{opt} of CEL lasers to be used this way is the same N as in equation (89). If we had more CEL lasers, i.e., $N' > N$, then it would be equivalent to dividing the measurement time into N' parts. But since $t_m/N' < \gamma_c^{-1}$ the phases still would not lock and we would have to include usual laser phase diffusion noise. Thus $N'_{opt} = N$. Since $\sqrt{N} = \varepsilon_{max}^{-1}$ as given by expression (83) or (85), the above detection scheme, which utilizes N CEL lasers, realizes the optimum enhancement of sensitivity.

In conclusion, we have shown that a detection scheme based on the correlated emission Hanle laser might, in principle, have a sensitivity superseding the usual quantum limit. The theoretical upper limit for the enhancement of the sensitivity is given by equation (83) for the detection of gravity waves and equation (86) for the detection of rotation. Furthermore, this device is free of the usual "dead band" frequency locking problem associated with the detection of ultrasmall signals, since the signal appears not as a frequency shift but directly as a phase shift.

7. DISCUSSION AND SUMMARY

Starting from a Hamiltonian model of a three-level system, where the upper two levels are the Zeeman sublevels of a nearly degenerate upper state and these levels are, in turn, coupled to the lower level via interacting with two modes of the quantized radiation field (see Figure 1), we have developed both the linear and nonlinear theory of the Hanle laser. In developing the linear theory we have used a Fokker-Planck equation approach and derived an explicit expression for the diffusion coefficient of the relative phase of the two modes. From this expression we have established physical conditions under which this diffusion constant vanishes. Then, in the next step, we have developed a nonlinear theory of the Hanle laser which is valid if the conditions for noise quenching from the relative phase, found in the linear theory, are satisfied. We were able to show in this way that noise quenching persists in the nonlinear theory. The key feature of our theory is the transformation given by equations (28)-(31) that leads to the equation of motion (32). In this way, instead of two upper states pumped simultaneously, we deal with one upper state that is pumped and another one that is empty. Under the noise quenching conditions this empty upper state is decoupled from the system, i.e., a population never builds up. This reduces the problem to a two-level system coupled resonantly to one quantized radiation mode. This quantized mode is, however, a linear combination of the initial "bare" modes. For example, if we prepare an initial coherent superposition ψ_+ of the two upper states ψ_a and ψ_b ($\psi_+ = \psi_a + \psi_b$) via the Hanle effect, then this coherent superposition is coupled to the ground state. This interaction selects $A = a_1 + a_2$ as the lasing mode. Since now both a_1 and a_2 can be expressed in terms of A, the spontaneous emission will be common to a_1 and a_2 and it cancels from the beat signal. The emerging physical picture is, thus, slightly different from a quantum beat laser. In that case the strong classical coupling establishes coherence of the upper two levels. In the Hanle laser atomic coherence is an initial value problem. It is achieved by pumping via the Hanle effect.

Therefore, the conditions for correlated spontaneous emission are different from those of a quantum beat laser. There, the conditions involved the coupling constant of the upper two levels and detuning while in the case of a Hanle laser these conditions specify the pumping (i.e., initial conditions). The question of stability can, at this point, easily be answered. We have formally reduced the problem of a Hanle laser to a one-mode laser problem. It is well known that the steady solution of this latter is stable above threshold. Hence, the correlated spontaneous emission operation in a Hanle laser is stable above threshold, as well. It is perhaps interesting to note here that line narrowing beyond the Schawlow-Townes limit has recently been observed experimentally[14] under the coherent coupling conditions of a quantum beat laser.

Finally, we have suggested that the correlated emission Hanle laser be used as a metric gravity probe. We have analyzed the noise performance of the system and established the following. Since the signal appears directly as a phase shift, the system is free of the "dead band" (frequency locking) problem of active systems. The sensitivity is limited only by the irreversible losses (absorption and detection) and we have established the theoretical limits of sensitivity for this system. We have also shown that by using a special detection scheme one can achieve a sensitivity superseding the standard quantum limit.

ACKNOWLEDGMENTS. It is a pleasure to acknowledge numerous discussions and conversations regarding different parts of this work with M. Orszag and L. M. Pedrotti. This work was supported by the Office of Naval Research.

REFERENCES

1. See W. HANLE, *Z. Phys.* **30**, 93 (1924) for the original experiment. Detailed accounts of subsequent work together with a theoretical explanation can be found in V. WEISSKOPF, *Ann. Phys. (Paris)* **9**, 23 (1931) and G. BREIT, *Rev. Mod. Phys.* **5**, 91 (1933).
2. M. O. SCULLY, in *Atomic Physics I*, edited by B. Bederson, V. W. Cohen, and F. M. Pichanick (Plenum Press, New York, 1969), p. 81.
3. C. M. CAVES, K. S. THORNE, R. W. P. DREVER, V. D. SANDBERG, AND M. ZIMMERMANN, *Rev. Mod. Phys.* **52**, 341 (1980).
4. W. W. CHOW, J. GEA-BANACLOCHE, L. M. PEDROTTI, V. E. SANDERS, W. SCHLEICH, AND M. O. SCULLY, *Rev. Mod. Phys.* **57**, 61 (1985).
5. See, for example, M. SARGENT, III, M. O. SCULLY, AND W. E. LAMB, JR., *Laser Physics* (Addison-Wesley, Reading, Mass., 1974).
6. M. O. SCULLY, *Phys. Rev. Lett.* **55**, 2802 (1985).
7. W. S. BICKEL AND S. BASHKIN, *Phys. Rev.* **162**, 12 (1967). *For a more update version see*: W. W. CHOW, M. O. SCULLY, AND J. O. STONER, JR., *Phys. Rev. A* **11**, 1380 (1975). Also M. O. SCULLY AND K. DRUHL, *Phys. Rev. A* **25**, 2208 (1982) and references cited therein.
8. M. O. SCULLY AND M. S. ZUBAIRY, *Phys. Rev. A* **35**, 752 (1987).
9. J. BERGOU, M. ORSZAG, AND M. O. SCULLY, Nonlinear Theory of the Quantum Beat Laser, submitted to *Phys. Rev. A*.
10. J. BERGOU, M. ORSZAG, AND M. O. SCULLY, Phase Noise Quenching via coherent Pumping and the Effect of Atomic Motion, submitted to *Phys. Rev. A*.
11. M. O. SCULLY AND J. GEA-BANACLOCHE, *Phys. Rev. A* **34**, 4043 (1986). See also J. GEA-BANACLOCHE AND M. O. SCULLY, in *Quantum optics*, edited by D. F. Walls (Springer, New York, 1987).
12. M. O. SCULLY, *Phys. Rev. A* **35**, 452 (1987).
13. See, e.g., W. SCHLEICH AND M. O. SCULLY, in *New Trends in Atomic Physics* (Les Houches, Session **38**, 1982), edited by G. Grynberg and R. Stora (Elsevier, Amsterdam, 1984) p. 997.
14. P. TOSCHEK AND J. HALL, Technical Digest of the XV International Quantum Electronics Conference (26 April–1 May, 1987, Baltimore), paper WDD2, p. 102.

INDEX

Asymmetric-top molecules, 90
Atomic constants, determination, 49–51
Average saturation parameter, 161

Beam-foil spectroscopy, 31, 57
Bennet holes, 154
BLA (broad line approximation), 129, 154
Bloch–Siegert effect, 142
Breit–Franken formula, 28, 88–89
Brossel–Bitter experiments, 96

Ca laser, 228
CEL (correlated spontaneous emission laser), 342
Chemical shift, 99
Chromosphere-corona transition region, 275–277
Coherence narrowing, 50
Collision induced Hanle resonances, 59–64
Collisional depolarization factor, 272

Delayed fluorescence detection, 52
Density matrix formalism for the Hanle effect, 17–27, 47–49, 243–254
Detection matrix, 89
Directional quantization, 6, 7, 41
Dressed-atom model, 19, 77–78

Effective absorption enhancement, 164
Effective emission enhancement, 164
Electric dipole interaction Hamiltonian, 134–135

Electric field level crossing in molecules, 100–102
Electric polarizability, 30
Electron–nucleus interaction in crystals, 299–304
Emission matrix, 22
Excitation matrix, 20, 89
Excitons, 309–329
Excitons in cubic crystals, 310–316
Excitons in hexagonal II–VI crystals, 316–317
Excitons in III–VI crystals, 317–323

Fluctuation-induced Hanle resonances, 64
Fluorescent scattering, 267
Forward scattering, 54–55, 72–74, 181
Four-wave mixing, 27, 59–64, 79
Franken equation, 28, 88–89
Free electrons in solids, 283, 284–309

Gaussian laser beam, 163–166
Geometrical depolarization factor, 258
Graded-band crystals, 296–299
Gravity probe, 359–365

Hanle effect
  and nonlinear optics, 76–77
  broad-band excitation, 17–23
  classical interpretation, 9–13
  collisional excitation, 56–57
  electric or second, 18

Hanle effect (*contd.*)
  in strong laser fields, general
    characteristics, 65-67
  laser excitation, 23-27
  optical, 74-75
  quantum mechanical interpretation, 13-16
  with fluctuating fields, 78
  with two-photon excitation, 75-76
Hanle laser, 79
  linear theory, 344-350
  nonlinear theory, 351-357
Hanle parameters, 250
Hanle phase matrix, 249
Hanle scattering matrix, 249
He-CdII laser, 201-206
He-Ne laser, 192-200
He-ZnII laser, 201-206
Heavy-hole bands, 285
Hertzian coherence, 20, 25, 141
Hidden alignment, 57
High-order coherences, 67-69
Hot excitons, 327-329

Impurity centers, 329-332
Induced emission, 18

Kippenhahn-Schlüter model, 262
Kuperus-Raadu model, 262

Landé g-factors in molecules, 98-99
Laser-level populations, measurements, 51
Level crossing, 8, 28-30
Level crossing in electric fields, 30
Level overlapping, 8
LIF (wavelength-resolved laser-induced fluorescence), 93
Lifetime investigations in molecules, 95-98
Lifetime of the excited state, 12
Light-hole bands, 285
Line crossing, 54-55
Local strains in crystals, 291

Macaluso-Corbino effect, 192
Magnetogram, 238, 242
Magnetometry, 55-56
Mueller matrix, 245

n-type crystals, 291-295
NLHE (Nonlinear Hanle effect), 123-229
  Doppler-broadened three-level system, 147-151
  homogeneously broadened three-level system, 139-147
NO_2, 104-118
Noble-gas Ion lasers, 206-218
Nonlinear Hanle effect, fluorescence detection, 69-71

OODR (optical-optical double resonance), 125, 180
Optical coherence, 19, 23, 25, 141
Optical radio-frequency double resonance, 94, 96, 99, 103
Optically pumped far-infrared lasers, 218-228
Optoacoustic detection, 129, 130
Optogalvanic detection, 57-59, 129, 130
Oxygen laser, 228

p-type crystals, 295-296
Parity violation, 51
Photosphere, 275
PIHE (Pressure induced Hanle effect), 27
Plasma diagnostics, 56
Polaritons, 324, 327-329
Polarization degree, 3, 12, 88, 288
Polarization spectroscopy, 74
Polarized luminescence in semiconductors, 285-288
Power broadening of the Hanle resonance, 67, 133
Power enhancement, 184, 186

Quantum beat laser, 342
Quantum beats, 16, 91-92, 96

Radiation narrowing, 50
Radiation trapping, 50
Radio-frequency coherence, 20
Ramsey structures, 76
Rate equations, 150, 155-163
Rate-equation approach, 129
Rigid asymmetric rotor, 90
Rigid symmetric rotor, 90
Rosette motion, 4, 35

Saturated linewidth, 132-139
Saturation intensity, 125, 132-139
Saturation parameter, 125, 137
Saturation resonances, 67
Scattering polarization on the solar disk, 262-269

Selective excitation of molecules, 92-95
Solar magnetic fields, 238-240
Solar prominences, 254-262
Spontaneous emission, 13, 18
Squeezed light, 79
Stark-Zeeman recrossing, 102-104
Stimulated level and mode crossing, 71-72
Stokes vector, 244
Subnatural linewidth effects, 51-54
Sunspots, 239
Symmetric-top molecules, 89

Time-resolved level-crossing spectroscopy, 29
Turbulent magnetic fields, 269-275

Uniaxially deformed crystals, 290-291

Xe laser, 200-201

Zero-field level crossing, 8
Zero-field saturation resonance, 168